ETHNOMEDICAL SYSTEMS IN AFRICA

ETHNOMEDICAL SYSTEMS IN AFRICA

Patterns of Traditional Medicine in
Rural and Urban Kenya

CHARLES M. GOOD

THE GUILFORD PRESS
New York   London

Printed in the United States of America

Last digit is print number: 9 8 7 6 5 4 3 2 1

Library of Congress Cataloging-in-Publication Data

Good, Charles M.
  Ethnomedical systems in Africa.

  Includes bibliographies and index.
    1. Folk medicine—Africa. 2. Healing—Africa.
  I. Title. [DNLM: 1. Health Services, Indigenous—
  Africa. 2. Medicine, Traditional—Africa.
  WB 50 HA1 G6e]
  GR 350.G6 1987   615.8′82′096   86-26998
  ISBN 0-89862-779-6

This book is lovingly dedicated to my three children

Charles III, Geoffrey, and Njambi

Acknowledgments

The idea for this book originated a decade ago. I deeply appreciate the hospitality and generous sharing of time, information, and life that so many people, both in Kenya and the United States, extended to me during field-work and in the preparation of this book. I regret only that the mere mention of names is so inadequate to convey my sense of indebtedness and gratitude to those who enabled me to undertake and finally complete the study.

Fieldwork in Kenya was funded by grants from the Geography and Regional Science Division of the National Science Foundation. Time away from my duties as Head of the Department of Geography at Virginia Poly-technic Institute and State University was made possible by a University Study-Research Leave. Much appreciation is reserved for the exceptional personal and leadership qualities of my friend and colleague Bob Morrill, who managed departmental administrative responsibilities so capably in my absence during 1977–1978. Since he became Department Head in 1982 Bob has consistently provided a positive, supportive atmosphere for research and scholarship.

Faculty and staff of the Department of Community Health in the University of Nairobi Medical School graciously welcomed me into their midst upon my arrival in Kenya and offered me a friendly, collegial atmosphere and base from which to study local systems of traditional medicine. Dr. F. John Bennett, then Head of the Department of Community Health, generously provided me with office space, secretarial assistance, access to the Department's Land Rover, and invaluable counsel throughout the year. I am also grateful to Dr. James Kagia, who succeeded John Bennett as Department Head, for his kind support of the research. The project would never have been initiated without the active support of Professor K. Thairu, then Dean of the Faculty of the Medical School. Dr. J. J. Thuku, Chief Administrator of Kenyatta National Hospital, offered a cordial reception when our research team arrived to conduct a KAP study of traditional medicine among patients in the filter clinics and wards in December 1977. Hospital staff were also most cooperative and tolerant of our intrusion into their busy daily routines.

Sharing an office and field experiences with Roy Shaffer, M.D., enabled me to observe and learn from a skilled physician and health educator who is

actively committed to community-based health care. The warm friendship
and hospitality Roy and Betty Shaffer and their family have continued to
extend to my family and me will always be remembered with great affection.

The case studies of rural and urban traditional medical systems that
form the heart of this book reflect a truly exceptional team effort. My
greatest debt is to my Reseach Associate Violet N. Kimani, Lecturer in
Medical Sociology in the Department of Community Health at the University
of Nairobi Medical School. We worked long and closely together in Mathare
Valley and in Murang'a District. Violet's effectiveness in presenting the
purposes of our research to administrators, traditional medicine practitioners
(TMPs), biomedical workers, and the general public opened many doors and
contributed significantly to building the rapport and trust that was so essential for our success as participant-observers of traditional medical practices.
She patiently helped me to better understand the social and cultural foundations of many of the behavior patterns and symbols we observed. She pointed
out many nuances that I would not have perceived otherwise and gracefully
tolerated my idiosyncrasies. Violet exemplified the skillful listener and interviewer, provided leadership and a role model for our research assistants, and
was adept at communicating across the different ethnic groups in our surveys.
She thus made extensive contributions to the research that made this book
possible. We also collaborated as authors on two papers published in 1980 in
the *East African Medical Journal* and *Anthropos*.

Our six research assistants applied their time and many individual
talents with dedication, insight, and good humor. B. W. Jacob Gacani, M.D.
(a medical student in 1977) and I worked side by side during the first 3
months of the Mathare Valley survey. His intelligence, empathy, and personal warmth helped to win the respect of TMPs and other informants and
enabled the research to proceed steadily with minimal delays. (Mrs.) Mar-
garet Matharia, a qualified Enrolled Nurse-Midwife, brought exemplary
personal and professional attributes to the project. Her deep curiosity about
traditional medicine, and her persistent hardwork, dependability, and hu-
maneness in her relationships with TMPs and patients proved to be invaluable assets. She and her family welcomed me into their home with frequent
Kenyan hospitality, including late afternoon teas which provided a much-
needed time of refreshment and reflection after a day's work in Mathare.
Richard Muthee was a member of our team for several months before joining
the Kenya Army as a nurse. Muthee's outgoing personality, interest in the
research, and keen sensitivity to the concerns of other people made a lasting
imprint on our work and on me. Penina Makobu provided capable assistance
in both Mathare and Murang'a District. David Mwaniki and I worked
closely together in the Kilungu Hills of Machakos District for 6 months and
frequently camped overnight at the Chief's Headquarters on the ridge that
affords the magnificent overlook at Ithemboni. Mwaniki introduced me to
many aspects of Kamba culture. He repeatedly and cheerfully hiked with me

across the steep hills of Kalongo and Musalala to interview many of the dozens of *andu awe, akimi wa miti,* and *asikiya* who had attended or had been identified at the initial *baraza* at Kalongo Market called by the Sub-Location Chiefs to allow us to explain the purpose of our research in Kilungu. Large sections of the material discussed in Chapters 4 and 5 reflect Mwaniki's diligence and cooperative spirit. The study also benefited from the multiple talents and first-rate assistance of Jane Lawry-White of Cambridge, England. Jane had just completed her B.A. degree in geography and anthropology at the University of Durham when we learned of each other's existence. After joining the project she applied her research skills at the Kenya National Archives and as a field assistant in Mathare and Murang'a District. She discovered important archival materials that inform key sections of Chapter 3 and also coauthored a paper (Good, Kimani, and Lawry, 1980) based on our field study of the initiation of a Kikuyu medicine man (*gukunura mundu mugo.*

Numerous government officers and private individuals provided assistance at crucial times during the research. In Nairobi, Mr. Ombogo, the District Officer at Kasarani whose responsibilities included Mathare Valley and Eastleigh, personally introduced us to the key Kenya African National Union (KANU) representatives in Mathare I, II, and III and created opportunities for us to explain the purposes of our work in appropriate public settings. Dr. Charlotte Neumann, M.D. from the University of California at Los Angeles offered invaluable contacts among health workers and assisted with the selection of the study area.

In Machakos District we received good help from Mr. Benson Kaaria, the District Commissioner; Mr. Shikombo, D.O.1, at Kilome; Kilungu Location Chief Joel Mutyandia, at Ithemboni; the Sub-Location Chiefs of central Kilungu; and from the staff at Nunguni Health Sub-Centre. Sister Goretta (Teresia Biberauer), the Sister-in-Charge at Kikoko Mission Hospital, and Mr. Mutie, the Clinical Officer in the hospital's outpatient clinic, graciously allowed us to observe their activities and provided data from patient records.

More than 100 TMPs and hundreds of their patients agreed to cooperate with us in this study, often at very considerable personal inconvenience and expense. This book belongs to them and exists only because of them. I have chosen to provide anonymity for the TMPs who appear in the case studies. However, I could not possibly fail to mention here the real names of several TMPs who made special contributions of surpassing value. They include Kaswi Kitheka, Kavivi, Mbiti Nguli, Esther Kimolo, Justin Mwangi, Muthunga wa Nzovi, Munene Kavai, Noa Kimbuku, and William Kimoja.

I wish to thank several colleagues who kindly read and helped to improve various drafts of the book. As author, I of course accept full responsibility for errors of fact and interpretation that Kenyans and others may detect in these pages. Larry Grossman provided an extensive and thorough critique of eight chapters and Bon Richardson made helpful comments on the first four chapters. Bob Stock exceeded anyone's reasonable expectations by reading eight chapters overnight during the 1985 medical geography symposium at Nottingham! Janet Crane, editor with Guilford Press, raised important questions and gave constructive criticism throughout the preparation of the book. Rennie Childress, production editor at Guilford, handled the many details of getting the manuscript into book form with great skill and patience. I also appreciate Ken Smith's timely assistance with computer analyses of the survey data while I was still in Kenya. Students Terri Fromwiller, John Beebe, Frances Baker, and Allan Cox provided cartographic assistance.

Jane Vyula, ably assisted by Vanessa Scott, typed at least two revisions of each chapter, offered useful editorial suggestions, and handled the book's long birth with patience and understanding. My appreciation of their extraordinary investments of time and skill goes far beyond my words.

My family encouraged and sustained me both in the field and in the writing task and I am deeply grateful for their support and indulgence.

Charles M. Good

Preface

Ethnomedical Systems in Africa is a study of the evolving patterns of indigenous therapies in Africa south of the Sahara with particular reference to comparative rural and urban trends. It is based on case study materials gathered during fieldwork in Kenya. The book represents a venture by a medical geographer into a field in which the key contributions have so far come mainly from anthropologists, sociologists, psychiatrists, and others in the social and behavioral sciences. The result is a study that is strongly interdisciplinary in scope, set within the locational and ecological framework of medical geography.

Indigenous African beliefs and practices concerning sickness and healing have long fascinated visitors from outside the continent. Early European explorers and travelers, missionaries, and colonial administrators (Hobley, 1910/1967; Tauxier, 1927)—many of them amateur ethnographers or natural-history enthusiasts—often recorded and published their personal observations and impressions of traditional medicine. Essays on "the medicine man," religion and magic, spirit worlds, and witchcraft and sorcery were already familiar features of the Africana literature of the early 20th century. Serious scholarly accounts that demonstrated the inseparability of traditional medicine and religious and moral culture were written principally by European anthropologists and often based on extensive fieldwork. Evans-Pritchard's studies, culminating in his classic *Witchcraft, Oracles and Magic among the Azande* (1937) of the southwest Sudan, and Field's *Religion and Medicine of the Ga People* (1937) on the Gold Coast, are outstanding examples of anthropological works of the pre-World War II era. Harley's *Native African Medicine* (1941), a study of the Mano people of Liberia based on material collected between 1925 and 1933, and Gelfand's *Medicine and Magic of the Mashona* (1956) and *Witchdoctor: Traditional Medicine Man of Rhodesia* (1964) are representative studies by expatriate physicians who lived and worked in English-speaking Africa during the colonial era.

Colonial rule and the spread of Christian missions in Africa posed an unprecedented challenge to the integrity and character of indigenous religions and their associated systems of traditional healing. Several generations of Africans educated in colonial government and mission schools and colleges were taught that indigenous religions and traditional healers were savage and primitive (Ademuwagun *et al.,* 1979). In Kenya, for instance, widespread adoption of Christianity was probably delayed because the missionaries— especially Protestants—demanded rapid, radical changes in African social and cultural practices, including the elimination of traditional religious cere- monies in sacred groves and consultations with traditional healers (Munro, 1975). Aided by the missions, colonial authorities vigorously repressed and officially banned some organized religious therapy systems. In the former Belgian Congo, for example, the destruction of secret societies and the discrediting of public priests appears to have seriously undermined tradi- tional authority systems of the Kongo people. The consequences included social disorganization and decay, which contributed to the emergence of messianic movements, such as Kimbanguism, and to a perceived expansion of witchcraft (*kindoki*). The Kongo people turned increasingly to personal- ized forms of witchcraft and sorcery protection procured from the surviving traditional healers (*nganga*). However, these tactics proved unsuccessful for the Kongo people because they neither arrested the erosion of traditional chiefly authority nor rid their countryside of devastating epidemics of disease (Janzen, 1978).

Despite increased interest in Africa and the widespread availability of more accurate knowledge about the continent's cultures and history during the last 40 years, the character of traditional medicine and the medico- religious functions of traditional healers remain subjects of widespread igno- rance and misapprehension. Even though it is fashionable in some intellectual quarters to give them "credit," indigenous medical traditions are often cast in a negative light. They are still viewed by most Westerners and many educated Africans, at least publicly, as the incarnation of a shameful legacy of pagan- ism, barbarism, and black magic. The activities of traditional medical practi- tioners are often held up as "examples of irrationality and malpractice, a mixture of superstition, deliberate deception, and ignorance. . . ." (Maclean, 1971 p. 13). These perceptions are reinforced by the still-powerful and perva- sive syndrome of witchcraft and sorcery beliefs that continue to form the basis for explaining much illness and misfortune in contemporary Africa.

Okot p'Bitek, the late Ugandan writer and humanities professor, was strongly convinced that most of Africa's contemporary social ills, and the primary sources of these problems, are indigenous. Avoiding the fashion of many among the African educated elite to look for external causes, p'Bitek (1978 p. 7) was highly critical of what he called "not discrimination by bloody white settlers or colonialists against Africans, but discrimination of Africans against Africans." Prejudice against the traditional African medical expert by his African counterparts in the medical schools and hospitals is just one case in point. He is treated with spite and "called a 'witchdoctor' (a misleading term, implying that he is a witch and therefore, a bad person), and his skill dubbed fetish. . . . [They] refuse to learn anything from him, or allow him to enter their hospitals" (p'Bitek, 1973, p. 7).

Personal experience confirms that many African doctors and other members of the official health establishment frequently refer to the "secrecy" factor in traditional medical practices and use it as a rationale for avoiding contact with the work of local healers. Such secrecy does exist in varying degree and to some extent probably does discourage interaction between traditional and biomedical practitioners. Ademuwagun *et al.* (1979, p. vii) attempt to explain the logic that contributes to this communication problem.

> Most African healing systems have not been formalized in print so that their principles could be open to outside scrutiny. Part of the ethics of many African healing systems is secrecy; this protects the society against the indiscriminate use of such medicines by certain individuals. Such secrecy also reflects the fact that the knowledge of indigenous medicines can be an index of one's power and influence in society. Just as Western practitioners of medicine guard their professions through tedious methods of registration and induction, so does the African traditional medical class obtain the same protection through secrecy. Unfortunately the success of that secrecy has resulted in a serious blow to the credibility of the entire system. Many people, including many urbanized West-ern-educated Africans, deny the efficacy—not to mention the existence—of indigenous African medicine about which they have often heard but which they have little formal knowledge.

In contrast, my fieldwork demonstrated that traditional healers do share professional knowledge *when respect for their integrity and a genuine desire to learn from them are communicated.* Other studies also underscore the significance of these preconditions for gaining access to valid information about *bona fide* traditional medicine. As Maclean (1971, p. 83) found in her work with 100 Yoruba healers in Nigeria, "there proved to be no difficulty whatsoever in obtaining very detailed accounts of remedies for a number of specific complaints. . . . *Babalawo* and *onishegun* alike readily offered their own prescriptions. . . ."

Because of the enormous burden of ill-health, and the continuing—even increasing—dependence of Africa's people on traditionally based therapies, it makes little sense for traditional medicine to remain compromised and discredited because the official health establishment denies its worth or scientific intelligibility. This low status is inconsistent with evidence emerging from the expanding literature on theoretical and applied aspects of traditional medical systems.

Studies of traditional medical systems in tropical Africa over the past two decades reflect increasing conceptual sophistication and sensitivity to the gains in knowledge and understanding that are possible through interdisciplinary approaches in research. Unfortunately, although the geographic perspective and methods are directly applicable to the analysis and interpretation of traditional medicine, only a few geographers have shown interest in the topic. Consequently, most contemporary work continues to ignore the locational dynamics of traditional medical systems.

The expansion of traditional medicine is a significant trend in African urban areas. Traditionally based therapies are thriving in towns and cities because they fill a void for services that are in demand, they are accessible, and they offer culturally recognized responses to illness. As confirmed by the Nairobi case study in this book as well as studies in other African cities (Chavunduka, 1978; International Development Research Centre, 1980; Maclean, 1971; Swantz, 1974), instead of crumbling, traditional medicine displays a remarkable resilience and capacity to adapt to the needs of town dwellers. Traditional medical practitioners are among the largest categories of self-employed persons in the vast "informal" urban economy. Together with their patients they represent a significant dimension of the economic and social geography of the African city.

The themes of the present study draw the reader into the borderlands of the social and behavioral sciences, the humanities, and the medical sciences. I can think of few realms of inquiry that demand a greater measure of commitment to the interdisciplinary perspective. A brief examination of the book and its bibliography will reveal the immense intellectual debt I owe to the many scholars whose work has produced most of the important ideas on the topic and stimulated my interest in African traditional medicine and in medical pluralism.

Anthropologist John M. Janzen's seminal study, *The Quest for Therapy in Lower Zaire* (1978), remains the most important book on African therapy systems. His articulation of the role of the "therapy managing group" and his method of analyzing individual courses of therapy sequentially, "case by case, episode by episode" through the universe of the four therapeutic systems in Lower Zaire are major contributions to our general understanding of illness management and health-seeking behavior in African ethnomedical systems. Janzen's study of assistance patterns highlights the persistence of a broad network of kin who are available to provide emotional and material support in times of illness and misfortune. At issue is whether this kind of therapy managing group is typical in view of the pluralistic and rapidly changing circumstances in Africa today. The African regional pattern points to diversity in the size and composition of TMGs.

Although Janzen's analysis and perspective are essentially aspatial and restricted to a rural milieu, his approach points the way toward discovering the *geography* of the therapeutic process. The fieldwork required as a precondition of such geographical analysis involves the extraordinarily arduous task of tracing the formation and activities of therapy managers, or "significant others" who act as brokers between the sufferer and therapy specialists, and following the movement of ill persons among the various specialists and the places where therapeutic activities are performed. Anyone who has attempted to develop case studies of individual courses of therapy in an African setting will readily appreciate the enormous constraints to obtaining comprehensive and accurate accounts of illness history, therapy-seeking, and actual treatments. These constraints include the logistics of perpetual readiness; the time frame (some cases extend over months and even years); the spatial range of therapy-seeking; considerations of subjects' privacy; availability and cost of personnel, and of transport, regardless of seasonal conditions; and the need for both social science and biomedical expertise on the research team. It is essential that fieldwork be adapted to these realities.

Among the many other scholars whose ideas and observations have guided my own thinking I would especially like to mention the following individuals and their work.

Godfrey Chavunduka is a sociologist whose book *Traditional Healers and the Shona Patient* (1978) is a landmark, population-based study of the organization of traditional medical practice and the process of defining illness and choosing therapy in an urban setting in pre-Independence Zimbabwe.

Lloyd Swantz is a sociologist who spent 5 years in the field preparing to write *The Role of the Medicine Man Among the Zaramo of Dar-es-Salaam* (1974). His observations on traditional medical practitioners (*waganga*), social adaptation, and health care patterns in the Tanzanian capital directed my attention to the many parallels, and some contrasts, in Nairobi and other African cities.

Una Maclean, a specialist in social medicine, wrote *Magical Medicine: A Nigerian Case-Study* (1971). This pioneering book, the first I read on the subject and the first ever written about urban ethnomedical systems in Africa before 1978, is a perceptive account of Yoruba healers, medicines, and sickness behavior in the old wards and new districts of Ibadan.

Arthur Kleinman, a psychiatrist–anthropologist, has written extensively on the comparative, cross-cultural, and behavioral aspects of health care systems. His *Patients and Healers in the Context of Culture* (1980) taught me a great deal about the internal structures and behavioral aspects of health care systems. My own conceptualization of an ethnomedical system as a structure that includes "the full complement of health care strategies available to some or all members of a community" (Chapter 1) grew, in part, from Kleinman's work on the generic features of health care systems, including the overlapping popular, professional, and folk sectors. I also found his concept of "Explanatory Models" (EMs) of patients and practitioners—the ideas about an episode of sickness and its treatment that are differentiated and used by all participants in the clinical process—to have potential applications in at least two key aspects of African health. First, learning about and operationalizing EMs should enhance the therapeutic effectiveness of biomedical practitioners. Second, exchange of representative EMs between traditional medical practitioners and biomedical workers will increase the likelihood of success of any efforts to promote formal cooperation between specialists in these different medical-cultural systems.

Historian Steven Feierman is a perceptive observer of the social forces affecting health in Africa. His lucid analyses of "Change in African Therapeutic Systems" (1979) and *The Social Origins of Health and Healing in Africa* (1984) sharpened my understanding of how African healing—as an integral element of household and local community relations—and disease have been part of a "single interrelated web of changes" affecting contemporary African societies.

The fieldwork for this book was undertaken in Kenya from June 1977 to July 1978, and again during the summer of 1979. Although I anticipated spending most of my time with urban *waganga* who live and work in the shanties of Nairobi's teeming Mathare Valley, it was evident early in the study that the city and rural domains of traditional medicine are not discrete entities. They are linked together by kinship ties and by interregional complementarity that features local and regional movements of practitioners, clients, and *materia medica*; information networks; and flows of cash and material goods. These transactions are interconnected with other socioeconomic sectors, including the informal jobs market, housing, and government and private biomedical services. In view of this pattern of linkage it was essential to select at least one rural area for comparative study. A plan of reconnaissance-type rural fieldwork was adapted as the most realistic procedure given the available time and resources and the logistical constraints of the project. I also had to choose an area reasonably near my residence and workbase at the Medical School at Kenyatta National Hospital in Nairobi, yet remote enough to qualify as a functionally rural setting (i.e., agricultural and beyond workers' daily commuting range of Nairobi or other major towns) where "traditional" values are more observable. The need to minimize daily travel time and field days lost because of impassable roads during the rains were additional considerations. For comparison I also wanted the Kamba people as the primary ethnic group in the rural study because the majority of *waganga* in the Mathare Valley survey in Nairobi population were Kamba. According to the Kenya Population Census of 1979 the total Kamba population was 1,725,569. They are the country's fourth largest ethnic group after the Kikuyu, Luhya, and Luo peoples.

Among the Kamba one person is referred to as an Mkamba. The Kamba people speak *Kikamba*, a term which also designates a custom or practice.

The ethnicity factor narrowed the choice for the rural study area to either Machakos District or Kitui District. These two large administrative districts are situated to the east of Nairobi and are collectively known as *Ukambani*, the Kamba homeland. Although the home villages of most Kamba *waganga* in the Nairobi study area were located in Kitui, this district is too remote for convenient day trips from Nairobi and many parts of it are frequently cut off during the rains. Conversations with knowledgeable Kamba acquaintances and others in Nairobi led to the selection of Kilungu Location in the spectacular Kilungu Hills of southern Machakos District as an acceptable alternative to Kitui. Valuable material was also obtained during numerous visits over several months to the rural homes of Kikuyu traditional healers and the clinics of some of the numerous town-based Kikuyu *waganga* in Murang'a District, the Kikuyu heartland situated about 80 kilometers north of Nairobi.

Five tasks occupied most of my field time in Kilungu. First, a census of TMPs was taken with the assistance of the Sub-Location Chiefs of Kalongo, Musalala, Ndiani, and Wautu. They called a *baraza* (meeting) of TMPs at Kalongo Market where David Mwaniki, my Kamba research assistant, and I introduced and explained the purpose of our presence in the area. Virtually all of the 25 TMPs who attended the initial session agreed to be included in the census, and many stayed on that first day for several hours until they could be interviewed. We recorded names and villages of TMPs who did not attend the *baraza* with a view to contacting them at their homesteads at a more convenient time. Ultimately, about a fifth of the TMPs in the census proved difficult to find—due more to poor communication and timing than disinterest or fear on their part. We spent hours and days walking up and down steep paths and through gardens and bush in order to interview a particular TMP and map his or her residence, only to find on arrival that the individual had "gone to Mbooni to visit her sick sister," was cultivating a plot of land in a distant village, or had not returned from a trip to the shops.

Second, I plotted the distribution of TMPs on a 1:50,000 topographic map in order to determine and evaluate their pattern, density, and location relative to each other and to the population at large. Third, we selected several TMPs for in-depth interviews concerning their own background as practitioners and the general organization of traditional medicine in Kilungu. Each of these conversations was guided by a formal interview schedule and varied from 3 to 7 hours in length. Fourth, detailed observation of a small number of patients in consultation and therapy with TMPs was undertaken, mainly in practitioners' homes. Finally, we observed both staff and patients at a government health center and at the daily clinic of a mission hospital, and consulted their official records for data on utilization patterns.

Contents

ETHNOMEDICAL SYSTEMS IN AFRICA

1

Medical Pluralism: An Interdisciplinary Perspective

INTRODUCTION

Coping with health problems occupies a large part of most people's lives throughout Africa. Yet for millions of people on the continent, access to basic biomedical (Western-type) health care remains a remote, elusive prospect. This inadequacy is not confined to Africa, but it is more acute there than in any other region of the Third World. Although most countries in Africa routinely allocate substantial proportions of their budgets to health services and related infrastructure such as water supplies, sanitary works, and roads, sustained improvements in community health status and increased accessibility to government and private health services have not materialized. Instead, health ministries find themselves preoccupied just with preventing the deterioration of existing "aspirin and bandage" services. Demographic projections offer little scope for optimism that this trend can soon be reversed. In Kenya, for example, the population is expected to double to 35 million by A.D. 2000. Even today, despite the fact that free outpatient services are constitutionally guaranteed to every Kenyan citizen, it is estimated that only about 16% of the population live within 2 miles of a health center, and only 15% to 20% ever see a doctor or a nurse (United States Agency for International Development, 1980).

In Kenya, the location of the case studies described in this book, health planners had proposed bringing 60% of the population within access to primary health care by 1983 (Republic of Kenya, 1978). However, the resources for accomplishing this anytime in the foreseeable future with the conventional biomedical approach of additional hospitals, health centers, dispensaries, doctors, and clinical assistants do not exist. The recurrent costs of the existing health services are great; and Kenya's reliance on imported fuels and volatile international agricultural markets subjects it to recurring financial pressure. Also, changes in the political climate (as in August 1982) can severely interrupt Kenya's lucrative flow of international tourists and thereby have a significant impact on the country's financial and employment prospects. Many other African governments face similar financial problems with accompanying ripple effects on the provision of health services.

1

In contrast to the limited reach of biomedical facilities, the influence of traditional medical practitioners (TMPs) is extensive in Kenya and virtually every other African country. After self-treatment, they are the main providers of health care to the rural poor. In Swaziland, for example, Green and Makhubu (1984) report that there are over 5,000 TMPs and thus a TMP/ population ratio of 1:120. At least 85% of the Swazi population is said to utilize their services.

Traditional medical practitioners also have powerful influence in the lives of town dwellers. They coexist side-by-side with and complement the biomedical facilities that are concentrated in the urban areas. Traditional medical practitioners are recognized by their own communities, and often by people from other ethnic groups (particularly in urban areas), as individuals who are competent to provide one or more types of therapy through the use of such methods as divination and psychotherapy as well as by using plant, mineral, and animal substances.

Traditional medical practitioner is a general term that encompasses a broad variety of roles such as religious-medical specialists, herbalists, mid-wives, and combinations of these skills. Traditional medical practitioners are the main providers and conservers of African traditional medicine. In this study, *African traditional medicine* refers to "the total body of knowledge, techniques for the preparation and use of substances, measures and practices in use, whether explicable or not, that are based on . . . personal experience and observations handed down from generation to generation, either verbally or in writing, and are used for the diagnosis, prevention, or elimination of imbalances in physical, mental or social well-being" (Bannerman, Burton, & Wen-Chieh, 1983, p. 25). Ideally this is an all-embracing system of healing that is deeply embedded in indigenous religious and sociocultural institutions. It reflects values and practices, both local and foreign, which have been incorporated and adapted over the centuries. Traditional medicine has public and private faces. Often romanticized, and publicly assailed as anachronistic and dangerous, it is privately patronized by people from all walks of life.

At its best, traditional medicine represents the authoritative delivery of a culture's cumulative experience with the healing arts. It relies on the re-sources of the past; yet to remain vital to contemporary society it must remain open to the future and in dialog with the total culture.[1]

Many basic principles and methods of traditional healing have a wide distribution throughout sub-Saharan Africa. But there are also important differences based on personal, ethnic, and regional features that create unique place-to-place expressions of the art of traditional healing. A strong factor contributing to these differences is that traditional healers (particularly the religious-medical specialists) tend to be highly individualistic in their use of cultural symbols and supporting paraphernalia, in their style of diagnosis, in recommendations regarding therapy, and in their dosage of remedies.

Most of our understanding of current health issues in Africa comes primarily from studies undertaken by or on behalf of biomedicine. What is known about well-being and disease, institutions serving health care needs, and therapy-related behavior of individuals and communities is almost always defined in terms of an orthodox Western medical science outlook and is focused on Western-type health services and facilities. Despite social and behavioral scientists' growing interest in Africa's flourishing systems of traditionally based health care, broad-based accounts of such systems are rare. Reliable and applicable information on traditonal medical systems is rarely available to those responsible for formulating realistic policies and programs for the health sector in their own countries. Indeed, the role of medical pluralism in Africa is typically highlighted as a matter of public concern only when the actions of traditional healers and other "unofficial" practitioners are alleged to have a negative impact on the strategies of biomedicine, as for example, in cholera control campaigns in Kenya and Tanzania (*Daily Nation,* March 3, 1978; *Nairobi Times,* 1978).

The first part of this study aims to develop a broad frame of reference concerning medical pluralism, including its implications for the creation and provision of access to appropriate forms of health care. The regional focus is Africa, although illustrations are drawn from the United States and other countries. Against this general background I compare and contrast the patterns of contemporary traditional medicine in rural and urban areas of Kenya. I gathered materials for these case studies during 14 months of fieldwork conducted during 1977–1979 in the Kilungu Hills of Machakos District, in Murang'a District, and in the shanty settlements of Nariobi's Mathare Valley. In these places, and for most of the 20 million people living elsewhere in the countryside and towns of Kenya, the traditionally based therapy systems continue to be a very important (and frequently the only) line of defense against sickness after self-treatment with home remedies.

Information about Kenya's traditional medical systems is scarce and difficult to obtain. The problem is compounded by the legacy of negativism toward many indigenous institutions that was inherited from the colonial era; by the impress of Western science, technology, and Christianity on society; and by the corresponding and understandable ambivalence many Kenyans have today toward traditional medicine and its place in a developing society. Since Kenya achieved its independence from Britain in 1963 the political elite has strongly and openly fostered a highly competitive, capitalistic style of modernization. This has led to a continuing, enthusiastic embrace of many Western values and institutions. One of the most visible manifestations of this process and the values that accompany it is a costly hospital-based, doctor-dependent, and urban-biased system of curative health services that remains not only beyond the effective reach of Kenya's rural majority but is also inadequate for its townspeople.

In view of these circumstances it is crucial to understand the primary alternative resources and strategies people depend upon when they need assistance in health matters. To this end, the present study is intended to provide a baseline of information about traditional medicine and to interpret the findings with reference to Kenya's medical pluralism, socioeconomic development, and unmet needs for health care. It brings the analytical and integrative perspectives of medical geography to bear on the changing character, role, and development potential of traditional medicine, with particular reference to its present relationship and possible formal cooperation with Kenya's Western-inspired biomedicine. I shall argue that medical pluralism is fundamentally a positive quality in Africa's health picture today. It is an underdeveloped resource that can be creatively organized in the service of better community health throughout the continent (Good, 1987).

Quite intentionally, I have chosen to examine the findings of the Kenya case studies in terms of ideas developed by scholars in a variety of disciplines rather than creating a single, all-embracing theoretical framework. This eclecticism still permits the study to reflect the distinctive approach of medical geography, and thus to emphasize the locational patterns and interconnectedness of traditional medicine with other subsystems of society. A geographic focus is essential because the present spatial organization of traditional medicine reveals many of the changes affecting patterns of seeking and providing health care as well as the nature of therapy itself. It also simultaneously reflects changes impinging upon traditional medicine from the wider society, such as increased mobility of people, competition, weakening of extended family cohesiveness, commercialization, marginalization of people, and the creation of rural–urban support networks.

It is essential to understand the origins and cognitive basis of the different societal models of health and disease. But I am less concerned with recording the "tradition" in African traditional medicine and more interested in examining generally how the indigenous systems are evolving. The changes underway reflect, in place and over time, (1) the inexorable penetration of urban influences, (2) contact and competition with biomedicine, and (3) the spread and deepening of inequality in society. These three factors emerge again and again in relation to the five main organizing themes of the present study. They are (1) ideas arising from religion and, lately, science which are the sources of fundamentally different paradigms of health, illness, and disease; (2) the character and social geography of the places people visit and the structures and institutions they rely upon for the management and treatment of sickness; (3) the choices and behavior patterns which give form and meaning to a community's therapeutic systems; (4) the nature, distribution, and ecology of disease and illness treated by TMPs; and (5) the potential of traditional medicine and biomedicine to further develop their complementarity for the benefit of community health.

The study's frame of reference is necessarily wide because health beliefs and practices cannot be separated from the totality of human experience. It is sufficient to recall that health systems "have broad ranging ties with people's cosmology, their philosophical outlook, with their socio-economic structure and . . . their entire way of life" (Twumasi, 1981, p. 151).

MEDICAL GEOGRAPHY

Medical geography's emergence as a viable and strongly multidimensional subfield in Western professional geography has occurred largely during the last 35 years. Broadly defined, medical geography applies the concepts and methods of locational, ecological, and sociocultural analysis to health-related problems (Hunter, 1974). Medical geography has a strongly integrative, holistic, multidisciplinary approach embracing the cultural, social, economic, and biophysical elements of human environments (Hunter, 1973; Meade, 1986). This breadth of scope is reflected in the two main areas of activity that distinguish current research in the field (Mayer, 1982). The first area encompasses disease ecology and the spatial patterning of morbidity and mortality. This includes (1) the mapping,[2] identification, and explanation of patterns of disease occurrence and their covariation with sociocultural, political, and biophysical factors in space and time; (2) the process of disease diffusion within and between areas and populations; and (3) understanding health and disease with particular reference to the interactions of demographic, environmental, and behavioral variables in and among specific places and populations.

Medical geography is thus firmly grounded in geography's tradition of spatial analysis. In addition, its coemphasis on human–environment relations links it closely with the interdisciplinary field of human ecology (Meade, 1977), to which important theoretical and substantive contributions have been made by medical anthropologists (e.g., Alland, 1970; Foster, 1982a; McElroy & Townsend, 1979); medical sociologists (e.g., Fabrega, 1974), medical geographers (e.g., Fonaroff, 1968; Hughes & Hunter, 1970; Hunter, 1966; May, 1961), and other social scientists (Turshen, 1984). Significant parallel contributions that illustrate the integrative approach of human ecology have also come from geographers whose primary interests are not medical but rather the "cultural ecology" of Third World agricultural societies (e.g., Grossman, 1984; Knight, 1974; Neitschmann, 1973; Porter, 1979; Richardson, 1983).

The second major research tradition in medical geography focuses on the spatial arrangement and utilization of the principal elements of health-care delivery systems, and the characteristics of the populations involved. This branch of research has been strongly infused with concepts and methods derived from the subfields of economic and social geography, and regional planning. It involves analysis of the location, planning, and utilization of health services and of the features of health systems that influence their effectiveness (Joseph & Phillips, 1984).

To date, most studies undertaken in either the disease ecology/geographic epidemiology tradition or the health services subfield have unfortunately tended to exclude consideration of the other. For example, patterns of morbidity or mortality are usually not considered in terms of their relationships to political philosophy as this may be reflected in the spatial organization of a health care delivery system. Such conceptual blind spots are not confined to a particular discipline, however; much health-related social science research is geared to narrowly professional, biomedical definitions of appropriate topics.

In general, research in medical geography has been restricted to problems defined within a biomedical framework. Consequently, medical geographers in Africa and elsewhere have ignored the existence of traditionally based medical systems and have overlooked medical pluralism—a global phenomenon—at all scales. Fewer than two dozen studies by geographers are partly or wholly concerned with indigenous medicine. These include a first attempt to address some of the key issues in the field (Good, 1977), articles on village India (Bhardwaj, 1975, 1980; Bhardwaj & Paul, in press), on the Chinese in Malaysia (Meade, 1976), on Calabar in Nigeria (Gesler, 1979a & b), on rural and urban Kenya (Good, Kimani, & Lawry, 1980; Good, 1980; Good & Kimani, 1980), and Kloos's work on Ethiopia (1973, 1976, 1977, and 1978). Other studies examine the interface of traditional and Western-type medicine in Africa (Good, 1987; Good, Hunter, Katz, & Katz, 1979) and health care strategies, geographical and health implications of disease etiologies, and the diversity of healers in Hausaland, northern Nigeria (Stock, 1980, 1981, 1982a & b, 1983).

Fortunately, scholars in related fields have given much greater attention to the health care traditions that coexist with biomedicine. Contributors from medical anthropology, behavioral psychology, and sociology continue to advance interdisciplinary ideas about medical systems and health behavior and are pointing to methodological issues that are relevant to many concerns in medical geography (Yoder, 1982a & b). Thus, medical geographers can also contribute to the development of research on medical pluralism. Their approach is valuable because it emphasizes the analysis of processes and relationships within places (site characteristics) and among places (relative location), over time, which create locational patterns. Such locational attributes are often crucial for illuminating the pattern of disease and sickness and for understanding the characteristics, functions, and landscape features of health systems and health care behavior. In combining locational and ecological approaches medical geographers generally highlight variables and connections that are more often than not treated as background material or taken for granted by their colleagues in anthropology, sociology, psychology, and epidemiology.

PLURALISTIC CONFIGURATIONS OF HEALTH CARE: A GLOBAL PATTERN

Pluralistic medical configurations are in no sense limited to the technologically less-developed countries. Despite the technological brilliance of scientific biomedicine, the behavior of people seeking health care in Western societies suggests that the scope for mutual understanding and fruitful cooperation among a variety of contemporary alternative therapy systems is possibly greater than ever before. Both East and West Germany, for example, have evolved unique solutions for reconciling the criteria of "science" and "efficacy" in biomedicine with the continued demand from their populations for alternative drug therapy provided by homeopathy, anthroposocial healing, and herbal healing (Unschuld, 1980). In the United States today, orthodox biomedicine does not singularly fulfill either the demand or need for health care. The preeminent medical model, of course, remains strongly wedded to molecular biology, to increasingly sophisticated and costly equipment, and to physician-dominated definitions and control of standards and access to health care resources. Yet there is considerable evidence that the boundaries of "legitimate" medical practice (as defined by medical associations, health insurance companies, and other centers of influence in the medical industry) are expanding once again to include systems such as chiropracty and psychological counseling. Other evolutionary trends, including modifications and additions to medical school curricula such as community health and social science, are also underway. Today, many laypersons and some medical professionals are reexamining the biomedical model—particularly the therapeutic limitations of the narrow, mechanistic view of the human organism that has so infused medical practices in most of the United States.

Whereas biomedicine continues to be widely perceived as the only legitimate basis of therapy in the United States, the revival of awareness and increasing acceptance of a wide variety of "folk" and popular health-care strategies is decidedly "medical chic" (Atkinson, 1979). There is a genuine interest in alternative medicine that runs deeper than passing fads or fancies, and it is increasingly political in outlook. Many concepts and practices have long historical roots, and it is evident that large numbers of people believe them to be potentially efficacious and beneficial to health. Whether the desired outcome is achieved through pharmaco-organic, physical, or psychological therapy, or some combination of these, may be of little concern to the person or family seeking relief from sudden, inexplicable, or persistent illness. Among the numerous examples of these alternative or competing therapies are botanical healing and various forms of psychotherapy among rural and urban black (de Albuquerque, 1979; Varner & McCandless, 1979), American Indian, hispanic, and other ethnic communities (Kunitz, 1981); transcendental meditation, yoga, and other forms of mysticism and techniques of mastering mind and body processes derived from Asian religions; the physical fitness movement, Alcoholics Anonymous, vitamin therapy, and special dietary regimes with restorative and preventive goals; and even such "deviant" belief systems as astrology and palmistry. We have recently witnessed the emergence of special centers in the United States and Mexico where persons chronically ill with diseases such as cancer and arthritis are treated with controversial and sometimes illegal medications (i.e., not sanctioned by the Food and Drug Administration or the American Medical Association) such as dimethyl sulfoxide (DMSO), Laetril, and snake venom compounds. Homeopathy, "Rolfing," acupuncture, Christian Science, naturopathy, community paramedical training, and midwifery are also receiving attention from many quarters. Not the least important is the spiritual healing movement which recently gained currency in American Episcopal and Catholic churches (Vecsey, 1978).

At present there is very limited systematic knowledge of the alternative therapies that are in fashion in the United States (Hulke, 1979). Little is known about regional variations in acceptance and patronage; locational features and spatial patterns created by the circulation of patients, therapists, and materia medica; or the extent of their complementarity and competition with other therapy systems.

MEDICAL PLURALISM AND COOPERATION IN THIRD WORLD SOCIETIES

Coexistence of various traditional medical systems, such as Yoruba, Unani-Tibbi, Ayurvedic, or classical Chinese, with each other and with biomedicine is a well-documented phenomenon in many parts of Africa, Asia and Latin America. Together these different modes of interpreting and responding to sickness form what Charles Leslie describes (in Janzen, 1978, p. xiv) as "more or less pluralistic, more or less integrated, and more or less syncretistic regional systems."

International awareness of the role and contributions of traditional therapy systems in African societies is primarily a postcolonial development. This emerging consciousness coincides with a growing realization that the kind of biomedical delivery system introduced by colonial administrations and subsequently inherited by independent African governments does not and cannot meet many of their societies' most pressing health care requirements. Rooted in the old colonial system, the Western-inspired blueprint for medical care is inefficiently designed, dependent on expensively trained personnel working in budget-absorbing hospitals, and too often grossly overwhelmed by the unrealistic expectations and demands placed upon it. In many ways it is inappropriate and of limited relevance to the conditions of life of people in developing countries (Dorozynski, 1975). It renders insufficient care even in central urban areas, and practically no care in large rural areas. Nevertheless, even where biomedical health services are physically present, prevailing social and cultural values and actual attendance patterns indicate that biomedicine is not preferred for many illnesses, nor for common events such as childbirth.

Although they are undeniably complementary, the pluralistic medical systems of the technologically less-developed countries are inequitably distributed and poorly interconnected. As is well known, biomedical health services are typically so thinly scattered that they are physically unavailable and inaccessible to rural populations that lack efficient motorized transport. In the Ivory Coast, Lasker (1981) found that the choice of therapy is more dependent on relative accessibility than on any traits associated with the individual patient. Similarly, a recent study from northern Nigeria (Stock, 1980) reveals that as distance from a health facility increases, people delay seeking institutional treatment, fail to complete a prescribed course of treatment, and are more prone to resort to alternative forms of therapy. Measures of this distance–decay relationship show that visits for the treatment of common symptomatic illnesses decline at an approximate rate of 25% per kilometer away from a dispensary. Thus "the effective range of the dispensary is only about four kilometres; at this distance the utilization rate is only about one-third of the zero kilometre value" (Stock, 1980, p. 426).

As the preceding example suggests, a large proportion of Africa's people suffer a critical lack of even the "aspirin and bandages" variety of Western health services. This situation is expected to continue well into the 21st century, regardless of incremental expansion of health personnel with biomedical training and the construction of additional physical facilities. The overall shortfall in biomedical health care is exacerbated by a shrinking supply of traditional medical practitioners in some rural areas; to some extent by a dehumanizing commercialization of both traditional and Western-type therapeutic services; by radical political, social, and economic upheavals of the sort Ugandans, Ethiopians, and Ghanaians have recently endured; and, as in Kenya, by unprecedented rates of population increase and diminishing budgetary resources for the health sector (Republic of Kenya, 1986, p. 30).

Until quite recently, there has been a widespread tendency in Western societies to subscribe to one of two fallacies about health beliefs and behavior indigenous to other societies. Steven Polgar (1962) labeled one of these the "empty vessel fallacy," whereby a territory and people are assumed to lack systematic health beliefs and a therapy system until Western medicine arrives on the scene and fills the void. Alternatively, the "fallacy of the separate capsule" recognizes the existence of indigenous medical resources and theories of disease, but maintains that they are inferior and not a viable alternative to Western professional health practices (Scrimshaw, 1979). Barbara Pillsbury, a medical anthropologist, offers an eloquent response to those who hold these stereotypes and suggests important theoretical and practical reasons for redefining and expanding the study of "medical systems" in the field. She observes that:

> Nowhere do people live in a health care vacuum. In all cultures people become ill; in no cultures do others stand idly by. Each culture has produced over the centuries its own adaptive methodologies for coping with illness. These embody an indigenous etiology, that is, a system explaining the occurrence of illness and disease based on the worldview and religious beliefs of the particular people in question. It is the underlying explanatory system, in interaction with features of the ecological niche of the population, that dictates local strategies for coping with ill-health. These are strategies that international medicine has all too long ignored (Pillsbury, 1978, p. 1).

"'Dead as the dodo.' That is what many health professionals and authorities would wish traditional medicine to be and what an even greater number of them believes it to be" (Vuori, 1982, p. 129). Yet it is evident today that traditional medical systems remain one of the major resources used by Third World populations to cope with their problems of poor health. Certainly in Africa, and much of Asia and Latin America, there is a huge number and striking variety of traditionally based healers including herbalists, diviners, midwives, fertility specialists, and spiritualists. According to the World Health Organization (W.H.O.) and other authorities they form the main source of assistance with health problems for at least 80 percent of rural inhabitants (Bannerman, Burton, & Wan-Chieh, 1983). However, it should be kept in mind that neither TMPs nor their various specialties are evenly distributed across the landscape. Traditional midwives deliver 60% to 90% of all children born in these regions. In addition, traditional healing continues to retain and expand its influence in towns and cities, side-by-side with biomedical services.[3] Assessing the current pattern of medical pluralism in India, for example, Dunn (1976, p. 155) states emphatically that "whatever the development of cosmopolitan health care" in that country, "there can be little doubt that popular traditional medicine will indefinitely survive." Recent studies by the geographer Surinder Bhardwaj clearly illustrate the complementary roles and competitive aspects of multiple therapy systems in South Asia (Bhardwaj, 1975, 1980; Bhardwaj & Paul, in press).

Ahmed and Kolker (1979) identify seven general functions, both psychological and sociological, that traditional indigenous medical systems perform independently of the local presence or adequacy of biomedicine. These are (1) relief of stress and anxiety created by the uncertainties of illness; (2) low cost (a questionable assertion addressed later) and convenience; (3) reinforcement of the patient through primary group involvement in diagnosis and treatment; (4) control of deviance; (5) minimization of the trauma of cultural change; (6) alleviation of personal stress in urban areas caused by social dislocation, uprooting, and anomie; and (7) promotion of ethnic identity. Nchinda (1976) has argued that traditional African medical systems survive because they satisfy the following four basic users' requirements (each of which has an important locational dimension): accessibility, availability, acceptability, and dependability.

At the urging of the W.H.O., biomedical opinion has recently begun to shift toward a more open stance regarding the validity and potential efficacy of elements of the traditionally-based medical systems (Akerele, 1984). Although heated controversy continues and opposition by biomedical practitioners is widespread, there is increasing interest in the potential benefits to be obtained from stimulating cooperative therapeutic activities between biomedically oriented and traditional-style healers (Neumann & Lauro, 1982). In Africa, given "the inescapable reality that Western medicine will not be available to the masses for years to come" (Miller, 1980, p. 98), it is imperative to ask: Can formal cooperation among practitioners of biomedicine and traditional medicine promote health care that is culturally more appropriate, more accessible, and more effective than is currently available without such collaboration?

In the mid-1970s the W.H.O. began to develop policy guidelines to assist Third World health ministries to conduct research, initiate pilot programs, and subsequently to implement cooperative training programs and treatment strategies. The World Health Organization's latest contribution is the publication of a handbook for health administrators and practitioners edited by R. H. Bannerman, J. Burton, and C. Wen-Chieh and entitled *Traditional Medicine and Health Care Coverage* (1983).

A synopsis of the strengths and weaknesses of African TMPs and traditional medicine is presented in Table 1-1. These will be reexamined in the final chapter in the light of evidence from the Kenya case studies. It will be noted that some of the "weaknesses" of traditional medicine are also applicable in varying degree to biomedicine.

While the concept of cooperation is attractive in the abstract to many, and numerous limited experiments have occurred—most often with traditional midwives (Pillsbury, 1982; World Health Organization, 1982)—few countries have actually taken the positive steps necessary to innovate and rationalize their health delivery systems within the framework of their own medical pluralism. Careful examination of the general health care situation in Kenya and elsewhere in Africa suggests that a formal policy decision to pursue specific areas of collaboration with traditional medicine could yield positive health benefits to both rural and urban populations (Good, 1986).

Despite their individual and collective strengths, probably no single

Table 1-1. Strengths and Weaknesses of African Traditional Medicine and TMPs

---

Strengths
1. TMPs perform positive psychological and sociological functions (e.g., stress reduction, social control) regardless of the local adequacy of biomedical services.
2. TMPs have acceptability through cultural continuity.
3. TMPs are generally adaptive and open to innovation.
4. TMPs collectively possess a large body of indigenous technical knowledge, ranging from individual and group psychology to the properties, actions, and applications of plant medicines.
5. TMPs treat a broad spectrum of physiological and mental ill-health, including conditions which are not diagnosed, misdiagnosed, and incompletely diagnosed by biomedical practitioners—or are not manageable according to biomedical standards.
6. Many TMPs recognize and will refer serious diseases and conditions such as malaria, tuberculosis, and trauma to biomedical facilities.
7. TMPs are available in all rural and urban communities.

Weaknesses
1. Knowledge of bodily diseases dependent on perceptible signs of dysfunction that are difficult to verify.
2. Limited repertoire of diagnostic techniques.
3. Much variation in length and quality of training of TMPs.
4. Procedures with patients are highly idiosyncratic and quality of care rendered varies greatly due, in part, to lack of standardized training.
5. Standards of hygiene and sanitation are low.
6. Conceptualization of illness etiologies in terms of witchcraft and sorcery is widespread among certain types of TMP. Despite possible value in social control, these "fear systems" are potent sources of mental stress and physiological insult.
7. Few mechanisms exist for systematically evaluating the outcome of therapy.
8. Recruitment of *bona fide* TMPs in rural areas is increasingly problematic.
9. TMPs lack cohesiveness, and exchange of technical information and practical experience is rare.
10. TMPs are ineffective at correcting negative stereotypes and false information about their roles, e.g., the "witchdoctor" caricature of "skins, beads, and backwardness"; traditional medicine as a game of "con artists" pursuing self-enrichment schemes; and the "unyielding secrecy" associated with the methods and medicines of TMP's.
11. Inarticulate/weak response to the paradigm conflict with biomedicine.

Note: Data taken from Good (1987).

therapy system, or combination of systems, is today adequately meeting even the most basic health care needs in the Third World. Separately or together, all fall short (Good *et al.*, 1979). In their quest for health, people attempt to compensate for these deficiencies by utilizing both traditional and biomedical services. Such joint use is made concurrently or sequentially for the same or different aspects of the same illness. In this sense the pluralistic medical systems are informally but rather haphazardly integrated. In the great majority of cases, however, the traditionally based therapies and practitioners continue to be ignored by their country's ministry of health and thus by its Western-trained medical and health planning establishment.

COMPETING PARADIGMS:
BIOMEDICAL AND PSYCHOSOCIAL

Any attempt to encourage formal cooperation between traditional and Western medicine and to promote their appropriate integration will encounter basic conflicts. Wherever scientific, profession-centered biomedicine has penetrated there is a confrontation between fundamentally different paradigms of health and illness. While I recognize the pitfalls of oversimplification, it may be said that the *biomedical* paradigm continues to dominate thought and behavior in Western or biomedicine. Disease and illness are reduced to essentially Newtonian, mechanical, or organismic states—as physical manifestations that can be successfully diagnosed and treated separately from a person's psychological condition and social milieu (Kleinman, 1978b). Consequently, throughout the Third World people usually assign biomedicine the limited function of treating organic disease or acute illness symptoms.

Rappoport (1980) observes that the biomedical treatment model, among other features, gives precedence to "technique" over "person": to the replication of scientific procedures and to "objectivity." Its value structure tends to ignore if not disvalue the therapist as an individual whose unique charismatic qualities and style, and capacity to communicate and generate hope, are crucial to healing.[4] In effect, the biomedical model also minimizes the length of time considered "necessary" for routine diagnosis and patient consultation. In practice, despite truly noteworthy exceptions, little premium attaches to really knowing a patient in terms of his or her social network and identity in a community, or the physical and psychological conditions of the patient's life space. Broad-ranging counseling of the sort that is typical, for instance, of *bona fide* traditional African therapy procedures, would be an extraordinary occurrence. Whereas in traditional medical practice time itself is virtually an organic part of therapy, in biomedicine time is measured and compressed by organizational and economic pressures. Biomedical clinical procedure is thus well adapted to processing a heavy patient load, a pressure that is particularly burdensome on medical assistants, nurses, and doctors. In the overcrowded and underequipped outpatient clinics of African hospitals and health centers there is usually little dialogue between clinician and patient, and the mean doctor–patient contact time is commonly less than 2 min (Family Health Institute, 1978, p. 90; Lasker, 1981, p. 161).[5] The biomedical model is thus inadequate to the clinical realities in such circumstances. This is particularly apparent if, as Dr. T. A. Lambo, Deputy Director General of the WHO asserts, 70% to 80% of patients "are not suffering from any discernible organic disorder" (Singer, 1977, p. 246).[6] Most importantly, few doctors and medical assistants have the theoretical and practical training in cultural analysis and psychology that is crucial to providing an appropriate diagnosis for most sick people they encounter. My own field observations suggest that the biomedical system of health care misses a broad spectrum of disease and illness. A significant proportion of the conditions presented in African hospitals, clinics and dispensaries are not recognized or are misdiagnosed and consequently not treated appropriately (Gatere, 1980).

In contrast, the conceptual frameworks of traditional medical systems that evolved outside the Euro-American realm emphasize psychological, social, spiritual, humoral, pharmaco-organic, and other biophysical phenomena and processes in varying degrees and combinations. "Person" and place of therapy both are considered crucial elements in successful treatment of illness. In most African societies situated south of the Sahara, for instance, what I term the dominant *psychosocial* paradigm illustrates the clearest and most pronounced area of conceptual and cognitive divergence from the biomedical paradigm. Its core features illustrate why there are significant constraints to achieving formal cooperation between traditional and biomedical systems.

In the *psychosocial* paradigm, a person's body, mind, and soul are conceived as an indivisible whole, and at any time the condition of one is mirrored in the others (Swift & Asuni, 1975). The etiology and symptoms of disease are not interpreted as isolated or simply probabilistic occurrences as so often happens in the biomedical approach; instead, their source lies beyond the ill person's physical being and reflect discord in his or her "social body" (Swantz, 1979) and a "rupture of life's harmony" (World Health Organization Regional Committee for Africa, 1976). Illnesses tend to fall into two general categories: *natural* or *God-given illnesses*, which are unrelated to human will or actions; and *illnesses of man*, caused by bad-will or hostile conduct of one person or group against another, the commission of various indiscretions, or the activation of supernatural forces (Janzen, 1978). Illnesses of humans are manifested in the phenomena of witchcraft and sorcery, magic, ghosts, and disturbances of the ancestral spirits (Mbiti, 1970), and as consequences of broken taboos (cursing a child and intergenerational sexual relations are examples) and unmet religious obligations. In popular culture the root causes of much illness and misfortune are directly attributed to conflicts and tensions in interpersonal relationships. These feature both *horizontal* (kinfolk, neighbors, co-workers) and *vertical* dimensions (relations with the "living-dead" ancestral kin) (Mbiti, 1970).[7]

Illness as it is conceived in the psychosocial paradigm thus has important moral implications. As a theory of causation, it integrates the spiritual with the physical universe. Quite logically, the appropriate therapy for "illnesses of humans" is rarely limited to brief bilateral consultation between a specialist and the sufferer. Instead, treatment frequently extends over several days or longer, may require meetings in more than one physical location (e.g., an urban migrant's rural homestead may be in need of special ritual purification or protective measures), and involve a supporting cast of kinfolk and other associates of the patient who act as a therapy managing group (Janzen, 1978). In this kind of system, illness and misfortune are essentially religious experiences, demanding a religious approach and a qualified specialist whose therapeutic techniques reflect the patient's cultural beliefs. The traditional healing process in Africa often evolves into a kind of psychosocial drama, with the goal of restoring the sufferer to a therapeutic community (Onoge, 1975). In coastal Tanzania, for example, the Swahili greeting *u mzima*, "are you whole, well?", captures the meaning of health interpreted as wholeness in the context of a community, a group of people related by bonds of kinship, neighborhood or common work, and interdependent with one another (Swantz, 1979).[8]

In northern Nigeria, the common Hausa proverb *lafiyar jiki arzikine* ("health is wealth") parallels the ideal definition of health espoused by the World Health Organization. *Lafiyar* symbolizes much more than the absence of disease. It connotes "a general state of physical and social well-being which incorporates such characteristics as good relations with family and social contacts, a settled religious and moral state, freedom from danger and fear, success at work and in personal affairs, and the absence of sickness" (Stock, 1980, p. 73).

PARALLEL OR JOINT USE OF DIFFERENT THERAPY SYSTEMS

As Leslie (1980, p. 193) points out, it is "the experience of illness, not the biological reality of disease, [that] causes people to consult others about their health." Despite the fact that many individuals hold strong ambivalent feelings toward it, biomedicine does hold a respected position in the Third World because of its greater effectiveness—relative to traditional medicine—in identifying and treating many clinical symptoms of disease, particularly acute ailments. Nevertheless, even if fully staffed and equipped biomedical facilities were placed within easy access of everyone, people would still make use of traditional healers and others available to them who are skilled at treating *illness* if not a particular disease (Good *et al.*, 1979; Leslie, 1980).

The cognitive linkage of people's beliefs about the etiology of disease and illness with particular and often sequential therapeutic options is a factor that should command attention from health professionals. Parallel use of biomedical and traditional medical specialists, as well as other sources of therapy, is often pronounced where large segments of the society conceptually separate the treatment of overt symptoms of illnesses from what are perceived as their underlying root causes. In Nairobi, for instance, a mother whose child experiences a sudden loss of weight and alertness may take him to the pediatric filter clinic at Kenyatta National Hospital to be seen by a doctor; and a man injured in an industrial accident will recuperate as an inpatient at the hospital. Assuming that both individuals are soon "cured" in a biomedical sense, neither the adults directly affected nor their families may consider the respective cases resolved until there is a satisfactory explanation as to *why* they were singled out for distress and injury. At this stage a diviner or other medicoreligious specialist is frequently consulted. In other cases, medical doctors are unable to identify or relate to the patient's complaint and send him or her away; or treatment is unsuccessful despite several return visits. Any of these three kinds of everyday experiences with the biomedical services serves to increase the flow of ill persons toward traditional healers and other kinds of therapists who are reputed to have skills in treating illness in its broader social context (Good, 1980b, 1987).

The necessity for cultural mechanisms to deal with the healing of illness is a universal characteristic of health care systems. There is no evidence that such devices are a vestigial stage to be "passed through" in the succession to a modern biomedical system. Indeed, as Kleinman (1978a, pp. 88–89) proposes, "we can look upon the legitimated role of social workers, psychiatrists, pastoral counselors, and patient advocates, as well as folk healers, in fully modern societies such as the United States as using a language of experience and treatment for illness which would otherwise go untalked about and untreated when sickness extends beyond the context of the family into the professional biomedical domain."

In view of the fundamentally different paradigms and behavioral correlates that distinguish traditional psychosocial and biomedical therapy systems, effective formal collaboration that actually enhances the health of individuals and communities cannot be expected to evolve on its own. Rather, cooperation will depend initially on the ethnomedical approach being cultivated, adopted, and infused into professional biomedical praxis—including a reformed academic medical curriculum. Because of shortages of both biomedical personnel and the health resources they require, bold and progressive initiatives are necessary to bring about a better allocation and utilization of those resources already available in the biomedical sector.

On the other hand, traditionally based practitioners are ubiquitous; those among them who are genuinely concerned about their patients' welfare have an unrealized potential to contribute more effectively to individual and community well-being. Nevertheless, serious reservations about the likelihood of a new role for traditional medicine persist. The orthodox biomedical system is, everywhere, intertwined and symbiotic with the established political and socioeconomic systems of a country. It is thus oriented toward serving the affluent, powerful, and mostly urban classes that are able to appropriate public resources. This political context of the health system will ultimately determine whether the "upgrading" and utilization of traditional healers in cooperative health care delivery will be of real value or if it will simply become another means of perpetuating so-called "second-rate" medical care to the already underserved masses. Might the price of legitimacy and other support for the traditional sector mean diminished service to the general population? The questions must be asked—who will benefit, and who will gain (McDonald, 1981)?

The health care systems of Kenya and other Third World countries are not static. Everywhere, improved access to appropriate health care depends less on huge injections of capital and more on innovative ideas for decentralizing and optimizing the use of health resources already at hand. Adoption of intensive experimentation with the concept and practice of ethnomedical analysis offers a realistic, feasible, and low-overhead approach to the formulation of rational policies for health service development in the 1980s and beyond. It holds out the prospect of linking multiple, spatially overlapping and interacting but poorly articulated therapy systems; and of facilitating better health care for the general population.

THE ETHNOMEDICAL MODEL: ITS RELEVANCE TO MEDICAL PRACTICES AND HEALTH

Despite scattered experiments with traditional birth attendants and traditional healers, such as in the Techiman District of Ghana (Warren, Bova, Tregoning, & Kliewer, 1982), formally structured collaboration among various types of practitioners in pluralistic medical environments of Africa is not yet a realistic expectation of national health services. Also, the cooperative ethic is unlikely to spread if Western medical science and technology continue to be perceived as omnipotent systems with benign ideological biases. Prospects for effective experimental programs of cooperation between TMPs and biomedicine are also unlikely if key decision-makers are reluctant to promote or to motivate real community participation in local health matters. Bureaucratic "turf" considerations and/or personal career and family considerations on the part of persons responsible for implementing health programs at district or subdistrict levels constitute a third stumbling block to cooperation. Finally, the changing political priorities of external funding agencies (e.g., U.S.A.I.D.) can also interfere with schemes to utilize TMPs (Pillsbury, 1982).

Leslie (1980) observes that even though biomedicine is almost always less scientific than it appears, health professionals throughout the world tend to translate proposals for cooperation with practitioners of alternative therapies as advocating the legitimization of quackery (Imperato, 1977). They contend such collaboration would require acceptance of procedures whose efficacy is at best difficult to evaluate according to conventional scientific criteria. These external, superficial interpretations of unfamiliar therapies (which may lead to conclusions of "quackery") and assumptions about their efficacy tend to screen out significant behavioral and semantic issues that are crucial to better professional communication and health care. At the same time, the difficulties of differentiating between genuine healers and those who have few skills apart from extracting their client's money must not be underestimated (International Development Research Centre, 1980).

In all societies, medical systems function as social systems that attach meaning to and structure the experience of sickness (Fabrega, 1977; Leslie,

1980). Medical anthropologists and most other social and behavioral scientists now recognize the analytical and practical value of distinguishing between disease and illness as two elements of *sickness*. *Disease* connotes the impairment of biological and/or psychological processes, whereas *illness* can be described as "the psychosocial experience and meaning of perceived disease" (Kleinman, 1980, p. 72).

Because most biomedical practitioners restrict the focus of their work to disease, the distinction between disease and illness is for them rarely more than a secondary consideration. Consequently, neither the patient's interpretations of particular sickness episodes nor the rationale and qualities of therapies that are available from other healing systems are routinely taken into account in clinical activities. This observation is especially applicable to the practice of biomedicine in Africa and other Third World regions where many fundamental and subtle cultural differences commonly separate patients and practitioners.

The work of Kleinman (1980), Leslie (1980), and others suggests that the purposes of biomedicine, insofar as these are directed toward individual and community health can be advanced by integrating the principles of the "ethnomedical" model into biomedical practice. Building upon recent developments in semantic anthropology, ethnomethodology, and comparative medical studies, the *ethnomedical approach* places particular emphasis on the "meaning contexts of sickness," and joins different systems of medical knowledge. Health, disease, and illness are interpreted in the light of the many ways these terms are perceived by patients, families, practitioners, and communities. In practice, the "ethnomedical analyst," whether medical practitioner or a social scientist, recognizes that people adopt sick roles when they do not have diseases, and experience diseases without being ill or taking on sick roles (Leslie, 1980). As already noted, the biological and psychosocial realities, while interdependent, are not the same thing; and their relationship is culturally defined and structured.

In the ethnomedical approach, episodes of sickness are interpreted in terms of Explanatory Models (EMs) that are used by patients and practitioners in all health care systems. Explanatory models are formed from a variable cluster of cultural symbols, experiences, and expectations associated with a particular category of sickness; and all EMs reveal sickness labeling and a cultural idiom for expressing the experiences of illness. They provide interpretations of sickness "to guide choices among available therapies and therapists and to cast personal and social meaning on the experience of sickness" (Kleinman, 1980, p. 105). *The nature of the exchange between patients' EMs and practitioners' EMs is therefore crucial* because it influences the entire procedure and outcome of health care.

Coping with sickness is a process organized for an ill person and his or her family within culturally composed networks of routine everyday meanings, symptom perception, and interpersonal relationships—what Kleinman (1980) calls the "semantic sickness networks." In patient and family EMs, these networks link a variety of experiences and beliefs about symptoms, causality, and significance. Quite often specific types of interpersonal problems and social tensions that are believed to trigger or accompany particular illnesses also form an important element of a network. These semantic sickness networks thus provide the basis for decisions regarding the selection and phasing of treatment options. Consequently, from the ethnomedical point of view, "it is the EM and the semantic sickness network it constitutes and expresses for a given sickness episode that socially produce the *natural history of illness* and assure that it, unlike the *natural history of disease*, will differ for different health care systems" (Kleinman, 1980, p. 107; emphasis in the original).

A number of culture-specific syndromes, identified in Western, non-Western, and transcultural ethnographic and medical anthropology studies, provide useful illustrations of "semantic sickness networks" (e.g., Hand, 1976). These include *susto* among selected Latin American groups; pathologies linked to an excess of "hot" or "cold" somatic qualities (such syndromes are nonsensical in the framework of a biomedical clinician's EM, who might in turn recommend a course of treatment the patient considers incompatible and even dangerous) (Kleinman, 1980); "rootwork" among southern Black Americans; "heart distress" in Iran; and *ching* ("fright") in Taiwan or Hong Kong. Concepts of infection (e.g., "evil eye" in North Africa and the Middle East) and specific "folk" therapy prescriptions (e.g., "feed a cold, starve a fever" in a middle-class English or American suburb) produce distinct lay patterns of choice among alternative therapies that may be utilized according to hierarchies of resort.

Contemporary American popular culture also illustrates the relevance of the ethnomedical approach. It contains a rich but largely unsurveyed variety of culturally constructed patterns of symptom perception, labeling, communication, and behavioral responses. For example, terms such as "nervous breakdown," "male menopause," "professional burnout," "shell-shock," "rape," "lump in the breast," "upset stomach, " and even "cancer" have profound personal, cultural, and social meanings that are crucial to establishing a basis for the provision of psychologically and culturally appropriate clinical care. They also reflect changing beliefs about the cause and severity of illness; yet there are few systematic accounts of popular American or other EMs and "networks" in the literature of social science and medicine.

Most functional social science studies of medical systems reflect definitions of health and sickness legitimated by the biomedical model and are concerned with these states primarily as they relate to professional and bureaucratic health care interests. Examples include defining disease and biological malfunction, an emphasis on technological innovation, and compliance with curative and preventive programs.

In contrast, ethnomedical analysis employs an interdisciplinary and interpretive methodology. It rejects the truncated, compartmentalized, and pre-selective approach to sickness in favor of examining a broad range of culture-specific and transcultural experiences in both clinical transactions and in the extraclinical domain of healing. Analysis is focused on the actual illness experiences of people and examines how they are perceived, labeled, communicated, and managed in interactions with family, social network, and therapists. In terms of this framework the social *geography* of therapy systems assumes much greater complexity, dynamism, and spatial range through the redefinition of "relevant" actors and through assessment of their locational behavior. For these reasons the ethnomedical model elaborated by Kleinman and others is an essential tool for nurturing successful collaboration between health-care traditions.

Extending well beyond the horizon of conventional biomedical practice, the ethnomedical approach offers several distinct practical advantages to health professionals, social scientists, health planners and, not least, the lay public.

First, analysis of recurring semantic sickness networks aids in understanding the logic of lay health resources and their utilization patterns. Such information can help detect key differences and areas of conflict between the various local folk, popular, and biomedical systems of health care. Health planners in Africa, the United States, and elsewhere will be better informed about the total environment of health-related behavior. They will also be able to mediate more realistically and effectively between the different systems when planning new programs or redesigning the configurations of existing resources for health.

Second, knowledge of semantic illness networks is directly applicable to the goals of disease prevention and health maintenance. Byron Good's (1977) analysis of a local Iranian explanatory model, for example, revealed information of great value to health authorities who were proposing to expand the use of oral contraceptives. Noncompliance with this program by village women was due chiefly to fear of "heart distress," which they perceived as a side effect of the pill. In Kenya, attempts by medical authorities to treat childhood malnutrition among the Digo people of coastal Kenya have often been unsuccessful because the ultimate cause is locally understood and treated in social terms. Rickets, marasmus, kwashiorkor, and loss of strength (*nguvu*) in children are given a signal label, *chirywa*, which the Digo attribute to parents' sexual misbehavior (Daily Nation, January 27, 1978; Gerlach, 1959).

Third, semantic network analysis of health-related concepts and transactions in any society can aid in creating a base of practical knowledge about typical sources of stress and the transition periods in the life of the individual which are popularly associated with changes in health status. It can also help identify diseases and illnesses that are perceived to be the most common and threatening (Pfifferling, 1975). Evaluation of usual family and social network patterns of coping with illness and maintaining health should help uncover which coping strategies are adaptive or maladaptive (Kleinman, 1978a).

Ethnomedical knowledge of this kind is probably at least as important to local health authorities interested in developing appropriate preventive and maintenance programs as is orthodox epidemological data on the prevalance and causes of various diseases. Indeed, Kleinman (1978b, p. 79), citing Kunstadter (1975), urges that "both sources of information are needed to specify epidemiological webs of pathogens, vectors, host susceptibility, and behavioral and environmental contingencies promoting certain sicknesses and resisting specific public health interventions."

There are at least three other factors that should finally convince social scientists (and biomedical professionals) of the usefulness of the ethnomedical approach. First, its applications are not restricted to traditional and "non-Western" medical systems. It is also appropriate to the analysis of both lay and professional health behavior in societies where biomedicine is dominant, co-dominant, or only a variant (as in Kenya).[9] Second, it serves to demystify and make more interpretable the otherwise exotic and perplexing aura (especially for Westerners) that typically surrounds the practices of both traditional and popular healers and their clientele (Rappoport, 1980). Finally, it should directly lead to more psychosocially *and* biomedically sophisticated models of health problems and therapeutic alternatives, and to more effective utilization of health services (Kleinman, 1978a).

DEFINITION AND CLASSIFICATION OF FOLK MEDICINE

Western biomedicine arrived in Kenya about 80 years ago. Today its appeal is widespread, but because of its restricted availability it is a practical alternative for only a small town-dwelling and periurban minority since nine out of ten Kenyans live in rural areas. As in many other technologically less-developed countries, biomedicine's bureaucratic structure, professionalization, and extensive resource requirements clearly inhibit its emergence as the dominant medical system accessible to the masses. Indeed, even where it is readily accessible, biomedicine often is not perceived by Kenyans as a comprehensive or final solution to health problems. Its diagnostic and therapeutic successes are limited because in practice its concepts, outlook, and institutional features lack the necessary "goodness of fit" with those of the local culture. Partly for this reason, a rich variety of traditionally based modes of therapy remain in use in Kenya and in virtually every other part of Africa (Lasker, 1981; Mullings, 1984). They are collectively and locally the dominant medical systems throughout Kenya today, cutting across both ethnic and class differences. Formalized, functional relationships between the local systems and biomedicine do not exist. Nevertheless, concurrent or serial utilization of traditional medicine and clinical medicine as treatment strategies is often more the rule than the exception.

Kenya's traditional and biomedical systems of healing are all undergoing change and adaptation from within and without in response to interaction with each other, and to the more general process of capitalistic modernization. One of the difficulties in analyzing and comparing these different therapy systems in Kenya, as in the rest of the world, stems from the imprecise terminology that is used to describe them in the literature and in general speech. For example, traditional medical systems are typically labeled "folk medicine," an ethnocentric term which has several different meanings, including: (1) any system other than Western biomedicine, (2) all unwritten systems, and (3) any "simpler" system at variance with Western biomedicine (Press, 1980). Press proposes a standardized definition of "folk medicine" that, by eliminating vague, stereotyped notions about its content and complexity, gives the concept sharpness and utility. Thus "folk medicine" is limited to "systems or practices of medicine based upon paradigms which differ from those of a dominant medical system of the same community or society" (Press, 1980, p. 48). According to this definition, which I adopt in this study, authochtonous medical systems such as Ayurveda, classical Chinese, pre-contact Navajo, or Kamba are not "folk," although any ethnic group's medical system can be so labeled if it is not the dominant practice of the community.

On the other hand, when folk medicine is freed from its non-Western or non-biomedical implications it becomes possible to "label and approach *Western biomedicine as folk medicine* in many tribal and peasant areas of Africa, Asia, and Latin America where it operates in the presence of a locally dominant medical system" (Press, 1980, p. 49; emphasis added). Thus in Kenya, biomedicine may be viewed as a "variant" or folk medical system because it is subordinate to and, as an operational model, departs in fundamental ways from the traditionally based medical sytems of the Kamba, Luo, Boran, Giriama, and others that remain dominant in terms of spatial coverage, accessibility, and frequency of utilization. A key distinction here in differentiating medical systems is their geographic sphere of influence. This dimension, which is seldom noted in the study of medical pluralism, concerns "the actual presence and availability of the constituent systems 'on the ground'" (Press, 1980, p. 47) in a specific locality.

THE NATURE OF ETHNOMEDICAL SYSTEMS

The total medical resources available to and utilized by a community or society, including popular, traditional and biomedical forms of therapy, can be said to constitute its *ethnomedical system*. An ethnomedical system evolves from a people's world view. It incorporates theories of illness and patterns of illness behavior that reflect socially constructed systems of meaning, values, norms, and interpersonal relations. Ethnomedical systems are maintained by an array of supporting social institutions and therapy specialists, including (in the technologically less-developed countries) newer roles such as medical assistants, nurses, and university-trained physicians. The dynamic geographic features of ethnomedical systems are observable in the spatial arrangement and interplay of resources and places where various phases and elements of disease, illness, and therapy are identified, acted out, studied, and administered. As conceived here, an ethnomedical system encompasses all of the acquired systematic knowledge, resources, organizational patterns, behaviors, and strategies (traditional, indigenous, scientific, imported) that are utilized by a community to promote the individual and collective well-being of its members.

Ethnomedical systems are socially accepted designs for labeling and managing illness that are both adjustable and adaptable to changes occurring in other social and economic subsystems. They are flexible wholes that can selectively accommodate and borrow from exotic medical traditions without losing their underlying conceptual coherence and integrity (Janzen, 1978). This idea may be compared with Press's (1980, p. 47) definition of a medical system, which he interprets as "a patterned, interrelated body of values and deliberate practices governed by a single paradigm of the meaning, identification, prevention and treatment of sickness."

An ethnomedical system also includes the *full complement* of health care strategies acceptable to some or all members of a community, *including* biomedicine as well as self-treatment and traditional practitioners. This model is compatible and nearly coincident with Kleinman's (1980, p. 50) model of health care, which he defines as "*a local cultural system composed of three overlapping parts: the popular, professional, and folk sectors*" (emphasis in original). An adaptation of this system which substitutes "biomedical" for "professional" and "traditional" for "folk" sector is shown in Figure 1-1.[10] The *popular* sphere of health care, which centers on self-treatment by the individual and family, is generally the first therapeutic choice for most people in most cultures. It is thus the largest, but also the least examined and poorly understood, part of any health care system. The popular sphere contains several levels, including individual, family, social networks, and community beliefs and activities. It is "the lay, non-professional, non-specialist, popular culture arena in which illness is first defined and health care activities initiated" (Kleinman, 1980, p. 50). In the case of the United States, for example, Kleinman (1980, p. 50) estimates that "70 to 90 percent of all illness episodes are managed within the popular sector."

In the technologically less-developed countries, traditionally based indigenous beliefs and practices generally serve as the "core" reference points in the ethnomedical systems for most of the population, although competition, complementarity and interdependence of different systems are typical wherever biomedicine is also available.

In summary, an ethnomedical system encompasses (Fabrega, 1977; Good, 1980b):

1. Folk knowledge and beliefs of laypersons with specialized knowledge.

2. Traditions, symbols, and values related to health, illness, and disease.

3. A society's causal theories and taxonomies of sickness.

4. Supportive social institutions and organizational arrangements (e.g., therapy managing groups, extended family, dispensaries, hospitals).

5. Recognized specialists (including traditional medical practitioners, traditional midwives, and more recent actors such as medical assistants, nurses, and biomedical doctors).

6. The spatial arrangement and interactions between various physical settings where various elements and phases of illness and therapy are determined, evaluated, and administered.

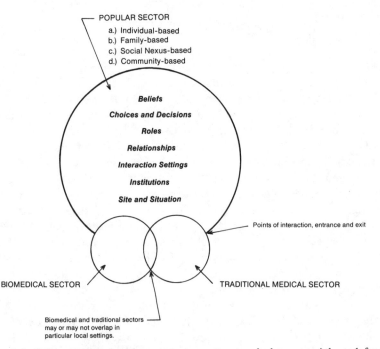

Figure 1-1. Ethnomedical health care system: structural elements. Adapted from *Patients and healers in the context of culture* by A. Kleinman, 1980, Berkeley: University of California Press.

THE CONCEPT OF COMMUNITY IN RELATION
TO HEALTH CARE

"Community," as referred to above, may sometimes signify an entire society or an ethnic group. In Kenya, however, and in every other new state, socioeconomic differentiation, complexity, and mobility are increasing. As a result the appropriate geographic and social units for the analysis of health-related behavior do not always coincide with a whole "society" because other bases of affiliation, such as class ("peasant," "unskilled laborer," "landless," "wealthy elite") and residential location (rural, urban), cut across ethnicity and promote differential access to "traditional" and "modern" medical resources. Thus, the concept of community used here is necessarily quite flexible. It cuts across and incorporates class as an additional basis for determining what people share, together with ties of kinship, neighborhood, common work, and other sources of propinquity. "Community" is thus also measured with a sliding geographic scale, and may simultaneously embrace rural and urban settings today because both are elements of a single socioeconomic system. In a political sense community means that a landless and jobless Kikuyu male, and a widowed and underemployed Luhya woman, both of whom reside in one of Nairobi's Mathare Valley shantytowns, are of a class apart from their more "successful" kinfolk who are coffee growers in the village or are ensconced in a ranked civil service job. Similarly, Kamba and Luo peasant farmers, although residing in widely separated areas of Kenya and differing in language, social organization, diet, and other dimensions of ethnicity, have a common bond in terms of the more restricted variety, quality, and accessibility of their health care resources relative to the urban elite in their respective societies.

In a very practical sense, the development and expansion of access to appropriate, basic health care resources for large numbers of rural and urban dwellers in the technologically less-developed countries will inevitably depend on how "community" is conceptualized by policymakers in the official medical and human services agencies. Innovative approaches to health services necessitate reforms that can come about only through difficult and courageous political decisions.

POLITICAL AND STRUCTURAL CONSTRAINTS TO
"HEALTH WITH THE PEOPLE"

A major barrier to formal collaboration between traditional and Western-trained medical practitioners arises from existing political and structural limitations of the biomedical sector. National approaches to health care are dominated by ideologies and strategies articulated by and for elite classes.

This phenomenon, an informal system which some have interpreted as part of a worldwide "medical cultural hegemony" (Elling, 1981) places little value on the health needs of the poor. The conventional technological, "bricks and mortar" approach to health care, which is also strongly urban in focus, obscures possible reforms that could lead potentially to more integrated, efficient, and socially equitable utilization of health, economic, and other resources. In this sense, equitable distribution of highly professionalized, physician-dominated, hospital-based health services, available to every village, is not feasible, sustainable, or desirable. Instead, as Gish (1979, p. 12) argues, "the task is to attempt to change the composition of health services— away from hospitals and towards primary care—through their more equitable distribution and to make them part of overall economic and social development."

Rural and urban communities could be mobilized and empowered to be less passive and dependent "receivers" of government administered services. Mechanisms are needed that will motivate politically disadvantaged communities to become more involved in, and responsible for, defining their own health care priorities and in shaping the kind of health services that function in their midst. Fundamental issues must be addressed, and their solutions cannot be found in rhetoric. Are health services to be designed "with the people" or "for the people"? Can cooperative strategies with traditional healers be developed when even private practitioners of "modern" biomedicine are so poorly articulated with the government-controlled health services in the Third World (Gish, 1979)? How are the basic health care needs of rural and urban people best identified, and how do they vary geographically in relation to cultural and class differences and ecological patterns? What fundamental reforms of the existing institutional and spatial structure of health care delivery might substantially increase, by 20, 30, or 50%, the small number of people who now enjoy access to low-cost primary care? These are crucial questions. Typically, those in government who could demand the answers do not ask the questions (Mburu, 1979). In this kind of conceptual vacuum, the evaluation of formal, effective cooperation between indigenous traditional healers and the official system of health services remains problematic.

Whether the setting is a country like Kenya or an old industrialized state such as the United Kingdom, the major obstacle to rational analysis and possibly more effective utilization of the so-called "alternative therapies" is the resistance of the biomedical health professions (Leslie, 1980). An example of this conservative posture of biomedicine is found in a review of medical beliefs in East Africa, whose author concludes that recognition of traditional healers by the official medical establishment would "undercut much of the painstaking progress made by Western medicine" (Weisz, 1972, p. 323).

Health conditions in the Third World highlight the nature of underdevelopment as a process, and are thus inseparable from considerations of political economy. Until political systems are reoriented and confront the inequalities perpetuated by social and property relations and the vested interests of professional biomedicine, most people will experience little further improvement in health status and medical services (Gish, 1979). This argument certainly does not appeal to most high-ranking biomedical practitioners and bureaucrats, but it is honest.

This first chapter describes the main themes of this book and examines the nature and scope of medical pluralism with reference to Africa's present and future health care needs. Attention has been drawn to the main features and functional implications of the different paradigms that characterize traditional medicine and biomedicine—the dominant therapies that give form to Africa's contemporary ethnomedical systems. An interdisciplinary approach is of great value if one wants to recognize the elements and interpret the functioning of such ethnomedical systems, and I have proposed that the holistic-spatial perspective of medical geography is well suited to those purposes. Finally, the ethnomedical model was presented both as a key theoretical idea about social networks, labeling, and information transfer in health care; and also as a sensible strategy by means of which patients and therapists who interact within and across the conceptual, clinical, and spatial boundaries of distinctive therapeutic traditions can improve communication with each other. Such a process enhances the prospects for effective healing. Evidence of the practical value of the ethnomedical approach in biomedicine is presented in later chapters.

In Chapter 2, I sketch the national picture of health conditions and health care in 20th century Kenya. I begin with the colonial period in order to show the continuity of past and present circumstances. Discussion of health determinants, regional patterns, and therapy systems in Kenya as a whole is essential background for interpreting the case studies of rural and urban traditional medicine.

NOTES

1. I am indebted here to John H. Leith for his enlightening discussion (developed in connection with a different theme) of "tradition," "the living faith of dead people"; "traditioning," the act of transmitting an integrated body of belief and practice; and "traditionalism," "the dead faith of living people." See Chapter 1 in J. H. Leith (1981).

2. Maps of varying scale are valuable tools in the research of medical geographers in both traditions. The purpose of such mapping is most importantly to discover clues to the source, ecology, and diffusion of disease and to develop hypotheses about the causes and possible means of containing or preventing a disease hazard. Maps are also essential for analyzing the organization, relative location, demand (utilization), and need for health services in a particular area, and for

determining the appropriate geographic allocation of such services at different areal scales.

3. This process tends to run counter to the conventional core–periphery acculturation model. In the latter view, there is a stage of transition from "traditional" to "modern" behavior in which urban elements of a population are the vanguard of the modernizing society. The traditionality of a population is seen as the main "barrier" to their utilization of modern health services. Accordingly, the means of promoting changes in peoples' health behavior is to improve their socioeconomic status and/or to centralize health services. See, for example, G. M. van Etten (1976), pp. 54–58.

4. This is not to deny that other (more limited) approaches are emerging in association with biomedicine, including a psychobiological model, which explains biological processes in terms of psychological factors; a biopsychological model; general systems theory; a relational mode; the health education approach; and family medicine. See A. Kleinman (1978b).

5. In Taiwan, the mean time for consultation between Western-style practitioners and patients reported in one study is 3 min 15 s, with mean times for clinical communication 1 min 18 s, and for time spent in explaining, 40 s. See Kleinman (1980), p. 303.

6. In this regard there is a great need for creating a systematic, epidemiologic approach to community morbidity in terms of traditional illness taxonomies and therapeutic practices. Little is known of the stressors associated with traditionally defined illness episodes, rates for traditional illness syndromes, relationships to demographic and independent variables, such as life cycle events, migration, fertility and infertility, climacteric, entrance into school, and so on. See J.-H. Pfifferling (1975).

7. Mbiti (1970, pp. 97–118) provides an excellent, brief treatment of "spiritual beings, spirits, and the living dead" in African societies.

8. I am grateful to Mr. Abdullah Makembe of Tanzania for a useful discussion of the *uzima* concept.

9. These are categories proposed by Press (1980).

10. This adaptation is presented by permission of the University of California Press. Kleinman (1980) characterizes the "folk" sector as nonprofessional and nonbureaucratic. It may include both secular (e.g., herbalists) and sacred (e.g., shamans, ritual specialists) elements, and the folk sector may integrate elements from the other two sectors—particularly the popular sector. "Folk" in Kleinman's scheme approximates the "traditional medical sector" in the ethnomedical model shown in Figure 1-1.

2

Population, Environment, and Health Care in Twentieth-Century Kenya

In this chapter I provide historical and regional overviews of Kenya's health problems and sketch the range of therapy systems that are available—in unequal measure—to the country's burgeoning population of 20 million. My aim is to present the context of the case studies of traditional medicine in rural Ukambani and in Nairobi that are developed in Chapters 5 through 8. Attention centers on the complex associations between human health and disease and Kenya's ecological, demographic, economic and sociopolitical systems. I argue that the country's pressing health problems in the 1980s are closely linked to the entrenched, inequitable model of biomedical health services inherited from British colonial days. This is an extremely costly and ultimately unsustainable model because it features hospital-based curative services in the towns and is dependent on a large supply of doctors. Although 87% of Kenya's population is rural, the distribution of health services reveals enormous disparities in favor of the small urban population. Despite the inclusion of free outpatient medical treatment as a constitutional guarantee since 1965, today "the shortage of services in most rural areas is so gross as to make the ideals of inexpensive or free care meaningless" (Mburu, 1979, p. 580). Despite substantial expansion of the physical infrastructure of health services, the parallels between the present pattern of health care and the system that prevailed during the colonial era are striking.

THE COLONIAL LEGACY OF DISEASE AND HEALTH CARE

Poor health is not an inevitable biological condition of mankind, nor a state that necessarily improves in proportion to the amount of budgetary resources directed against it. Moreover, factors in the physical environment are often necessary for a disease to occur, but they are rarely if ever sufficient in themselves to explain the presence of disease. Instead, as the record from the late precolonial to the postcolonial world of the 1980s illustrates, disease is largely a consequence of human agency. It arises from within a complex web of social, economic, and political processes that reflect life-styles and unequal power relations based on class, race, and gender. These relationships, together with ignorance and miscalculation of the consequences of interventions in the physical environment, produce what Turshen (1984) aptly terms "the unnatural history of disease." In other words, the causes of ill-health (i.e., the disease process) are situated in the polity rather than in "natural environment" or in the individual as a biological organism. This perspective is an essential corrective to conceptualizations of disease and health services provision in an historical and sociopolitical vacuum—that is, as primarily biomedical and technical problems exacerbated by resistant cultural attitudes.

As in the Americas and Pacific, East Africa offers abundant evidence of the unnatural history of disease. During the late 19th and early 20th centuries, when the health of Europe's population was continuing to improve, devastating epidemics of jiggers (*Tunga penetrans*), rinderpest, Gambian sleeping sickness, nagana, smallpox, relapsing fever, and syphilis—which often triggered or compounded famine conditions—debilitated and nearly destroyed some African societies in parts of Kenya, Uganda, and Tanzania. These disasters were intimately connected with the breakdown of isolation and the unprecedented upheaval associated with the European conquest, partition, and transformation of the region's economic and social life. Military activities, the long-term effects of the Arab-Swahili slave and ivory trades, the disruption and reorientation of trade routes, the displacement of traditional economic life, and the redistribution of population are among the many factors that upset the ecological balance. These changes facilitated the introduction of new pathogens and forms of disease that confronted many populations with new and grave risks. The result was a deteriorating picture of African health in Kenya and other parts of East Africa at the beginning of the 20th century (Dawson, 1979; Good, 1972, 1978; Hartwig & Patterson, 1978; Kitui District, 1898–1912; Kjekshus, 1977; Turshen, 1984; Ukamba Province Annual Report, 1908–1909).

When officials in the colonial medical services in Africa tried to explain the often shocking standards of African health, few appreciated the profound health effects of the sociopolitical, economic, and ecological impacts that accompanied European overrule. Instead, "conditioned by the same capitalist forces that shaped colonial governments," medical authorities "looked for proximate sources of tropical diseases and found them in the poor hygiene of Africans, their lack of immunity to diseases they had no prior experience of, and a poor diet that lowered general resistance to disease" (Turshen, 1984, p. 14).

Poor health was an omnipresent reality for the African population in colonial Kenya. Many diseases showed distinctive foci and distributional patterns produced by the sensitive interplay of human ecology and habitat features such as temperature, rainfall, and vegetation. Although Kenya straddles the equator (Figure 2-1), its geological history and climatology combine to create dramatic ecological gradients and contrasts of epidemiological significance. Environments range from the productive Highlands (2,000–3,000 meters) in the south central and southwest of the country, the Great Rift Valley, and the ice- and snow-capped volcanic peaks of Mt. Kenya (5,199 meters) to the vast expanses of scrubland and semidesert in eastern and northern Kenya (Figure 2-2). Under colonial rule Kenya's higher, wetter, and

Figure 2-1. Kenya reference map.

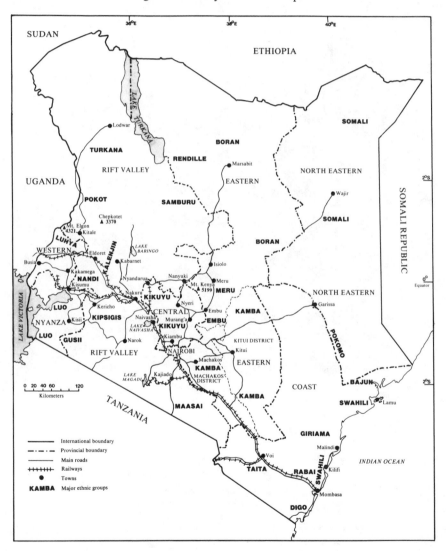

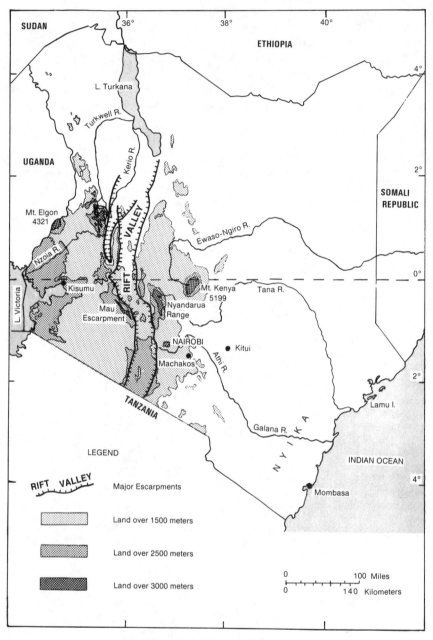

Figure 2-2. Kenya relief map.

more productive plateaus and mountain slopes experienced increasing population pressure in response to gradual reductions in mortality rates and a shrunken land base due to land alienation to European settlers.

Human mobility often served as an efficient mechanism for the dissemination of disease from one ecological zone to another. Malaria, for example, was indigenous and endemic in precolonial Kenya, but it was apparently unknown at altitudes above 1,500 meters until after the Europeans arrived (Diesfeld & Hecklau, 1978). In coastal and upland areas of Kenya malaria is transmitted by anophelines either on a year-round or seasonal basis (depending on rainfall and the fluctuating density of the vector), and it proved to be a major source of general morbidity and child mortality throughout the colonial period. Other vectored diseases such as plague (particularly in the early decades of colonial rule), human sleeping sickness (which spread in several epidemic waves from Uganda beginning in 1902), *kala azar* (leishmaniasis), tick-borne relapsing fever, and various filarial infections such as onchocerciasis (river blindness), elephantiasis, and hydrocele were also endemic in particular places.

Shortly after Kenya achieved its independence from Britain, Fendall and Grounds (1965) published a series of articles that highlighted the incidence and epidemiology of diseases during the colonial era. Their analyses show that then, as now, children formed a group at very high risk to numerous disorders. Chronic undernutrition—described as "nutritional poverty in the midst of plenty" and thus unnatural—was perennial and widespread, particularly at the coast and in Western and Nyanza regions. Malnutrition was exacerbated by periodic famines of local or regional occurrence, and was a contributory factor in half of all the many child deaths attributed to gastroenteritis and dysentery.

Respiratory infections were the commonest cause of mortality in children under 6, with incidence and prevalence greatest in the higher, colder, wetter regions. Otitis media and other ear diseases took a toll, while bacterial meningitis and encephalitis were frequently implicated in deafness among African children. Virtually every child contracted measles. Whooping cough was the fifth leading cause of child death behind respiratory infections, gastroenteritis, malnutrition, and malaria.

Localized outbreaks of smallpox, which was usually imported across Kenya's land frontier, accompanied periods of stress caused by civil disturbances, war, and famine (Dawson, 1979). Urban populations were most at risk, and children 10 years old or less accounted for 61% of all smallpox cases. Highly localized epidemics of cerebrospinal meningitis occurred with about 20% mortality. These outbreaks also correlated with times of stress such as the Mau Mau Emergency of 1954–1959, when some 4,000 cases were reported. Once again children represented about 70% of all cases.

While polio was not a major cause of death in colonial Kenya, its crippling sequelae presented enormous difficulties of management and rehabilitation for poor communities. Polio was characterized by high endemicity after World War II, and an epidemic in 1954 initiated a pattern of 3-year epidemic cycles. Unvaccinated children experienced an attack rate five times higher than in the vaccinated population.

Tetanus was the fifth leading cause of death among all hospital inpatients in the Fendall and Grounds (1965) survey. Children aged 5–9 accounted for the largest number of admissions. The incidence of neonatal tetanus was apparently greatest along the coast and among the Kikuyu-speaking peoples in central Kenya. In the latter region neonatal tetanus was linked to the traditional practice of dressing the severed umbilicus with earth obtained from the floors of dwellings that were made of hard-packed earth and cow dung.

Many other diseases debilitated the colonized African population, including syphilis (Nyanza accounted for 70% of the total reported cases in 1925), yaws, leprosy (particularly in areas bordering Lake Victoria), and enteric fever. Tuberculosis, introduced to Kenya at the beginning of the colonial era, rapidly became endemic. By 1948 the tuberculosis prevalence rate for the overall population (all age groups) was estimated at 11 per 1,000 population.

At the time of Kenya's independence an estimated one million people, or at least 11% of the total population, harbored schistosomiasis. This debilitating disease was endemic in seven major foci ranging from the coast to Lake Victoria, including several irrigation schemes. A large proportion of the population was also parasitized by intestinal helminths such as roundworms and hookworms. Tapeworm predominated in the pastoral zones, and hydatid disease—with dogs serving as the heavily infected link between livestock and humans—was endemic among Turkana nomads in the northwest desert zone. The high prevlence of hydatid disease in Turkana was related to the custom of using dogs to clean the face and anus of babies who had vomited or had diarrhea (Fendall & Grounds, 1965).

The disease profile for colonial Kenya described above provides only a partial accounting of the true variety and extent of morbidity and mortality. It also represents a narrowly biomedical perspective on disease and thus does not, for example, address the issue of what diseases were perceived as important within the scope of African traditional medicine. Nevertheless, this overview of disease in British Kenya establishes a context in which to examine the colonial response to African needs for health care.

In colonial Kenya the provision of basic medical care for the general African population "was delayed again and again," and "only in late 1945 was a Ten Year Programme for the development of health and hospital services recommended with a financially generous endowment" (Diesfeld & Hecklau, 1978, p. 26). This neglect of African needs for health care grew out of the distinct, racially based social and spatial structure that rapidly grew up in

Kenya under the British administration. Europeans, Asians from the Indian subcontinent, and Africans formed a three-tiered hierarchy in descending order of status. These classes also characterized the system of health care, with separate European hospitals, Asiatic Wards, and African hospitals.[1] By 1949, 15 years after Kenyan Africans were admitted to train at Makerere College in Uganda for a University of London medical degree, only 15 had graduated as doctors. Following their training, Africans had to serve as "assistant medical officers" under European and Asian Medical Officers and were discouraged from further professional advancement (Mburu, 1981).

Colonialism thrived as long as race and class differences could be exploited by one ethnic group capable of dominating the others. As early as 1903, the Colonial Office established separate and unequal policies and objectives for the medical department to pursue vis-à-vis the three main races in colonial Kenya. Protection of the European community's health—and particularly that of government officials—was first and foremost. Second, it was necessary to ensure that the native and Asian labor force remained healthy enough to work. Finally, control of common communicable diseases and environmental sanitation in towns and rural markets was also viewed as a priority. The social and locational priorities inherent in these measures, and the corresponding allocations of financial resources, strongly favored the European and largely urban Asian communities (Mburu, 1981).

Although the idea of government-sponsored rural health services for Africans was introduced in Kenya after World War II, it is an understatement to say that under the colonial administration such services experienced "a serious lack of financing" and suffered from the government's "inability to train a sufficient number of medical aides for every center" (Beck, 1981, p. 22). Indeed, I present evidence in Chapter 3 that shows that the burden of financing, building, equipping, and staffing rural health services in colonial Kenya was shouldered by the local native councils (LNC), which became known as African district councils (ADC) after World War II. Given the "pay as you go" fiscal policy of the Colonial Office in London, and its power to dictate the budget for the local medical department, the colonial administration in Kenya never seriously pursued the concept of support for basic health care for Africans. Although the district medical officer typically went on "tour" in his district once or twice during the year, he spent most of his time practicing curative medicine at the district hospital which was located in a town. If Africans wanted dispensaries in areas beyond the reach of the widely scattered health services sponsored by Christian missions, they had to finance them by assessing additional taxes on their own communities. This policy remained in effect until 1969 (6 years after Kenya achieved its independence), when the Local Government Transfer of Functions Bill shifted responsibility for health, education, and other services from the ADC to the central government (Hartwig, 1979).

While the majority of the African-initiated dispensaries suffered from a lack of trained staff and supplies, they nonetheless had strong support from the local population. By 1962, the year before Kenya's independence, the number of ADC dispensaries showed dramatic growth. While expansion was credited to the dispensaries' popularity with the African population, "the Ministry of Health also took credit for their growth from nil in 1946 to 140 in 1962" (Beck, 1981, p. 22).

In many parts of colonial Africa Christian missions were the first to introduce Western medicine on a permanent basis (W. D. Foster, 1970; Hartwig, 1979; Rotberg, 1965). It was impossible for the missionaries to ignore the heavy burden of disease within the African population; thus as a practical matter the provision of curative medicine was inseparable from the goals of evangelization and education.

In Kenya medical care provided by the colonial administration was essentially restricted to the towns and other employment centers, the military, and curative services for the relatively large European community. Christian missionaries provided the bulk of the medical services available in the African "reserves" (those rural areas that did not form part of the 31,000 square kilometers alienated to Europeans in the "White Highlands"). From 1907, beginning with the Church Missionary Society (CMS), and later the Church of Scotland Mission and Africa Inland Mission, Protestant mission groups pioneered the establishment of medical outposts in remote areas. Protestant and (from 1940) Catholic missions gradually built hospitals where there was evidence of a potentially larger and stronger following (Mburu, 1979). The different missions exercised denominational spheres of influence in order to diminish excessive competition among themselves.

But the missions certainly did not enjoy unqualified acceptance by the African populations. In 1912, for example, the British provincial commissioner for Central Province, referring to the Kiambu District adjacent to Nairobi, reported that

> it seems that the Missions are not popular in the reserve. The native has not settled in his mind that they are purely for the good of the Kikuyus, and their constant interference with the tribal authorities, the [. . .] of non-converted natives living on their land, and the encroachments made on native land, have resulted in an unconciliatory attitude on the part of the natives (Central Province, Annual Report, 1911–1912, p. 19).

Thus it has been argued that Africans accommodated the missions, above all else, because of the (primarily curative) medical services (Diesfeld & Hecklau, 1978; Mburu, 1979). Even medical activities were not easily accepted by all. In Nyeri District, for example, disease control was considered problematic because the Kikuyu people had "a profound dread" of contact with any place, such as a hospital, where deaths had occurred, or with doctors and dressers who had touched dead bodies. Such places and people were considered *thahu*, or ritually unclean (Nyeri District Annual Report, 1915–1916).

There was some antipathy on the part of the British colonial administration toward missionaries. Divisions arose over such issues as appropriate standards for medical doctors and the fallout from efforts in the 1920s by the Church of Scotland Mission to have the practice of female genital mutilation abolished. However, the government did come to depend extensively on the medical activities of the missions and made small grants available to them after World War II. However, up until World War II most of the medical missions, aided by their overseas contributors, remained self-sufficient and independent of the colonial administration. Regardless of inadequacies in the quality of care and the limited extent of coverage, missionary medical activities in Kenya came to represent what Mburu (1979, p. 524) has interpreted as "the single most important attempt to affect Africans, in their own environment, outside the prevailing colonial structure."

African traditional medicine is inseparable from the whole of African cultures. It continued to coexist—openly and covertly and side-by-side—with mission and government medical services in colonial Kenya. It remained the most readily available and frequently used system of health care within increasingly pluralistic ethnomedical systems. In contrast to territories such as the Belgian Congo (Janzen, 1978) the colonial regime in Kenya made no systematic efforts to prohibit the practice of traditional medicine. However, Christian missions often demanded the cessation of traditionally sacred observances and consultations with diviners and other religious-medical specialists (Chapter 3). Individual colonial administrators had a strong curiosity about African beliefs and medical practices, as evidenced by their often detailed descriptions of "medicine men," legal ordeals, and religious customs in official reports (e.g., see Machakos District Political Record Book, Vol. 2, up to 1910). In the main, however, in the early decades of colonial rule the administration in Kenya paid little attention to traditional medicine and its practitioners "as long as they did not disturb the peace" (Beck, 1981, p. 62).

After 1930 African traditional medicine received somewhat greater attention, apparently because of its relationship to the increasing visibility of witchcraft practices and the perceived threat they presented to public order in several areas of Kenya. The colonial administration had initially shown its concern about this phenomenon by instituting an ordinance that was intended to help control witchcraft activities (The Witchcraft Act, Cap. 67, 1925; rev. 1962, see Mutungi, 1977). English common law proved to be incompatible with many traditional African beliefs and practices. In the case of witchcraft, it was (and has been shown to be to this day) "absolutely impotent of inducing change among its subjects" (Mutungi, 1977, p. xv).

Many British colonial officers went to considerable lengths to explain the difference between the practitioners and methods of witchcraft, or the harmful "black arts" (Swahili = *uchawi*), and the beneficial or "white medicine" (*uganga*) practiced by "the harmless medicine man" (Machakos District Annual Report, 1933). Some administrators, including the District Commissioner of Machakos in 1955, emphasized the "undoubted good of white magic." Nevertheless, few Europeans had the inclination to understand the holistic nature of African approaches to disease and healing. Many missionaries, administrators, medical staff, and other Europeans who lived in Kenya surely saw little reason to credit a system that was to them so obviously "inferior" to Western medicine. Whether intended or not, from an African perspective they successfully denigrated the local population by stereotyping traditional medicine as an irrational assemblage of "native superstition," black magic, and witchcraft. European characterizations of traditional medicine as shameful and uncivilized, whether communicated from the pulpit, by staff in the mission hospital or government dispensary, by administrators, or by teachers in the schools, contributed to a "colonization of the mind" among the African population (Mburu, 1977, p. 168). Because the beliefs and thought patterns associated with traditional medicine are indivisible from African traditional religion and culture as an entire system for living, many colonized Africans found themselves caught between the premises and requirements of their own heritage and the competing pressure to appear "modern." As I noted in Chapter 1, the result was and is a strong ambivalence toward traditional medical practitioners (TMPs) on the part of many Africans in Kenya and elsewhere on the continent. Individuals will admit that "herbal medicine" has some value. Publicly, however, most people project an attitude of incredulity toward traditional medicine and disparage or joke about TMPs as charlatans and perpetrators of witchcraft. In spite of the carryover and inhibiting effect of these strong prejudices—which can be interpreted as an adaptive response by African peoples forced to accommodate the intrusive European culture—it is well known that TMPs are privately consulted by people from all social and economic backgrounds in modern Kenya (Mutungi, 1977).

HUMAN ECOLOGY AND HEALTH: KENYA IN THE 1980s

Medical geography focuses both on understanding spatial variations in disease patterns and health factors and on analyses of health problems which emphasize human ecology. Health is viewed as adaptability instead of the "absence of disease" (Meade, 1977, 1986). This positive conceptualization of health received strong impetus from the work of Audy (1971), who defined health as "a continuing property potentially measurable by the individual's ability to rally from insults, whether chemical, physical, infectious, psychological, or social." In this view, "disease" is not just a biological process or state, but rather "a measure of the maladaptive interactions among the

. . . triad of population, environment, and culture" at particular places and times (Meade, 1977, p. 382). These interactive elements of medical geography are surveyed below in their general relationship to present health conditions in Kenya. The country's health problems also mirror its prevailing sociopolitical system. As noted earlier, the retention (since independence in 1963) and maintenance of an inequitable, costly "colonial" model of biomedicine is today the paramount sociopolitical factor responsible for retarding progress toward "health for all" in Kenya.

The Biophysical Environment in Relation to Health

Kenya's land area is 582,646 square kilometers (approximately the size of Botswana or France), and it extends 4°21'N and 4°28'S of the Equator in East Africa. Tanzania, Uganda, Sudan, Ethiopia, and Somalia are border states (see Figure 2-1). Southeast Kenya has more than 500 miles of coastline along the Indian Ocean, while the southwest includes part of Lake Victoria. The national territory is administratively subdivided into eight Provinces, including Greater Nairobi, and these are further subdivided into districts, locations, and sublocations in descending order of scale.

Kenya's physical landscapes and ecological zones are renowned for their variety and often unexceeded beauty. One measure of the resource value of this natural heritage is the 19 national parks and game reserves in Kenya, and the excellent infrastructure that has been developed to accommodate tourism. However, despite this environmental diversity agriculturally productive lands constitute less than 20% of the total area. Arable land is at a premium outside the densely populated highland enclaves in central and southwest Kenya—the agricultural core—and the coastal zone along the Indian Ocean.

Arid and semiarid lands comprise over 80% of Kenya's land area. These "drylands" are utilized primarily by pastoral populations such as the Maasai, Turkana, Boran, and Somali. One in five Kenyans and half of the country's livestock are found in these zones (Bernard, 1985). While they are by no means uniform, the drylands are generally below 1,372 meters (4,500 feet) in elevation (Figure 2-2). Rainfall is unreliable and markedly seasonal, with 7–10 arid months. Rainfall totals are less than the 762 millimeters (30 inches) normally required to sustain rain-fed agriculture. Effective moisture availability is greatly reduced by high evapotranspiration rates. Vegetation varies from dry woodlands and acacia savanna in the more humid margins of the drylands to thornbush, scrubland, and sparse grasses in the desert zones in northern Kenya. Soils are generally thin and rivers are markedly seasonal.

The critical relationship of water to the health of humans and livestock is magnified in the drylands. Water is not only scarce, but potable supplies are often difficult to find and maintain. The close association and mutual dependence of human communities, vectors, and pathogenic agents on the same water sources in the drylands is an epidemiological factor of major signifi-cance. Although water supply schemes remain a national development prior-ity, the Republic of Kenya's *Development Plan for the Period 1984 to 1988* reports that the present population enjoying improved water systems ranges from 4% of the population in the nomadic pastoral lands of North-Eastern Province to 20% in the densely settled Central Province (Republic of Kenya, 1983, pp. 36, 154). Only one of six rural families obtains water from a borehole or piped supply.

Many of the diseases of poverty that are widespread among Kenyan populations are closely associated with deficiencies in water supply or sanita-tion. Health effects related to the scarcity of readily available water for domestic use are particularly well illustrated in the case of the pastoral Maasai situated on the savanna plains near Amboseli and Mt. Kilimanjaro. In some Maasai communities the prevalence rate of trachoma—a "water-washed" infectious disease whose incidence decreases as the volume of avail-able water increases—is nearly 100% (personal observation).

Schistosomiasis is also a significant health problem in Kenya. Among the country's several state-organized irrigation schemes, including Mwea Tebere south of Mt. Kenya, the Ahero and Kano II schemes near Lake Victoria, and those on the Tana River at Galole and Bura, only the remote Perkerra project in the Rift Valley south of Lake Baringo is free of schistoso-miasis (Diesfeld & Hecklau, 1978).

Although harsh, agriculturally limited, and famine-prone, the arid and semiarid lands have become the migration destination for thousands of marginalized Kenyan farmer families who lack access to adequate arable and grazing lands in the overpopulated higher-potential areas of the country. Campbell (1981) has identified 12 areas in Kenya's drylands that are the targets in this migration process. The migrations are both spontaneous, for example, in eastern Kitui District and in the foothill zone of Mt. Kilimanjaro on lands formerly used by Maasai herders as dry-season pastures; and government-sponsored, as in the Bura and Hola irrigation schemes along the Tana River. A struggle between herders and farmers for access to and control over resources is underway along the wetter margins of the drylands and in the river valleys and around the edges of swamps. The areas selected for settlement in these locations "represent an interface between the sedentary farming economies of the higher-potential lands and the pastoral and wildlife economies of the semi-arid rangelands" (Campbell, 1981, p. 40).

While variable soil and water conditions result in an uneven pattern of settlements and land-use conflicts along the interface of the high-potential lands and the drylands, in some areas population growth is two to ten times the national average (Bernard, 1985). This movement of people "down the ecological gradient" into the arid and semi-arid lands has been underway for nearly a generation as one response to Kenya's soaring demographic pres-sure. Such mobility necessarily exposes the incoming human and livestock populations to the threat of food shortages, the psychological shock of social dislocation, and to new infectious and parasitic diseases. To date there has been little systematic examination of the risks associated with interzonal mobility and resettlement in Kenya (cf. Roundy, 1976). The specific hazards connected with such movement may include exposure to leishmaniasis (the main foci of this disease is the upper Tana River basin in northern Kitui, Embu, and Meru), schistosomiasis, malaria, trachoma, dysenteries, meningococcal meningitis, tuberculosis, and anthropozoonoses such as anthrax. The resident drylands populations are also exposed to diseases imported by the new settlers. This situation is analogous to the epidemiologic risks of rural-to-urban migration and rural-return mobility—which continue to be the dominant forms of population movement in Kenya.

In contrast to the drylands, the zones of greatest ecological potential and densest rural populations outside of Kenya's coastal strip are situated above an elevation of 1,372 meters (Figures 2-2 and 2-3). These include the Nyandarua highlands (Aberdares), which extend north and northwest from Nairobi (1,676 meters) through Central Province to the volcanic soils of Mt. Kenya (5,199 meters); the high plateau situated on the western side of the Rift Valley; the well-watered "tea" highlands of Nandi, Kericho, and Kisii; and the densely populated lands of Western and Nyanza Provinces adjacent to Lake Victoria (Figure 2-1). Outlying regions of high potential include smaller zones of highlands or "hills" above 1,500 meters in Machakos District, such as the Mua and Kilungu Hills; the plateau around Kitui; and the Taita Hills.

Greater moisture, altitude, cold, soil and groundwater chemistry, and population density are among the important factors in the ecology of human health in Kenya's zones of higher agricultural potential. Respiratory infections, including bronchitis, pneumonia, tuberculosis, and measles as well as polio, viral hepatitis, gastroenteritis and intestinal helminths are widespread in the higher arable zones. Endemic goiter is also extensive. Prevalence rates range from 15% to 72%, with highest levels concentrated in the highlands of Central, Rift Valley, and Western Provinces (Diesfeld & Hecklau, 1978).

Dental fluorosis, a disease preventable through defluoridation of drinking water, is endemic in many areas of Kenya. A World Health Organization (W.H.O.)/Food and Agriculture Organization (F.A.O.)/United Nations International Children's Emergency Fund (U.N.I.C.E.F.) survey of 19,000 individuals conducted in 1968 found 44.1% with dental fluorosis (Nair, Manji, & Gitonga, 1984). There is also evidence that high levels of fluoride in groundwaters are widespread throughout the country, and particularly in association with volcanic soils of the Rift Valley and Highlands zones. A more recent survey (Nair et al., 1984) based on 1,286 groundwater samples (boreholes and wells) found the highest fluoride concentrations (ppm) in Nairobi Province (30.2, with 32% of the samples over 8.1 and 80.3% over 1.0 ppm); Rift Valley (57.0, with 10% of the samples over 8.1 and 70.3% over 1.0 ppm); Eastern Province (19.3 in Machakos District, with 13% of the provincial samples over

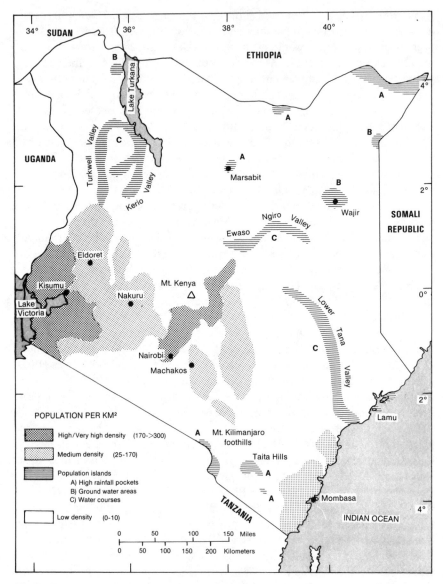

Figure 2-3. Kenya population regions. Adapted from Ojany and Ogendo (1973) by permission of Longman Group Ltd.

8.1 and 65% over 1 ppm); and Central Province (22.0, with 8.3% of the samples over 8.1 and 53% over 1 ppm).[2] These provinces account for approximately 60% of Kenya's population. At present the greatest need is for studies to determine the extent to which the location of water supplies actually consumed by the human populations, and the pattern of fluorosis, covary with the high levels of fluoride found in groundwater. Defluoridation of water supplies might greatly benefit the health of the population in many areas of Kenya.

Whereas most areas above 2,000 meters are more or less permanently free of malaria due to insufficient warmth for *Anopheline* development, human population mobility and periods of exceptional rainfall in drier areas can and do promote epidemic outbreaks among locally non-immune populations. In addition to abnormal rains, which may promote rapid breeding of mosquitoes and increased malaria transmission where conditions favor *Anopheles*, projects designed to enhance agricultural production and nutrition frequently have unintended (but avoidable) health consequences (Hughes & Hunter, 1970). In Kenya's Nyanza Province and Kisii District, for example, thousands of ponds for *Tilapia* fish farming were dug in the 1950s in an effort to boost available protein. However, over 80% of these fish ponds became infested with *A. funestus* or *A. gambiae*, and the increased mosquito density caused an upsurge in malaria transmission and morbidity (Desowitz, 1980; Diesfeld & Hecklau, 1978). In the 1970s a large irrigated rice farming project on the Kano Plains near Kisumu in western Kenya promoted the expansion of *A. gambiae* from 1% to 65% of the local mosquito population (Chandler, Highton, & Hall, 1976).

Malaria is transmitted for up to 6 months annually in Machakos District, in eastern Central Province, and on the plains to the south and east of Mt. Kenya (Figure 2-2). It is hyperendemic (age-specific spleen rate of 50–74% for children aged 2–9) or holoendemic (>75%) in coastal Kenya and in much of the Western and Nyanza provinces. Furthermore, *P. falciparum* resistance to chloroquine is now evident in approximately 30% of new malaria infections in areas of Siaya District in Nyanza Province (personal observation, Saradidi Rural Health Project, July 25, 1986). High sickle-cell trait (HbS) rates (20–40%) also occur among many of the ethnic groups settled in these areas, but it is not known whether the presence of this genetic mutation has lowered the death rate from cerebral malaria (Foy & Kendall, 1974).

Other important vectored diseases found in some of the arable zones include human sleeping sickness, with foci mainly in the Lambwe Valley in Nyanza Province and Samia Location in Western Province near the Uganda border; and onchocerciasis, with a small endemic focus in northern Western Province. This disease has been eradicated from several large areas of western Kenya (Diesfeld & Hecklau, 1978; Onyango, 1974).

The Population and Cultural-Behavioral Dimensions

Demographic composition and structure, population distribution, and host characteristics such as genetics and immunological and nutritional status are well-known variables affecting health status and exposure to disease. This population dimension of medical geography interacts with environmental features and with a broad spectrum of cultural behavior.

The cultural-behavioral dimension in medical geography focuses on all attitudes, beliefs, and practices—including specific elements of livelihood systems—that maintain, promote, or ameliorate exposure to health hazards. Conditioned by ethnicity, class, place, age, sex, technology and many other factors, the relationship of behavior to health is inseparable from the environmental and demographic dimensions. Human behavior produces a differential and often predictable exploitation of the physical and cultural resources of the environment. It transforms the healthfulness of habitats and moves environmental elements such as irrigation water, vectors, and pathogens from one place to another. Behavior determines in part who will be exposed to which insults, in which locations, and for how long (Meade, 1977).

Kenya's African population doubled from 5.2 to 10.7 million in the 21 years between 1948 and 1969. Population totals and density in Kenya's seven provinces and Nairobi, based on the last national census in 1979, are shown together with the density ranges for districts in Table 2-1. Even at this level of generalization one can see the association between areas of high ecological potential and high population density, that is, Central, Western, and Nyanza provinces (Figure 2-3). Districts in other provinces with high population densities include Machakos (the location of the rural case study in this book), Embu, and Meru in Eastern Province, and Kericho, Trans-Nzoia, Nakuru, and Uashin Gishu districts in Rift Valley Province. In terms of physical regions the most densely settled areas of Kenya are the eastern slopes of both the Nyandarua Range and Mt. Kenya; the Kikuyu Escarpment, immediately west and north of Nairobi; the Kakamega Hills in Western Province; the Kisii Highlands in Nyanza District, and the Mua Hills of Machakos District to the southeast of Nairobi.

Kenya's current (1986) population of over 20 million—at least 50% of whom are children under age 15—is increasing at the rate of 4% annually and will double in 17 years (A.D. 2003) if growth remains unchecked. The crude birth rate of 53 per 1,000, perhaps Africa's highest, and the relatively low crude birth rate of 13 per thousand, have produced a demographic crisis (Population Reference Bureau, 1984). President Moi's recently announced national campaign to impose sanctions on couples who exceed a completed family size of four children (*Popline*, 1985) recalls China's recent population policies. It bears testimony to the magnitude of this population growth process that began to manifest itself in Kenya over 25 years ago. A new and as yet unpublished national survey indicates that at least in a few areas, such

Table 2-1. Kenya Population and Density By Provinces and Districts, 1979

Province	Population	Area (km²)	Density (km²)	
Nairobi	827,775	684	1,210	
Central (5 districts)	2,345,833	13,173	178	66–280 (range)
Coast	1,342,794	83,040	16	
Mombasa (5 districts)	341,148		1,622	2–34 (range)
Eastern (6 districts)	2,719,851	155,759	17	1–96 (range)
Northeastern (3 districts)	373,787	126,902	2	2–3 (range)
Nyanza (4 districts)	2,643,956	12,526	211	143–395 (range)
Rift Valley (13 districts)	3,240,402	163,884	19	2–161 (range)
Western (districts)	1,832,663	8,196	223	
Kenya (total)	15,327,061	564,162	27	

Note: From *Population Census, 1979* (Table 12) by Republic of Kenya, 1979, Nairobi Government Printer.

as densely populated Nyeri District near Mt. Kenya, the idea of family planning and the adoption of modern contraception are showing significant expansion (personal communication, Alan Johnston, Ministry of Planning and National Development, Nairobi, August, 1986).

At 8.0 births per woman, Kenya's total fertility rate is apparently the highest in Africa and may be the world's highest (Population Reference Bureau, 1984). These extraordinary levels of childbearing can be interpreted in part as an increasingly maladaptive response to conditions of social insecurity and economic marginalization among an impoverished majority. Negative impacts on maternal and child health include anemia, low birth weights, malnutrition, and increased vulnerability to infectious diseases and stress. In addition, existing health agencies are strained to provide the curative and rehabilitative therapies required to deal with the ill-health that stems, directly and indirectly, from high fertility.

Kenya's population is organized into at least 40 African ethnic groups ranging in size from the Kikuyu (3.2 million), Luhya (2.1 million), Luo (2 million), and Kamba (1.7 million) to the Sakuye (1,824) and El Molo (538) (Figure 2-1). Some 32,500 Asian, 18,660 Arab, and 4,445 European citizens were also recorded in the 1979 Population Census, although there are also thousands more Asians and Europeans who reside in Kenya without Kenyan citizenship. By the end of the colonial period in 1962 the European population had expanded steadily to 55,759, while Asians numbered 180,000 or just over 1% of the total population (Diesfeld & Hecklau, 1978).

Swahili, a Bantu language indigenous only to certain coastal peoples, and English are both linga franca and official languages in Kenya. Most African ethnic groups have their own language, although ethnic and linguistic units are not always comparable (Whiteley, 1974). The African languages are organized into three major language groups, including Bantu (over 15 languages, including Kikuyu, Luhya, and Kamba, spoken by approximately two-thirds of the national population), Nilotic (13 languages, including Luo, Nandi and Maasai, spoken by three of ten Kenyans), and Eastern Cushitic (about 12 languages, including Galla and Samali, spoken by less than 4% of the population.

For virtually all Africans bilingualism is a prerequisite for individual economic and social advancement. It is also a necessary but often overlooked communication tool that can strongly influence the effectiveness of those who provide health services in urban and rural communities. For example, doctors and clinical officers who do not speak the language of the people who inhabit the areas where the government has posted them are at a distinct disadvantage as diagnosticians and health promoters.

Diabetes is an example of a disease that appears to exhibit marked differences among Kenya's ethnolinguistic groups and thus in its spatial pattern. Returns for 95 hospitals show the highest recording rates from 12 hospitals in three areas: Kericho in the highlands of western Kenya (49 cases per 100,000 inhabitants); the Central Province (30–45); and the Indian Ocean coast (20–50). In the remaining 83 hospitals reports of diabetes are negligible or nil (Diesfeld & Hecklau, 1978). The extent to which these data reflect the true geographic prevalence, as opposed to an artifactual distribution, cannot be determined in the absence of a more detailed study (Hunter & Arbona, 1985).

Kenya's urban population, estimated at 2.8 million or 15% of the total population, has increased rapidly since independence due to both natural increase and rural–urban migration.[3] If the present growth rate continues, 30% of the total population will be urbanized by the year 2000 (*Popline*, 1985). Demand for urban infrastructural services such as water, sewerage, housing, health services, and schools already is well beyond the government's capacity to provide. Nevertheless, in recent years urban areas have continued to consume a huge share of the national resources for services such as health care. For example, an estimated 80% of the government's recurrent expenditures for health is spent in Nairobi, Mombasa, and Kisumu. Kenyatta National Hospital (KNH) in Nairobi, the country's major teaching and referral center, is seen by the public as "the mainspring" of the national health services and consumes as much as 25% of the Ministry of Health's annual budget. Kenyatta National Hospital primarily serves residents of Nairobi (about 5% of the population), plus referrals. In addition, approximately 90%

of all doctors serve in urban areas (Family Health Institute, 1978). Despite the concentration of biomedical resources, there is no hard evidence that the urban poor are healthier than their rural counterparts.

The urban population is distributed among 91 urban centers (Obudho, 1983). Nairobi, the cosmopolitan national capital, is the major city with an estimated 1.3 million people or over 35% of the total urban population (Republic of Kenya, 1983). Its population is growing at estimated rates of 8–13% annually. Situated at the southeastern margin of the highlands, Nairobi is East Africa's largest city and the region's paramount center of transportation, commerce, banking, industry, tourism, and relatively sophisticated biomedical services. Other important urban centers include Mombasa (est. 510,000), the major ocean port for Kenya and Uganda; Nakuru (est. 110,000), a center of commercial agriculture in the Rift Valley which was once a major European enclave in the "White Highlands"; Kisumu, (est. 175,000), a growing commercial and industrial center at Lake Victoria; and Eldoret (est. 75,000), on the Uasin Gishu Plateau west of the Rift Valley.

Health problems in Kenya's urban centers include communicable diseases, social pathologies, and hazards associated with changes in life-styles. Much ill-health is linked to the enormous difficulties and lack of success at providing adequate and low-cost housing, safe water, sewage and refuse disposal, appropriate health services, and social services to rapidly expanding populations characterized by limited employment opportunities. Malnutrition, much of it apparently hidden from those who prepare official health statistics, is also a significant urban problem—in large part because the poor move into a money economy and can no longer grow the bulk of their own food.

During the last 20 years in Kenya "the estimated accumulated shortfall of [urban] dwelling units needed to be built has increased phenomenally" (Republic of Kenya, 1983, p. 165). A large proportion of the population in the largest cities such as Nairobi and Mombasa is housed in extremely crowded shanties, speculative slum barracks, and other makeshift housing which lack readily accessible, safe water and adequate sanitary facilities (*African Business*, 1985). In Nairobi, for example, large numbers of the poorly housed population depend on streams and rivers for water for drinking, brewing, cooking, washing clothes, recreation, and personal hygiene. Transmission of *Schistosoma mansoni* occurs along most of these water courses and surveys of children attending nearby schools have found prevalence rates for this infection ranging from 20% to 94% (Diesfeld & Hecklau, 1978; Highton, 1974). Although schistosomiasis is endemic in Nairobi, maintenance of the transmission cycle is undoubtedly linked in part to infected individuals who migrate from other endemic regions such as Machakos and Kitui Districts and the Lake Victoria borderlands.

Many additional health hazards emerge from the process of in-migration and mixing of tens of thousands of people from other places and disease environments. Infectious diseases such as measles, gastroenteritis, gonorrhea, and tuberculosis spread rapidly, and stress-related pathologies are intensified, in the urban nexus. Moreover, there are also large streams of people moving from urban-to-rural and urban-to-urban environments (Chapter 6). Unfortunately, despite the relatively dense urban networks of health service points such as the dispensaries and clinics operated by the Nairobi City Council, there is a dearth of readily accessible, pertinent, and reliable data on the ecology of urban health and the geography of risk in the capital. This deficiency continues to obscure the dynamics and true profile of many urban health problems and discourages appropriate interventions.

Approximately 85% of Kenyans are rural and economically dependent on agriculture. Since biomedical services are concentrated in the towns and periurban areas, access to dispensaries, health centers, and hospitals is difficult and costly for the large majority of Kenyans in both the high- and low-potential regions. Only an estimated one in five people are able to utilize the existing biomedical services provided by the government. Despite earlier government intentions to expand health service coverage to 60% of the population by 1983 (Family Health Institute, 1978, p. 75), population growth and government policies affecting the size and allocation of the health budget have prevented significant progress toward this objective (Republic of Kenya, 1983).

A REGIONAL APPROACH TO HEALTH FACTORS

In their excellent study entitled *Kenya: A Geomedical Monograph*, Diesfeld and Hecklau (1978) developed a system of 15 regional divisions of the country that are based on a synthesis of ecological, economic, social, medical, and cultural-historical conditions. The interrelationships of these factors show the conditions of life and provide the basis for a geomedical synopsis of health services and disease patterns. A condensed adaptation of the Diesfield–Hecklau classification is included here (by permission of the authors) in order to highlight the locational variability of Kenya's health problems, underscore the need for regional approaches to the planning of health services, and sketch the national setting of the case studies developed in Chapters 4–8. The two major socioeconomic criteria in the classification distinguish a Kenyan population divided between "African peasant and pastoral societies" and "class societies."

African Peasant and Pastoral Societies

1. *Peasant societies in areas of rainfed agriculture over 1,500 meters above sea-level.* Situated in the Highlands in southwestern Kenya, these societies include the Kikuyu, Nandi, Kisii, Luhya, and the Kamba of the hill

country in Machakos District. The lands they occupy are generally the most ecologically favored areas of rural economic development (Figures 2-1 and 2-2). Mean annual temperature maxima are below 26° C and mean annual minima can fall below 14° C. Rainfall is biomodal and generally adequate for two crops. Fertile volcanic soils are widespread and almost everywhere the population density exceeds 100 persons per square kilometer. In most areas the density ranges from 200 to 500 persons per square kilometer, but this density is surpassed in parts of Meru, in southern Kikuyuland, and in Kakamega District.

Intensive hoe cultivation is the dominant agricultural technique at 1,500 meters above sea level, with maize and legumes being the main staples. Ecological conditions often approach the ideal for cash and export crops such as tea, arabica coffee, pyrethrum, wheat, a variety of temperate zone vegetables and fruits, and grade dairy cattle. The latter are integrated with other agricultural activities and produce income from milk sales. Traditional, extensive animal husbandry has declined due to the need for and rapidly shrinking base of arable land relative to population growth and to agricultural modernization. The process of consolidating fragmented plots and registering individual titles to land, begun in the 1950s under the Swynnerton Plan, is now complete throughout large areas of the highlands.

Climatic conditions in the Highlands eliminate or greatly reduce the occurrence and distribution of animal vectors or the intermediate hosts of endemic tropical diseases. The intensity of malaria transmission also declines markedly above 1,500 meters. Tsetse flies and phelbotomes—the vector of leishmaniasis—are absent, and water temperatures are too low to support the snail hosts of *Schistosoma masoni* and *Schistosoma haematobium*.

Poor hygiene in dwellings, a high residential density, and coldness and dampness (favoring respiratory infections) promote diseases in the Highlands. Intestinal helminths, primarily roundworm (*Ascaris lumbricoides*) and tapeworm (*taenia*), are widespread in the population. Hospitals in selected locations including Meru (an area of high population density and land scarcity), Kericho, Londiani, Molo, and Naivasha (which include large-farm and plantation areas with a large population of agricultural laborers and their families) report a high frequency of kwashiorkor together with a medium or high measles incidence. Nasopharyngeal cancer is frequent among the Nandi and Kipsigis peoples as well as other highland dwellers. In addition to the likelihood of genetic factors in nasopharyngeal cancer (Linsell, 1974), carcinogenesis is possibly associated with irritation caused by long exposure to continuously smouldering wood fires in smoky and poorly ventilated dwellings.

The highlands above 1,500 meters tend to have the best infrastructure and the highest density of biomedical services, particularly church-related services. A remarkable diversity of TMPs and their imitators also operate in the villages and towns alongside biomedicine.

Peasant settlement schemes above 1,500 meters were initiated by the colonial administration of Kenya and continued after independence in order to alleviate the land shortage in the densely populated highlands. The largest resettlement program, known as the "Million Acre Settlement Scheme" (1962–1969), involved the purchase of 400,000 hectares of agricultural land from European farmers for redistribution among 35,000–40,000 African families. Both high- and low-density schemes were allocated in accord with settlers' net income goals. Credit, agricultural extension services, and grade dairy cattle were supplied to most settlers.

Health risks among the transplanted peasant families include psychosocial stresses and a worsening nutritional status associated with a production system that emphasizes cash crops. Biomedical services are generally inadequate and ignore mental health needs. Traditional systems of social security and health care have been weakened by population relocation and by fundamental changes in settlement, nutrition, livelihood patterns, and values. Increased infant and child mortality is a major threat in these circumstances. Contact with new disease reservoirs and reciprocal exchanges of such diseases as polio, tuberculosis, and hepatitis may occur among population groups already living in areas of in-migration.

2. *Peasant societies in areas of rainfed agriculture between 1,000 meters and 1,500 meters above sea level.* Ecological circumstances vary markedly in this zone. Climatic conditions generally do not favor production of higher value cash crops such as arabica coffee, tea, pyrethrum, potatoes, and temperate fruits and vegetables. The major areas of settlement are the densely populated but agriculturally less-developed plains of southwestern Kenya—the Luo and Luhya homelands—where maize, sorghum, legumes, millets, cassava, and sweet potato are cultivated; and Machakos and Kitui Districts in Eastern Province, the homeland of the Kamba people, where hoe cultivation of maize and legumes provides the staple diet. Population densities range from 100 to over 300 persons per square kilometer in Western and Nyanza Provinces, but in Machakos and Kitui Districts densities rise above 200 persons per square kilometer only in high potential "hill" zones.

Severe poverty and drought-related crop failure are marked in Machakos and Kitui below 1,500 meters (Porter, 1979). Cotton, sisal, and castor oil are grown for market in Machakos and Kitui, while sugar is a cash crop in Kisumu District adjacent to Lake Victoria. Lower population densities and longer fallow periods give rise to extensive bushlands in both the Kamba areas and southwestern Kenya. Although farm families keep more large livestock than in the areas above 1,500 meters, except in the Luo area near Lake Victoria where semipermanent cultivation with ox-ploughing is widespread livestock are not integrated with other farming activities. Grade cattle are poorly adapted to the climate and diseases of this zone.

In contrast to the Highlands, the warmer plateaus between 1,000 meters and 1,500 meters favor several infectious human diseases. Malaria is transmitted for up to 6 months in the Machakos-Kitui settled zones and practically year-round in southwestern Kenya. A zone of sleeping sickness persists in south Nyanza Province. Both southwestern Kenya and Machakos-Kitui are zones of endemic schistosomiasis, and western Kitui has endemic leishmaniasis. Hookworm and roundworm are favored by climatic conditions, poor sanitation, and ignorance of preventive meaures. Busia, Siaya, and Kisumu Districts in southwestern Kenya also have the country's highest levels of endemic leprosy (73–95.5 cases per 100,000 population).

Biomedical services in the densely settled Luo and Luhya-speaking areas in western Kenya are rated "good" in relation to area and "defective" or "moderate" in relation to population (Diesfeld & Hecklau, 1978). Traditional medicine is widely available, and among the Luo this is provided by the *ajuoga* (diviner), *jathieth* (who treats disease associated with natural causes), and the *jachoho* (birth attendant).

3. *African and Afro-Arabic peasant societies in areas of rainfed agriculture in the coastal lowland.* The human habitat along Kenya's palm-fringed "Swahili coast" is separated from the highlands and areas between 1,000 meters and 1,500 meters by the Nyika—a broad belt of arid, low-potential thorn-bushland that is sparsely populated (see Figure 2-3). The one major route across the Nyika connects the ancient Afro-Arabic port city of Mombasa with Nairobi and the Highlands. This corridor spans some 500 kilometers and comprises a railway, an exceptionally busy main road, and an air route (see Figure 2-1).

The narrow coastal zone varies in width from 30 to 60 kilometers. Its western periphery marks the moisture limits of rainfed cultivation and forms the margins of peasant settlement. Although the seasonal distribution of precipitation is not uniform in this zone, May is the month with the most rain at both Mombasa (235 millimeters) and Lamu (340 millimeters). The southern half of the coastal zone experiences eight to ten humid months (monthly rainfall of at least 50 millimeters) making double-cropping possible. Rainfall is increasingly seasonal toward the north and at Lamu the dry season extends from August through March.

Ecological potential in much of the coastal zone is not yet fully exploited. In most areas, population density is under 150 persons per square kilometer, rising to over 1,600 persons per square kilometer in the environs of Mombasa. From Malindi to the Somali border the density is generally less than 20 persons per square kilometer. Centuries of Arab and Swahili influence along the coast have contributed to a strong but varied imprint of Islamic culture among local populations. Shifting cultivation and bush-fallowing are the most widespread systems of agriculture, although in many areas these are interspersed with plantation systems of coconut, sugarcane, and sisal production. Maize and cassava, in the wetter and drier areas, respectively, are the staple foods. Cotton, coconuts, and cashews are important cash crops for smallholders. Tick vectors of East Coast Fever, mainly

Rhipicephalus appendiculatus, and tsetse flies limit the development of cattle keeping.

In addition to agriculture, many Africans at the coast are employed in the area's substantial tourist industry. Mombasa receives direct charter flights from Europe as well as tourists who come through Nairobi. This tourism is of epidemiological significance in that it potentially contributes to the importation of diseases, the transmission of local pathogens among the resident and visitor populations, and the rapid dispersal of pathogens to other parts of the world. Mombasa's port and other industries make it a major target for migration flows from all over Kenya and East Africa.

Malaria is endemic along the hot and humid coastal zone, with year-long transmission from Kilifi south to the Tanzania border and mainly seasonal transmission along the more lightly populated northern half of the coast. Bancroftian filariasis (*Wucheria bancrofti*), manifested as elephantiasis and hydrocele, has long been endemic along the entire coastal strip of Kenya. Urban and rural forms of filariasis (transmitted by *Culex fatigans*, and *A. gambiae* and *A. funestus* mosquitoes, respectively) have been distinguished. In several foci the "filariasis index" (microfilaria rate and density, and sign rate of hydrocele or elephantiasis formation) in males over 14 years old has exceeded 50%. *S. haematobium* schistosomiasis is also endemic on the coastal plain, with urine tests showing prevalence rates ranging from 22% to 67% of populations studied. Hookworm infestation is also high (45% to 95% of hospital populations) and predominates over roundworm except at Lamu. Infectious hepatitis occurs with exceptional frequency at the coast, registering the highest incidence (47.8 per 100,000 population) of any area in Kenya. Settlement schemes and irrigation projects established in the coastal hinterland have contributed to the spread of filariasis, schistosomiasis, and other intestinal helminths. The south coast is also the lesser of Kenya's two main foci of leprosy (the other is located in the Lake Victoria region), with a clinically diagnosable infection rate of about 1%.

The Coast Province also has the highest incidence of tetanus infections. Neonatal tetanus is closely related to the widespread local practice of sealing the navel with a mixture of earth, cow dung, and ash.

Biomedical services and transport links are relatively well developed along the immediate coast. There are 24 hospitals and 106 lower-order health service facilities in the four districts that comprise the coastal zone (*Weekly Review*)/Nairobi, February 6, 1981, p. 9). No systematic studies of African traditional medicine or other systems of medicine found at the coast, such as Unani ("Arab") and Ayurveda ("Hindu"), are available. However, the Digo and other coastal peoples continue to rely extensively on the *mganga* (s.)/*waganga* (pl.)—which roughly translates from the Swahili as "medicine man"—for assistance with health problems (Gerlach, 1959; personal observation, 1986). Peoples such as the Giriama (see Figure 2-1) are widely reputed to have a repertoire of powerful medicinal and poisonous plants. Kabwere, one of Kenya's most famous and awe-inspiring traditional healers, is head-quartered at Malindi.

4. *African peasant societies in marginal areas of rainfed agriculture.* Marginal areas form an extensive transition zone between the Kenya High-lands and the wide, arid strip (the *Nyika*) separating the coastal region from the highlands. Smaller, isolated zones of marginal agriculture are found on the Kavirondo Gulf at Lake Victoria and in the Cherangani Hills north of Eldoret in western Kenya.

Population density in the marginal farming zones varies with local moisture availability and soil quality. Overall, the mean annual rainfall is only 500–700 millimeters. In the environmentally disadvantaged zones within Machakos, Kitui, Meru, and Embu districts the population density is often below 50, and in some areas is less than 20 inhabitants per square kilometer. Shifting cultivation predominates and features drought-resistant species of cassava, millet, legumes, and, in favored wet areas, maize. Farmers and herders in eastern Kenya have evolved a wide variety of adjustments designed to cope with drought and reduce subsistence risks. Lacking suitable cash crops, many people keep cattle and sell them in order to meet their cash needs. Overall, the economy of the marginal zone is distinctly peripheral to and controlled by the wealthier highlands farming regions (Porter, 1979).

Malaria is mesoendemic in the marginal, semiarid zones of agriculture. Elevations between 500 meters and 700 meters form a natural focus of leishmaniasis (kala azar), with low endemicity characteristic of West Pokot, the northern Rift Valley, the Kerio Valley, northern Samburu, and Isiolo District. As already noted, the main focus of kala azar includes parts of northern Kitui and southern Meru Districts, and the middle Tana River basin.

High-risk agriculture and food shortages pose major threats to nutrition and health in the marginal areas. Since a large proportion of the able-bodied males of working age migrate outside the region to find jobs, women (Fergu-son, 1986), children, and the elderly are particularly exposed to nutritional deprivation (Porter, 1979). These areas of lower human population density also have few biomedical services.

5. *African pastoral societies in predominantly and semiarid lands.* Ke-nya's herding peoples are mainly subsistence pastoralists. They are distrib-uted at low densities (see Figure 2-3) over approximately three-quarters of the national territory. Extensive regions within this zone receive under 500 millimeters of precipitation annually, usually concentrated in a period of less than 3 months. Habitats are fragile and easily degraded. Human and live-stock carrying capacities are very low under subsistence herding and unra-tionalized management. Despite varied herding strategies, including differ-ences in stock combinations and mobility patterns, high losses of cattle associated with prolonged drought and overstocking make pastoralists ex-tremely vulnerable to food shortages. Cultivation is restricted to widely

scattered mountain and hill zones and is at best a supplemental food source against the herders' "extraordinarily critical nutritional position" (Diesfeld & Hecklau, 1978, p. 86).

Ethnically, Kenya's pastoral population is predominantly non-Bantu. The camel-herding Somali, Boran, and Rendille, and the Samburu and Galla peoples occupy much of northern Kenya and portions of the interior Coast Province (Figure 2-1). The Turkana are found in northwest Kenya between Lake Turkana and the Uganda border, while the Kenya Maasai—who once controlled a vastly greater territory—are confined to Narok and Kajiado Districts in southern Rift Valley Province (Figure 2-1). Human population density varies from four to ten persons per square kilometer throughout the zones of pastoral economy, but the lack of water and tsetse infestation over large tracts of bushland places great pressures on available resources.

Health conditions in the pastoral regions are at best poorly known and the density of the health services network is extremely low. Hospitals are found only at a few widely dispersed administrative centers such as Marsabit, Lodwar, Maralal, Garissa, Kajiado, and Narok. Health centers and dispensaries are confined to small, scattered market settlements and the quality of statistical records is poor. Since the population is both sparse and highly mobile, epidemiological observations based on health facility records may mean little. Systematic reporting of epidemics such as cholera and smallpox, spreading from neighboring and similar regions in the adjacent countries of Ethiopia and Somalia, did not begin until the 1970s.

Nomads such as the Samburu of north-central Kenya rely on diviners (s. *laibon*; pl. *laibonok*) who practice ritual medicine and prepare counter magic potions made from specific species of shrubs and roots. Herbal cures for various diseases are "public knowledge and accessible to everyone" depending on their location and personal knowledge of the plant and its uses (Fratkin, 1975; Spencer, 1973).

In general, much of the morbidity and mortality in Kenya's pastoral regions is never observed by biomedical personnel. Moreover, pastoralists such as the Boran camel herders of Isiolo District may by-pass biomedical services even where they are immediately accessible (personal observation, Garbatula, 1978). The pastoral populations do not appear to increase significantly over time, but they do decrease in places due to high mortality or emigration.

Infectious eye diseases such as trachoma and others that may cause blindness are endemic among most herding communities. Although Kenya lies south of Africa's "epidemic meningococcal belt," which extends into the Karamoja District of northeast Uganda, smaller outbreaks of cerebrospinal meningitis do occur in northern Kenya. Occasional fresh cases of yaws are also found among some of the northern nomads whose hygienic conditions make them particularly susceptible. The Maasai, many of whom have drifted into Nairobi and other towns in southern Kenya in recent years, are reportedly experiencing sharp increases in sexually transmitted diseases.

The pastoral regions are typically malarious near water. Exceptional rainfall may greatly increase the number of mosquitoes that survive from one rainy season to the next. This development together with the low immune status of the population favors localized epidemics. Other vectored diseases include leishmaniasis, which is reported sporadically in the north and northwest.

Unusually high rates of hydatid disease occur among Turkana children and adult women in northwest Kenya (11.6 cases per 1,000 population in the Lodwar catchment area). Transmission is between dogs infected with the canine tapeworm *Echinococcus* and infants within the homesteads. *Echinococcus* frequently occurs in association with brucellosis and anthrax in nomadic cattle herding communities where dogs are common and in close contact with people. Lodwar District Hospital, a small, desert medical oasis in Turkana District with an architecture not unlike a French Foreign Legion outpost, is a magnet for women with advanced cases of hyatid disease. The hospital is typically administered by some expatriate medical officer who is willing to trade the locational hardships and professional isolation for a few months or a year of "experience." On any day several Turkana women with huge hydatid cysts can be seen sitting or lying about the hospital's veranda awaiting their turn for surgery. Such operations are regularly performed by surgeons with the Flying Doctor Service, an agency of the African Medical and Research Foundation (AMREF) based in Nairobi.

Among Kenya's pastoral populations the beef tapeworm (*Taenia saginata*) is the leading intestinal helminth, followed by roundworms. Tuberculosis is also a significant health problem in several of the pastoral districts, whereas leprosy appears to be considerably lower among the nomadic peoples than it is among the sedentary farming populations.

Class Societies

Kenya's six "class societies" are associated with several plantation-style systems of agriculture (Diesfeld & Hecklau, 1978). The origins of these class societies is directly traceable to the colonial economy in which Africans provided cheap labor and Europeans controlled the land and provided capital and managerial skills. This pattern continues today, albeit with significant social modifications. The most important change is that since independence wealthy Africans have purchased former European farms and ranches. Consequently, Kenyanization of the former "White Highlands" by African farmers will soon approach 100%. The term "Afro-European class society" must therefore be seen in its proper context. It describes a structure undergoing rapid change that "is only valid within certain historic limits" (Diesfeld & Hecklau, 1978, p. 89).

Five Afro-European "class societies" are identified by Diesfeld and Hecklau (1978): those (1) engaged in a pastoral economy in arid areas; (2) on large farms in areas of rainfed agriculture; (3) on coffee plantations; (4) on tea plantations, and (5) on sisal plantations. "Afro-Asian" class societies on sugar plantations represent a sixth form of economic and social organization. Although these six class societies constitute only a small fraction of Kenya's total population, they account for a disproportionate share of the country's commercial agricultural production and highlight the marked inequalities that characterize contemporary Kenyan society as a whole.

Class societies involved in *ranch systems of livestock production* are found mainly high bushlands and grasslands above 1,800 meters that are tsetse-free. Crop production is generally impractical in these areas because rainfall is low (700–800 millimeters) and the dry season is at least 9 months long. The greatest number and acreage of ranches are found on the Laikipia Plateau, north of Nanyuki in central Kenya; the north side of Mt. Kenya below 2600 meters; and the areas north and south of Lake Naivasha in the Rift Valley. Human population densities are low, ranging from under 10 to less than 20 persons per square kilometer over extensive tracts. In many areas at least 4 hectares are required to support one cattle unit. Living standards for the African stockmen and their families are low and sanitary conditions are not well-developed. Health services are poor, and the vast distances within the areas makes access to biomedical care very difficult. Although elevation eliminates endemic malaria and other warm climate infections from most of the large-scale ranching areas, communicable diseases such as tuberculosis, polio, and hepatitis are present. Also, inadequate food production by African worker families engaged in a life-style increasingly integrated into the cash economy increases the risk of malnutrition among their children.

Areas characterized by *"Afro-European class societies on large farms"* form the core of the former "White Highlands." They include much of Nakuru District, the high plateaus of Uasin-Gishu and Trans-Nzoia situated west of the Rift Valley; and the high foreland of the northern Nyandurua range and Mt. Kenya. Large-scale mixed farming featuring both grain production (wheat, barley, and maize) and animal husbandry (cattle and sheep) attains its highest elevations on the Mau Escarpment west of Nakuru, where the cultivation limit approaches 3,000 meters. Vector-borne diseases are comparatively insignificant in these temperate zones of large-scale mixed farming. Infectious diseases predominate, including respiratory ailments. Hospital returns indicate the presence of roundworm eggs (*Ascaris lumbricoides*) in at least 40% of the populations surveyed. The density of health services in most of this region is characterized by Diesfeld and Hecklau (1978) as "moderate" to "medium" in relation to area and population.

Kenya's *coffee and tea plantations* are also situated in productive highland areas between 1,500 meters and 2,400 meters above sea level. Estate coffee production occurs mainly on the volcanic soils of the eastern and southern margins of the Nyandarua Range (Aberdares) near Nairobi. Tea plantations are concentrated in well-watered zones (over 1,300 millimeters of rain) on slightly acid soils in Kericho and Nandi Districts in southwest Kenya. Labor for coffee production is recruited seasonally from the adjacent, densely settled Kikuyu homelands. Workers on tea estates typically live with their families in company-owned housing. While the density of health services in the estate coffee- and tea-producing regions is considered "good" in relation to population and area (Diesfeld & Hecklau, 1978), there is little published information concerning the health profile of laborers and their families.

Approximately 25,000 laborers are employed on some 50 *sisal plantations* in Kenya. The main producing areas include the drier and agriculturally poorer sections parts of the Coast Province such as Taita District and parts of Kilifi; the eastern, lower margins of Kiambu and Murang'a Districts north of Nairobi; and northern Nakuru District in the Rift Valley. Once again, there is little systematic knowledge of health and health service problems on the large-scale sisal plantations. Diesfeld and Hecklau (1978) suggest only that there are "severe" deficiencies. "Massive schistosomiasis infection" is among the problems known to exist in the sisal estates of Coast Province, particularly in Taita District around Voi and Taveta.

Kenya's *sugar plantations*, the locus of rural *Afro-Asian* class societies, are situated in two areas: the Miwani sugar belt situated to the east of Kisumu near Lake Victoria, which stretches for some 35 kilometers; and in Kwale District on Kenya's south coastal plain near the Tanzania border. Of the nearly 30,000 hectares devoted to cane, over half is owned by large sugar factory companies, another third is on large farms located near the sugar factories, and the rest is produced by African smallholders who, together with the large farms, supply sugar to the factories. Processing of the cane sugar is controlled by a few large enterprises owned by wealthy Indian families such as Miwani, Madhvani, and Metha. Environmental conditions in both of the main producing areas favor intense malaria transmission, schistosomiasis, and roundworms. Mass administration of chemoprophylaxis to stem a high level of malaria morbidity has been successfully conducted by Kenya's Department of Vector-Borne Diseases since 1966 on the Ramisi Sugar Estates in Kwale District at the coast. Little else is known about the specific health profile of African workers on the Asian-owned sugar estates.

Despite wide gaps in the preceding overview of health conditions in Kenya (several of these shortcomings are addressed in the case studies in Chapters 4–8), the preceding discussion underscores the fundamental importance of the ecological-regional approach to the analysis of health problems.

An understanding of how environmental conditions, social structure, political economy, cultural behavior, and health are dynamically interrelated—and why and how these associations vary spatially—are crucial concerns which should guide the development of appropriate interventions for community health.

CORE DISEASE PATTERN: A SUMMARY

At the national scale Kenya mirrors the basic core disease pattern associated with a large group of technologically underdeveloped, urbanizing Third World countries. The core diseases comprise five main categories: malnutrition, diseases related to deficiencies in water supply or sanitation, contact and airborne infections, psychosocial stress, and accidents. These are not static categories. As Kenyan society continues to evolve and becomes more urban the process of "epidemiological transition" (Omran, 1977) from infectious diseases to chronic and degenerative diseases should be anticipated.

Protein-energy malnutrition (PEM) is considered Kenya's principal nutritional problem (Haaga, *et al.*, 1986). Approximately 24% of preschool-age children reportedly have mild PEM, while 2% have severe PEM. Kenya is currently working to develop a nutritional capability that will provide a per capita average of 2,557 calories daily by 1988 and at least 2,360 calories for each adult (Republic of Kenya, 1983, p. 186). A study of malnutrition undertaken as a part of Kenya's Integrated Rural Survey found that in the age group of 1–2 years only 36.6% of the girls and 56.5% of the boys were normal in regard to both height and weight (Balcomb, 1978).

Diseases related to deficiencies in water supply or sanitation include four groups (Bradley, 1974). *Water-borne* diseases, where water acts as a passive vehicle for a pathogen, include the classic "common source" infections such as typhoid, cholera, and infectious hepatitis. *Water-washed* and fecal-disposal diseases diminish in reponse to increasing the volume of water available to people. They include intestinal helminths such as roundworm (ascariasis), hookworm (typically *Necator americanus* except *Ankylostoma* on the coast), trachoma, scabies, dysenteries and enteroviral diarrheas, poliomyelitis, and leprosy. Schistosomiasis, whose transmission cycle requires aquatic snails as intermediate hosts, is the classic *water-based* disease in Kenya. Significant *water-related vectored* infections—those spread by insects that breed in or bite near water—include, first and foremost, malaria; also trypanosomiasis, bancroftian filariasis, and onchocerciasis. [Visceral leishmaniasis, or *kala azar*, transmitted by sandflies (*Phelbotomus spp.*) is also a very important vectored disease in Kenya but its ecology is related to anthills rather than water].

Contact and air-borne diseases are usually transmitted from person to person. The main infections in Kenya are tuberculosis; measles; sexually transmitted diseases, particularly gonorrhea; leprosy; influenza; whooping cough; diphtheria; and meningococcal meningitis. Several of the diseases have been targeted for control as part of Kenya's Development Plan for 1984–1988 (Republic of Kenya, 1983).

Diseases of *psychosocial stress*, including alcoholism and hypertension, but especially depression, neuroses, and psychoses, are apparently widespread and greatly underestimated among Kenya's urban and rural populations. The philosophy and structure of the biomedical services—which emphasize curative medicine and labeling of sick persons according to the International Classification of Diseases—makes them poorly equipped to identify and cope with mental and psychosomatic ill-health. The broad scope and importance of this area of need in health care is soon evident upon examining the activities of TMPs and their patients and the nature of the illnesses the latter present (Chapters 5–8).

Accidents, including traffic accidents and occupational exposures, are an increasing source of morbidity and mortality on the Kenyan scene. This category can be expected to account for a larger proportion of the core disease pattern as the Kenyan population urbanizes and mechanizes, and as the use of toxic substances in agricultural and industrial processes expands.

A final category of diseases that have particular geomedical relevance include several discussed earlier such as endemic goiter, fluorosis (Nair, Manji, & Gitonga, 1984), and diabetes (Diesfeld & Hecklau, 1978). Cancer is among the important diseases in this cluster, including nasopharyngeal carcinoma and cancer of the esophagus. Esophageal cancer has a roughly tenfold male predominance and is most common in the Western Province near Kisumu and throughout the Rift Valley highlands. Attempts to find ecological associations of this cancer with such factors as raw alcohol consumption have not been rewarding, and it is extremely rare in Kenya outside the areas mentioned (Diesfeld & Hecklau, 1978; Linsell, 1974).

As in Uganda and Tanzania, cancer of the penis is found almost entirely among Kenyan ethnic groups that do not practice circumcision. These include the Luo, Samia, and Turkana, among whom this cancer is three times higher than expected. Although cancer of the cervix is the most frequent cancer in Kenyan females, according to Linsell (1974) it is not associated with the pattern of distribution of penis cancer.

Liver cancer is common in Kenya and shows "significant variations in incidence . . . over short distances" (Linsell, 1974, p. 388). Possible etiological associations include chronic liver damage, "nutritional" cirrhosis (more typical in Africa than alcoholic cirrhosis), viral infections, and aflatoxins produced by fungi on stored cereals in rural areas.

Kenya's core disease pattern is in transition. Visible progress has occurred in the form of reduced infant mortality (the rate fell from 120 to 86 per

1,000 births between 1963 and 1982), although the country's relentlessly high fertility level surely depresses the quality of life for many children who survive. Longevity has also increased, from 40 years in 1963 to 54 years in 1982 (Republic of Kenya, 1983, p. 35). Nevertheless, preventable diseases and malnutrition continue to account for the larger share of ill-health. The persistence of these problems underlies the fact that poverty and political impotence, particularly within the rural population, are the key factors that block satisfactory progress toward healthier communities in Kenya and many other areas of Africa. It remains to be seen how Kenyans cope with ill-health in the future, including the growing burden of mental and stress-related illnesses that are currently treated primarily by TMPs—or not at all. In the discussion that follows attention is focused on the various therapies currently available, and for whom they are practical alternatives.

Kenya's biomedical services are the only officially recognized system of health care in the country; yet this system is accessible only to a favored fraction of the population. A description of the physical infrastructure of biomedical services is included below because it is this model—dependent on hospitals, doctors, and a heavy burden of recurrent expenditure on "bricks and mortar"—that largely explains why biomedicine is more a promise than an alternative for most of Kenya's citizens.

THE RANGE OF THERAPEUTIC OPTIONS IN KENYA

As noted in Chapter 1, the *popular sector* of health care is where illness is first perceived and acted upon, either by individuals directly or with the assistance of their close associates. Therapy begins with self-awareness and, depending on the problem at hand, with self-treatment using household remedies and consultations with a cluster of kin and friends. With few exceptions (e.g., see Maclean, 1971), the popular sector has received little attention from social and medical scientists in Africa or elsewhere. Although it is also not a main focus here, it is essential to recognize that the popular sector is typically the first line of resort for the Kenyan population, as it is throughout the world. Moreover, the practices adopted in the popular sector often directly influence the outcome of therapies that may be resorted to later—for the same or a different illness—in the "biomedical" and "traditional medical" sectors (see Figure 1-1).

The primary position and importance of the popular sector in health care among the population of Kenya is shown in Figure 2-4. In this diagram, as in Figure 1-1, the popular sector is the link between the various strategies that together comprise the potential range of therapies found in the local ethnomedical system. The inferred interaction of the popular sector with the biomedical and traditional sectors is a crucial point. Kleinman's (1980, p. 51) argument is pertinent:

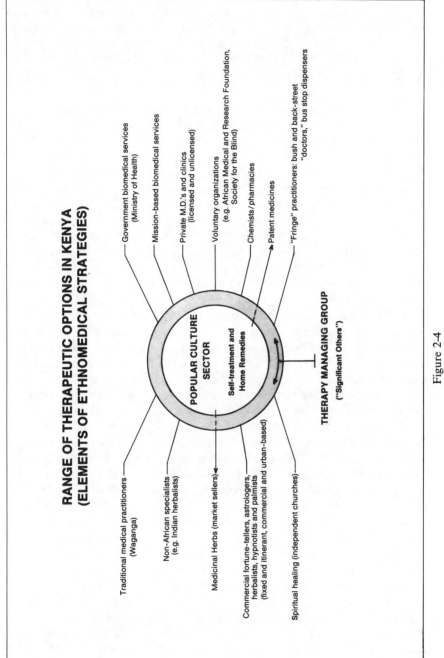

RANGE OF THERAPEUTIC OPTIONS IN KENYA
(ELEMENTS OF ETHNOMEDICAL STRATEGIES)

Government biomedical services (Ministry of Health)

Mission-based biomedical services

Private M.D.'s and clinics (licensed and unlicensed)

Voluntary organizations (e.g. African Medical and Research Foundation, Society for the Blind)

Chemists/pharmacies

Patent medicines

"Fringe" practitioners: bush and back-street "doctors," bus stop dispensers

POPULAR CULTURE SECTOR

Self-treatment and Home Remedies

THERAPY MANAGING GROUP ("Significant Others")

Traditional medical practitioners (Waganga)

Non-African specialists (e.g. Indian herbalists)

Medicinal Herbs (market sellers)

Commercial fortune-tellers, astrologers, herbalists, hypnotists and palmists (fixed and itinerant, commercial and urban-based)

Spiritual healing (independent churches)

Figure 2-4

61

The customary view is that professionals organize health care for lay people. But typically lay people activate their health care by deciding when and whom to consult, whether or not to comply, when to switch between treatment alternatives, whether care is effective and whether they are satisfied with its quality. In this sense the popular sector functions as the chief source and most immediate determinant of care.

"Significant others" (Janzen's therapy managing group) in the lay community play a varied but often central mediating role in the process of choosing and evaluating therapeutic options for ill persons.

The 12 therapeutic options shown in Figure 2-4 are representative but do not exhaust the strategies that may be available to individual Kenyans. Moreover, as the case studies in Chapters 5–8 demonstrate, during an illness people often resort to a wide variety of combinations of these strategies in their quest for healing. In theory, and assuming cash resources are sufficient, persons who live in larger urban areas such as Nairobi Mombasa, and Kisumu have direct access to all or most of the strategies shown in Figure 2-4. In contrast, rural folk have a much narrower range of options from which to choose unless they are willing to travel and absorb the related expenses to obtain the health services they prefer.

In the case studies of rural and urban ethnomedical systems presented in Chapters 5–8, attention is centered on the much-neglected and poorly comprehended *traditional medical sector* and its linkages to the biomedical sector. In both the rural and urban settings the findings are drawn from practitioner- and facility-based participant-observation and interviews. The activities and locational patterns of TMPs, of *waganga* (Swahili), and their patients form the main framework of analysis.

A complementary survey designed to assess knowledge, attitudes, and practices (KAP) concerning traditional medicine was also conducted among 212 outpatients at filter clinics and 119 inpatients at the Kenyatta National Hospital (KNH) in Nairobi—Kenya's huge referral center and the teaching hospital for the University of Nairobi Medical School. The survey results undoubtedly underreported the current influence of traditional medicine because many people have been conditioned through contact with Western institutions and attitudes to camouflage and deny any association with it. Nevertheless, the degree of contact actually recorded suggested that traditional medicine is a significant resource in health care throughout Kenya.

Some 22 ethnic groups were represented among the 331 patients questioned in the Kenyatta National Hospital KAP survey.[4] Only 11.7% of the patients ($N = 25$) had a current residence outside the immediate Nairobi metropolitan area. Of these, 21 lived in the adjacent Kiambu District. As expected, Kikuyu patients ($N = 159$) predominated (48% overall and 54% of outpatients) since Nairobi's African population is at least 60% Kikuyu and the city is contiguous with the southern edge of the Kikuyu homeland. Luo

($N = 56$) formed the second largest group represented (17% overall and 17% of outpatients), followed by 46 Kamba (14% and 9%) and 34 Luhya (10% and 11%). These are also the four largest ethnic groups in the national population.

Over two-thirds of all patients in the Kenyatta KAP survey and 96% of the Kamba—the group focused upon in the rural and urban case studies— reported that TMPs practice in their own (rural) home locations (Table 2-2). The corresponding figures for the other major ethnic groups were: Luo (88%), Luhya (88%), and Kikuyu (48%). Only 21% of the hospital patients admitted that they had ever consulted a TMP. However, the responses differed among ethnic groups. Over 41% of the Kamba patients said they had visited a TMP, compared with the Luo (32%), Luhya (15%), and Kikuyu (12%). One in five patients said they knew of at least one extended family member who had been treated by a TMP. Among ethnic groups the response rates for family members treated by a TMP were Luhya (34%), Luo (32%), Kamba (30%), and Kikuyu (10%). Some 23% of the patients also said they

Table 2-2. Aggregate Responses to Selected Questions in Kenyatta National Hospital KAP Survey on Traditional Medicine (Patient $N = 331$)

	Yes		No		No answer or don't know	
	No.	%	No.	%	No.	%
1. Have you ever consulted a traditional healer for any illness?	79	24	248	75	4	1.2
2. Do *waganga* work in your home area?	223	67	93	28	15	4.5
3. Is any member of your extended family a traditional healer?	42	13	287	86	2	1.0
4. If you previously consulted a *mganga*, were you satisfied with the treatment you received?	39	59	27	41	—	—
5. While consulting a *mganga* did you concurrently attend a modern medical service for the same problem?	19	30	44	70	—	—
6. Do you know of any person who was cured by a *mganga* after modern medical treatment had "failed"?	74	22	254	77	3	1.0
7. Apart from yourself, has any member of your extended family ever been treated by a traditional healer?	68	21	226	69	37	10.0
8. Are there illnesses that *waganga* treat more successfully than modern medical personnel?	98	30	143	43	90	27.0
9. Have you ever used herbal medicine?	105	32	223	67	3	1.0
10. Should traditional healers be licensed by the government?	100	30	94	28	137	42.0

Note. From Hospital survey, December 1977.

had personal knowledge of cases in which a TMP had cured an illness after efforts in the biomedical sector had failed. The personal use of herbal medicine in self-treatment was reported by 32% of the KNH patients.

Remarkably, 13% of the KNH hospital patients reported at least one TMP in their extended family. Rates for ethnic groups varied as follows: Kamba (28%), Luo (21%), Luhya (11%), and Kikuyu (5%). Although it cannot be assumed that the hospital patients represented a true cross-section of the total population of Kenya (14.7 million) in 1978, extrapolation of the 13% rate yields 127,400 TMPs![5] This figure may be 50% too large; yet even 63,700 TMPs produces a crude rate of 1 TMP for every 155 Kenyans.

Patients' attitudes about extending official government recognition to TMPs were surprisingly positive (Table 2-2). Only 28% said TMPs should *not* be licensed to practice, while 30% approved of licensing and 42% said they "don't know." These findings suggest a degree of public support for traditional medicine that was larger than anticipated. Such attitudes may have been influenced in part by the up-beat feature articles on traditional medicine in Africa that appeared with some frequency in local newspapers and magazines during the late 1970s. This climate of receptivity also punctuated Kenya's *Development Plan for the Period 1979 to 1983* (Republic of Kenya, 1978), which devoted more than five paragraphs to traditional medicine. In contrast, traditional medicine is not mentioned at all in the *Development Plan for the Period 1984 to 1988* (Republic of Kenya, 1983), although a traditional medicine unit with a pharmacology emphasis was established in the Kenya Medical Research Institute.[6]

Overall, the Kenyatta KAP survey reinforces the evidence that traditional medical systems are functioning extensively in Kenya. Although there are marked regional variations in attitudes toward and in the use of traditional medicine, TMPs are readily available in most rural areas and a significant proportion (30%) of the patient population specifically supported legitimization of TMPs through some form of licensing. Among the major ethnic groups surveyed, the Kamba exhibited the strongest identification with traditional medicine.

In most instances TMPs can properly claim to be the legitimate bearers of indigenous medical traditions. The other therapeutic options shown on the left side of Figure 2-4 are, with the partial exception of those who collect or sell medicinal herbs, comparatively recent additions to Kenya's medical pluralism.

Although admittedly their distinctiveness from urban TMPs is often blurred, *commercial fortune tellers, astrologers, hypnotists, palmists, and "herbalists"* are individuals who purvey expensive advice, talismans, and "medical" substances to gullible and often desperate people in the towns. While no published studies of such practitioners have been unearthed for Kenya or other parts of Africa, few individuals in these categories appear to represent the health interests of the public. They should not be confused with *bona fide* TMPs. Many of them operate out of rented hotel suites or central office buildings. They advertise in the major daily newspapers under such banners as "Kenya's World-Famous Astrologer, Herbalist, Palmist, Evangelist and Healer," or "Professor of Medicines," and publish preposterous claims about their healing and life-changing powers. Self-promotion of this sort by TMPs would be anathema under traditional rules of conduct (Chavunduka, 1978). Some among this newer breed of entrepreneur are itinerant, moving every few days within a circuit of places such as Nairobi, Eldoret, Kisumi, Mombasa, Dar es Salaam, and other urban centers in Kenya and East Africa. This category of practitioners, whose main tools appear to be legerdemain and the vulnerability of anxious clients, represents the least known of the many health-for-profit schemes found in Africa today.

The Central Government, through the Ministry of Health (MOH), has primary responsibility for *biomedical services* (Figure 2-4). Mission-based services are also of great significance, particularly in rural areas, while private doctors (including many Asians) have located their "surgeries" and clinics primarily in the larger towns and cities.

Government biomedical services in Kenya are of two basic types. Hospital services are distributed according to politicoadministrative units and special-purpose functions. Kenyatta National Hospital, with over 1,800 beds and a variety of specialist services, is the apex of the system. It mainly serves the population of metropolitan Nairobi, or about 5% of the national population excluding referrals (Family Health Institute, 1978). Kenyatta National Hospital employs over 3,000 workers including approximately 350 doctors and 1,100 nurses. In addition, the Nairobi City Council operates a maternity hospital, 7 maternity centers, 21 health centres and dispensaries, a venereal disease (VD) clinic, and a chest clinic. Nairobi also has a number of nongovernment, mission, and company-run dispensaries, hospitals, and maternity facilities. Nairobi Hospital, a private facility that catered to Europeans before Independence, now serves the privileged elite of all races. Overall, Nairobi has 12% of all hospitals and 19% of all hospital beds (604 beds per 100,000 population) in Kenya (*Weekly Review*, 1981).

Provincial hospitals are referral and training centers for Kenya's seven provinces and are located in the main towns of each province (Figure 2-1). These are Mombasa (Coast Province), Garissa (North Eastern Province), Machakos (Eastern Province), Nyeri (Central Province), Nakuru (Rift Valley Province), Kakamega (Western Province), and Kisumu (Nyanza Province). There is a district hospital in each of Kenya's 40 districts excepting the seven districts in which the Provincial hospitals are located. A number of smaller general hospitals are also found at the subdistrict level. Mathari Hospital, Kenya's main psychiatric facility, is situated in Nairobi adjacent to the shantytown of Mathare Valley.

Provincial and district hospitals—which provide the "curative services"—are seen as referral points for the health centers and dispensaries that form the main infrastructure of "rural health services." The formal responsibilities of these hospitals include the coordination and strengthening of community-based health care, teaching, and administration of rural health services (Republic of Kenya, 1983).

Rural health services are organized according to the concept of a rural health unit (RHU). Each RHU—there are more than 250 of them in Kenya—is based on a region that ideally serves 50,000 to 70,000 people. The RHU service region may include two or three locations—the administrative units that are the building blocks of the districts. Health centers (HCs) are usually the headquarters of a RHU to which are linked a number of satellite sub-HCs or dispensaries. A clinical officer has 3 years of training and is in charge of a RHU. He leads a RHU team that includes "enrolled" community nurses, a public health technician, and family planning educators. Each RHU team is responsible to the district medical officer of health (DMOH). The central government operated 233 HCs in 1982, and another 38 were owned by missionary organizations. Including municipal (2) and private (1) facilities, from Independence in 1963 to 1982 the number of HCs increased from about 160 to 275, or by 71% (Republic of Kenya, 1983). Nevertheless, the regional distribution and physical access to HCs shows marked inequalities. In 1978 the national average was reportedly one HC per 72,000 people, but the range was 1:10,000 in Lamu District on the coast to 1:166,000 in Turkana in Kenya's arid northwest (Republic of Kenya, 1978).

Ideally, HCs are intended to provide a broad range of preventive, promotive, and curative care. This includes inpatient services for emergencies and patients requiring observation plus onward referral for patients requiring a doctor or hospital; curative and rehabilitative care in daily outpatient clinics; low-risk deliveries, and integrated maternity, child health, and family planning services in daily clinics; consultations and laboratory investigations for refractory cases referred from sub-HCs and dispensaries; regular school health services; and mobile services to dispensaries and other centers in the RHU.

In 1982 the central government operated 802, and missionary organizations 232, sub-HCs and dispensaries throughout Kenya. In contrast, only a handful of dispensaries existed at Independence in 1963. Today they form the base of the rural health services. They offer outpatient and referral services but are not equipped to perform deliveries or accommodate inpatients.

In newly independent Kenya the *mission medical services* had become an indispensable element in the national system of biomedicine. By 1966, missions accounted for 46% of all hospitals and 23% of all hospital beds (Diesfeld & Hecklau, 1978). In 1982 missionary organizations still operated 21.3% of all health facilities in Kenya, including 39% of the hospitals and 20% of the dispensaries. During the 1984–1988 Development Plan Period the church hospitals are expected to operate at a cost amounting to only 4.4% of the Ministry of Health's recurrent and development expenditures on curative services (Republic of Kenya, 1983).

In 1979 Hartwig noted that the relationship between the church-run medical sevices and the Kenya government remained largely unchanged from that which prevailed under the colonial system. In other words, the church institutions suffered from "benign neglect" and some were approaching bankruptcy. They had been "expected to be essentially self-supporting financially, with the modest government grants frozen at decade-old levels" (Hartwig, 1979, p. 125).

Although Kenya's church-related health institutions have reportedly received little guidance from the Ministry of Health over the years, they are likely to remain essential for many decades to come. The churches are also major providers of health services in many other African states. They account for approximately 45% of the total available health services in Tanzania, 40% in Malawi, 34% in Cameroun, and 29% in Ghana (Hartwig, 1979). It is noteworthy that in Kenya's *Development Plan for the Period 1984 to 1988* the government projects increases in allocations to church hospitals of 53% for recurrent expenditures and 19% for development expenditures (Republic of Kenya, 1983). Despite the construction of many new HCs, dispensaries, and training facilities since Independence, and much official rhetoric about the need to expand primary care to the agrarian masses, rural health services in Kenya have consistently lost out to hospitals in the competition for resources. The philosophical and structural changes that must occur to move Kenya toward the World Health Organization (W.H.O.) target of "health for all" by the year 2000 have not materialized (Mburu, 1979, 1981; Republic of Kenya, 1983; World Health Organization, 1978). Kenya's official biomedical system—which is usually heavily patronized by the minority who have regular access to it—is still hospital-based, doctor- and technology-dependent, and profession-oriented. The roots of this entrenched system extend to the early colonial period. Because it has become so resource-consuming—in terms of both physical infrastructure and expensive labor—it is not a sustainable system within the limits imposed by Kenya's demographic growth and budgeting capabilities (Family Health Institute, 1978).

Kenya's *Development Plan for the Period 1984 to 1988* offers hopeful signs of changes that could bring about an increase in and an extension of appropriate biomedical services to the needy rural populations. This plan specifically recognizes the urban bias of the current system and pledges support for a new distribution of health care resources.

> Development of rural health infrastructure has lagged behind, in part, because of budgetary constraints. Public spending to maintain and extend costly urban-based hospitals will be contained, and the bulk of savings from the slow-down of capital projects in urban areas will be redirected towards smaller-scale projects at the district and sub-district levels (Republic of Kenya, 1983, p. 152).

Analysis and comparison of the national health budgets for 1978–1983 and 1984–1988 reveals an increase of 65.3% in *development* expenditure for rural health, that is, HCs, dispensaries, and training, in the current plan. The allocation for hospitals during 1984–1988 is actually 4.6% below the 1978–1983 level before adjusting for erosion of purchasing power. However, because the *recurrent* expenditure required to maintain the operation of existing hospitals continues to increase (the 1984–1988 expenditures are 91% above the 1978–1983 level), there is reason for concern about whether the Kenya government can be committed to an effective expansion of rural health services without dismantling some elements of the hospital-based part of the national delivery system. In fact, the actual *overall* allocation of funds for the rural services shows a decrease from 34% of the health budget in 1978–1983 to 26% in 1984 (Republic of Kenya, 1978, 1983).

Kenya also has branches of or is the main field for numerous *private voluntary organizations* (PVOs) that provide health services. These include the Family Planning Association, St. John's Ambulance Society, Kenya Society for the Blind, and the Kenya Society for Deaf Children. The African Medical and Research Foundation (AMREF), with a staff of some 400 based in Nairobi, is among the largest PVOs in Kenya. Its field of operations also includes Tanzania, Uganda, Malawi, and the southern Sudan. The African Medical and Research Foundation's programs are generally designed to complement and provide research support for the government's health programs. While the East African Flying Doctor Service, founded in 1957 and based in Nairobi, is perhaps the best known of AMREF's functions, during the last decade much emphasis has been given to strengthening the rural and primary health services. Activities include a surgical nursing and radio service to scores of rural hospitals, training and support of community health workers, research into health behavior and health education, including traditional medicine (see Githagui, June, 1985; September, 1985) and the publication of technical and nontechnical materials about and for local health workers (see Shaffer, 1984). Research on and treatment of eye diseases, hydatid disease, and leprosy are other concerns of this important organization (African Medical and Research Foundation, 1980).

In Kenya's towns, *chemists* (pharmacists) perform important roles that overlap with the biomedical and popular sectors (Figure 2-4). They are often primary consultants for the sick or their representatives, who often wish to avoid the cost or inconvenience of visiting a doctor or an outpatient clinic. In addition to dispensing patent (over-the-counter) medicines, they give advice about sickness and they prescribe and sell numerous drugs that are available in the United States only with a physician's signature. Thus in Kenya as in other African countries, chemists must be seen as established health care authorities who have the power to influence the outcome of their patrons' illnesses in both positive and negative ways. Research concerning the role of pharmacies has yet to be undertaken in Kenya. A new study from Addis Ababa by Kloos and co-workers (1986) points the way.

Marginal, self-styled, and unscrupulous practitioners who operate on the fringes of both biomedical and traditional healing are also an increasingly important force within the spectrum of therapeutic options in Kenya (Good & Kimani, 1980; Wasunna & Wasunna, 1974) and in the rest of the Third World (Van der Geest, 1982) (Figure 2-4). Attracted by opportunity for financial gain from a vulnerable public desirous of "modern," quick cures, these "bush doctors" (*Daily Nation*, 1977, 1978) and their urban counterparts—the "back-street doctors" (*Sunday Nation*, 1977) and "bus-stop dispensers" (McEvoy, 1976)—manipulate the "popular sentiments flowing from traditional . . . and modern medicines" (Conco, 1972). Many rural and urban dwellers who exercise the right to choose are drawn to these untrained, poorly informed entrepreneurs whose practices often have lethal outcomes. These *"fringe"* practitioners administer injections of liquid chloroquine and dispense antibiotics and other controlled-type drugs obtained through black market channels—often with tragic results for entire families. In Nairobi and other urban places the "bus-stop dispensers" and "doctor-boys" or "back-street doctors" hustle antibiotic capsules, abortifacients, and other illicitly acquired drugs to anxious clients in crowded city streets, bus depots, and markets. The consequences of this indiscriminate "pushing" of unlabeled synthetic drugs without concern for their pharmacology, interactions, and additive potential are varied and difficult to quantify. McEvoy (1976, pp. 193–194) has commented on this growing problem.

> At the health centre clinical problems of management and treatment arise from delayed presentation, and a clinical picture obscured or complicated by previous polypharmacy. The efficacy of first-line drugs is unnecessarily compromised by the emergence of resistance, hypersensitivity, and tolerance. Furthermore, injectibles are frequently diluted with unsterile fluids and administered with poor technique, resulting in local pyogenic infection, septicaemia, and serum hepatitis. . . . The environmental effects . . . are a danger to the community. At present penicillin, the cheapest and most readily available antibiotic in Kenya, is virtually useless against most common infections. The economic implications in the case of anti-tuberculous drugs are formidable, and we may soon see chloroquine resistance in falciparum malaria become widespread in Africa.

McEvoy (1976) recommends that the problem of unethical drug trafficking be approached first through interdisciplinary research in order to identify and understand the sources, distribution, and retail channels of the substances. He sensibly cautions against precipitous actions that could undermine the increasing efficiency of biomedical auxiliaries in health care, or policies that might discriminate against *bona fide* traditional healers.

The main purpose of this chapter has been to outline the general geographical pattern and structural features of Kenya's health problems, and also to describe the therapeutic options people may resort to in their efforts to cope with ill-health. This information provides a context in which to compare and interpret the evidence from two case studies presented in later chapters: the rural ethnomedical system in a part of Machakos District in Kenya's Eastern Province, in which both biomedicine and traditional medicine are examined; and urban traditional medicine in the national capital. The precolonial and colonial background of the rural case study, set among Kenya's Kamba people, is presented in Chapter 3.

NOTES

1. Individuals and groups within several of the diverse Asian communities eventually established separate hospitals, such as the Aga Khan Memorial Hospital in Nairobi in 1958 and the Pandya Clinic in Mombasa. These were open to patients of all races and "were financed by a pound-for-pound matching agreement with the government" (Hartwig, 1979, f.n. 15, p. 126).

2. The standardized healthful level of fluoride concentration is 1 ppm.

3. The estimated annual rate of urban population growth from 1969 to 1979 was 7.6% (Republic of Kenya, 1983, p. 145).

4. The Kenyatta National Hospital KAP survey was conducted during the first 2 weeks of December 1977. Operational conditions in the filter clinics made it very difficult to achieve the goal of a strictly controlled systematic sample of outpatients. In lieu of this I opted for roughly proportional sampling in each of the three clinics (male, female, and pediatric). Each patient was interviewed in private by a trained research assistant prior to being seen by a doctor. Each questionnaire contained 46 questions, and required 15–20 minutes to administer. During the survey period 5,769 *outpatients* were seen in the filter clinics, including 2,155 in the pediatric unit (37%) and 1,724 and 1,890 in the male (30%) and female (33%) clinics, respectively. Within these groups, 3.2% of the adults who brought children to the pediatric clinic and 4.5% and 3.5% of the outpatients at the male and female clinics, respectively, were interviewed. As a percentage of all outpatients interviewed in the survey, the proportions were: pediatric (32.5%), male (36.3%), and female (31.31%).

The ward patients in the KNH survey ($N = 119$) included 39.4% ($N = 39$) of those in four medical wards, 85% ($N = 53$) of the women in two gynecological wards, and 66% ($N = 27$) of those in the chronic sector (Ward 25).

5. This is based on an arbitrary extended family size of 15 members. Given a total population of 14.7 million in 1978, there would thus be 980,000 families. If 13% of these families included one TMP, that would mean Kenya had 127,400 TMPs.

6. The *National Guidelines for the Implementation of Primary Health Care in Kenya* (Bennett & Maneno, 1986) was published as the present book went to press. The *Guidelines* are intended to reflect and inform a major policy shift in Kenya toward community-based PHC strategies to achieve "help for all" by 2000. They outline a significant role for TMPs as bonified health workers cooperating with the official health services.

3

Habitat, Social Change, and Health Care in Precolonial and Colonial Ukambani

Kenya's traditional medical systems have had direct and subtle forms of contact with biomedicine and its accompanying cultural and technological "baggage" for some 80 years. This continuing encounter has provoked numerous incremental and irreversible changes in the character of traditionally based health care, including its style, content, and the outlook of its practitioners and clients. Pharmaceuticals provide a significant example of the consequences of such cultural exchange. Antibiotic pills, capsules, and injections spread rapidly within the towns and outwardly along and beyond the main roads in the countryside. They are by now widely and easily available, for a price; but as Bledsoe's West African work (1983) reveals, they are also often used inappropriately and for reasons (such as the meanings assigned to their particular colors and shapes) that have no relationship to their intended pharmacological function. In addition, peoples' health attitudes and behavior have been influenced in both positive (e.g., hygiene and sanitation) and ambiguous (e.g., erosion of sorcery power) directions through personal experiences with government and mission medical services and with independent biomedical practitioners-for-profit. Contact with the language of biomedicine, too, has naturally expanded the popular vocabulary for describing states of health and sickness and, consequently, peoples' perception and related behavior.

It is possible and valid to describe the character of today's traditionally based medicine minus its historical roots. However, only a longer view can provide the essential element of continuity, highlight the richness of detail, and reach toward the social and locational insights that give meaning to contemporary health care practices. Thus the present chapter provides an historical overview of the habitat features, social change, and evolving patterns of health care in precolonial and colonial Ukambani.

PHYSICAL AND ECOLOGICAL PATTERNS IN UKAMBANI

The Kamba people are Bantu-speaking agriculturalists with a strong tradition of cattle-rearing. They are ethnically related to the Kikuyu, Embu, and

Meru peoples who live in adjacent territories to the west and north. The Nilotic pastoral Maasai, historic adversaries of the Kamba in cattle raiding, occupy the grassy plains south of Ukambani.

Physically, Ukambani is situated just south of the equator within the Athi and Tana River basins. It comprises the districts of Machakos, with a population of over one million people occupying an area of about 15,000 square kilometers; and Kitui, where nearly a half-million Kamba are unevenly distributed across some 30,300 square kilometers[1] (Figure 2-1). This region forms the southern half of Kenya's Eastern Province, and has about the same area as Sierra Leone or Ireland. The land rises from east to west and forms a series of vast, open peneplains with often abrupt transitions to distinctive hill zones that locally exceed 2,000 meters in elevation. Most of the eastern plains, situated at 200–500 meters above sea level, are dry savanna and thorn-bushland dominated by *Commiphora* and *Acacia* spp. on red latosols. This arid zone receives less than 500 millimeters of rainfall annually in a markedly seasonal pattern, most rain occurring in April and November. Today the human population density only occasionally exceeds one person per square kilometer over extensive stretches of this dry eastern region of Ukambani, and the carrying capacity of the land for humans and livestock is very near the maximum possible without inducing significant land deterioration (Bernard & Thom, 1981). Extensive grazing,[2] primitive manufacture and trade of charcoal, wildlife management, and wildlife poaching in and around the huge Tsavo National Park are the principal economic activities in this region. Cultivation is restricted to isolated "islands" of favorable topographic, soil, and moisture conditions, and overall the ecological potential is low.

Most Kamba live in central and western Ukambani. These are areas of higher, wetter plains above 1,000 meters elevation, and steep-sided massifs that rise to about 600 meters above the surrounding plains and attain heights of more than 2,000 meters above sea level. Two rainy seasons are typical here: the long rains of March through May and the short rains of October through December. The flow of most rivers and streams also shows strong seasonal fluctuations, and even in the higher elevations many dry up completely shortly after the rains. Climate ranges from locally humid and agriculturally favorable conditions in the Mua, Iveti, Kangundo, Kilungu, and Mbooni Hills to dry subhumid and semiarid conditions over much of the Yatta Plateau, Makueni, and the high grasslands of the Athi and Kapiti Plains that lie closer to Nairobi. Agriculture is risky and food shortages are commonplace in the semiarid areas surrounding the hills (Porter, 1979).

Population densities and population pressure are both high in Machakos's western hills and in Kitui's central hills. Locally, densities in the Machakos hills exceed 450 persons per square kilometer, and it is here that rainfed agriculture reaches its greatest potential. Maize, pulses, cassava, millets, and sweet potatoes (in the wettest zones) are the major subsistence crops, while coffee, maize, vegetables, and fruits are the crops of greatest cash value. Cultivation, grazing, and fuelwood collection in the hills have left little of the natural vegetation undisturbed, apart from the familiar fig trees whose large dark-green canopies often mark the location of sacred groves (*Mathembo*). Present ecoclimatic zones and plant cover indicate that the original natural vegetation was probably mixed forest on the higher hills, merging into *Combretum* spp. woodland and savanna on the lower slopes, and *Acacia* spp. bushland and savanna on the lowest plains (Moore, 1979, p. 420).

EVOLUTION OF SOCIETY AND ECONOMY

Traditions of most Kamba clans and lineages generally agree that the earliest Kamba settlers in Ukambani came from the plains south of Mt. Kilimanjaro in present-day Tanzania. Northward migration of Kamba groups began from the Kilimanjaro area in the 15th–16th century when they were still hunters and pastoralists (Figure 3-1). These migrations took them through Kenya's Chyulu Hills and into Kibwezi, Makueni, and Nzawi in Ulu (the precolonial name for the area of Ukambani south of the Athi River). Kamba groups had adopted agriculture by the time they arrived in the Mbooni Hills in the mid-17th century. According to tradition and place names, the fertile Mbooni massif is the hearth area from which many of the 25 patrilineal Kamba clans originated and later spread across Machakos and Kitui Districts.[3] From Mbooni the Kamba expanded first across the Athi River into Kitui in the 18th century. A second movement out of Mbooni resulted in the settlement of the Kilungu Hills. This was followed by dispersals into the other hill districts of Ulu, including Kalama, Iveti, and Mukaa.

The 18th and 19th centuries are characterized for the Kamba as a period of "incessant wanderings of small groups in search of land and livelihood" (Spear, 1981, p. 104). New colonies were established in a wide variety of physical environments differentiated largely by elevation, temperature, and rainfall. These variations provided opportunities for people to make new ecological adaptations and to develop distinctive livelihood patterns. These ranged from intensive agriculture based on maize, eleusine (finger millet), bananas, and root crops in the well-watered highlands of central and northern Ulu; crops with greater drought-resistance such as millets, sorghum, and cassava in the lower, drier areas, grading into an increasing reliance on pastoralism; and hunting along the Taru Desert margins. Despite the steady increase in the importance of cultivation, livestock, particularly cattle, continued to play a major role in Kamba subsistence and social life (Bernard & Thom, 1981; Spear, 1981).

By the 19th century Kamba society had acquired a distinctive "frontier" quality characterized by mobility, a flexible social system marked by fission-

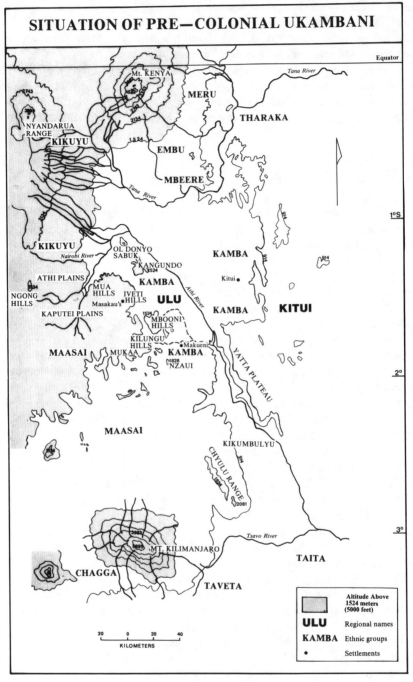

Figure 3-1

ing into small-scale population and territorial units, shallow lineages of three to four generations, and highly decentralized political institutions. Dispersed homesteads (*musyi*, pl. *misyi*) organized as *utui* (pl. *motui*), or neighborhood groups, formed the primary settlement unit. Individuals, families, and lineages in an area, which on occasion might include neighboring *motui*, or *ivalo*, were joined by propinquity, intermarriage, and an oath to cooperate with each other in activities such as cattle management and defense. Ecological variation and contrasting subsistence activities encouraged a symbiosis between the Kamba of the hills and the plains (Bernard & Thom, 1981) which created important differences among *motui*. This encouraged a degree of specialized production and the exchange of craft articles and foodstuffs at marketplaces (*ing'anga*) which were commonly located at the junction of hill and plain (Munro, 1975). Extensive regional trade also evolved, conducted by *utui* associations of hunters and warriors. Regional associations of *motui*, headed by influential entrepreneurs known as "bigmen" (*andu anene*), emerged to develop and control the caravan trade with the coast.

The renowned mobility of the Kamba continued to increase during the 19th century, stimulated in part by recurrent drought and famine,[4] particularly in the more precarious habitats of Kitui. Kamba traders and small colonies of Kamba migrants known as *avikilambu* (rain-seekers) moved out of the stricken areas and came into direct contact with Afro-Islamic towns and rural communities along the Indian Ocean coast. Kitui Kamba intensified their participation in commerce and established dominance over the trade in ivory and cattle between the eastern highlands and the Kenya coast until the 1860s, when Arab-Swahili merchants succeeded in capturing control of the caravan business (Munro, 1975; Spear, 1981).

Because of their trade and colonization activities numerous Kamba came into contact with Islamic and other African (e.g., Giriama, Rabai, and Digo) traditional medical practices. Selected elements of these foreign systems, such as reading and recitation of verse in divination proceedings (apparently inspired by the example of the Koran) and belief in *majini* spirits, were adapted and incorporated by the Kamba. Today, both practices can be witnessed frequently in local healing rituals, with *majini* often cited as causal forces in sickness.

Some Kamba traditional healers who functioned as religious-medical specialists (*mundu mue*, s.; *andu awe*, pl.) also played key roles in the evolving patterns of and responses to economic activity and social change in eastern Kenya during the 19th century. Kivui wa Mwendwa of Kitui, for example, was one of these *andu awe*. He rose to prominence beginning in the 1820s and 1830s as the most renowned of the Kamba ivory traders. Originally a medicine man of considerable repute who operated a vigorous trade in amulets and charms, Kivui later acquired much prestige and a very large following as a "big man." He was well known at Mombasa on the coast and in 1849 and 1851 personally guided Ludwig Krapf, an emissary of the Church

Missionary Society, on some of his travels to the east of Kenya's Rift Valley (Krapf, 1960; Munro, 1975; Spear, 1981).

Kitui's dominant position in the lucrative long-distance caravan trade declined after 1860 as Swahili commerce continued to expand into the hill country of the Machakos Kamba. New commercial patterns and military realignments occasioned by the decline of Maasai strength in the southern interior combined with drought, famine, disease, and expanded cattle raiding to challenge the established social order within and around Ukambani (Munro, 1976). Kamba prophets (*ndoto*), who held status as traditional healers (*andu awe*), "provided a source of new authority which replaced that of the *andu anene*, and they attracted large followings" (Spear, 1981, p. 129). A woman named Syokimau, who with Masaku is considered one of the two most famous Kamba prophets of the late 19th century, predicted the coming of Europeans and the railway to Ukambani. In contrast, Masaku's reputation as a prophet was built upon his remarkable ability to predict the rains and the location of elephants and Maasai cattle. This kind of "intelligence service" operated by Masaku is seen by one historian to represent "a changing emphasis in prophetic function under the stimulus of demand for cattle and ivory" (Munro, 1975, pp. 27–28).

Masaku's extensive brokerage activities attracted Swahili merchants, whose campsite near Masaku's home at the base of the Iveti Hills in western Ulu became a market known as "Masaku's". In 1889, Masaku's became one of the principal up-country stations of the Imperial British East Africa Company (IBEA), with the now Anglicized name of Fort Machakos; and in 1895 it became the District headquarters for the Ulu-Kitui area under the new British East Africa Protectorate. Machakos soon grew into a township and "a permanent center of Swahili influence" (Munro, 1975, p. 100). Although it was bypassed to the south by the Uganda Railway, Machakos remained the undisputed commercial center of the district. Today it is a vigorous market town and the largest settlement in Ukambani.

Kamba culture has experienced continuous change since its origination. It has been typified as a way of life marked by flexible rules that reflect "a looseness of basic structural orientation"; by a strong individualism, manifested in a "shallow cultural commitment" and aversion to coercive conformity; and by an acceptance of change and a "surprising willingness to absorb new ideas" (Machakos District Annual Report, 1924; Oliver, 1965). Because Kamba culture remained "quite literally a culture on the move" (Oliver, 1965, p. 425), it was characterized by a high degree of adaptability to new environments. This mobility is highlighted at a local scale in a report by one of the early British district officers (DOs) stationed in Kitui who wrote that he "could count numberless villages which have moved within the last two years; in fact, there is a constant moving about of the Akamba" (Dundas, 1913, p. 506).

Despite this mobility and mixing of the Kamba peoples, from the mid-17th century onward the movement of pioneer settlers (*Athaisu*) out of Machakos and into the relatively empty bushlands of Kitui initiated a pattern of physical isolation of the Machakos from the Kitui Kamba, forming cultural differences that persist today. Variations in language, clothing, and personal adornment, and the addition of a third circumcision ceremony (*nzaiko*) for men in Kitui (the latter also occurs in Kilungu)[5] exemplify some of the differences that emerged between the two sections (Lindblom, 1969; Munro, 1975). Allegedly through the influence of their Tharaka and Giriama neighbors, Kitui Kamba also gained a reputation for possessing more powerful medicines and poisons, and greater notoriety for witchcraft, sorcery, and countersorcery than most other Kenyan peoples (Kitui District Annual Report, 1936, Kitui District Annual Report, 1956; Ukamba Province Provincial Commissioner's Inspection Reports, 1910).

Social and cultural differences within Ukambani were not, however, restricted to the relatively benign contrasts between Machakos and Kitui. Subregionally, a well-remembered enmity characterized relations between people of the Kilungu Hills and the other Ulu Kamba long before the colonial era. While documentation of their now-muted differences is scarce, even today the *Akilungu* (it has been suggested that *Kilungu* means "wrong-doers")[6] remain a people apart in the mind's eye of other Kamba. The origin of the cleavage is apparently rooted in various differences in dialect and customs, and particularly in a social memory of conflict that emerged quite early in relations of the *Akilungu* with other Kamba settled on and about the neighboring massifs. *Akilungu* are remembered as people who waged ruthless wars and raided other Kamba, whom they called *evaau* (nephews), for livestock and women, thus provoking counterattacks. They developed a reputation for bad temper, thievery, and witchcraft, and as a people of extraordinary wildness, rapaciousness, and brutality who subjected prisoners to extremely cruel forms of torture (Lindblom, 1969; Munro, 1975; Owako, 1971). Raids by the Akilungu on settlements in the adjacent Mukaa hills and at Nzaui massif continued until a few years after the establishment in 1895 of Crown rule over the new British East Africa Protectorate.

The Akilungu generally "held aloof" from relations with the local administration headquartered at Machakos (Figure 3-2). Meanwhile, other Kamba with whom they had long been feuding seized an opportunity to retaliate and terminate the conflict with their adversaries by successfully appealing to the nascent colonial administration for support from British military units. In February 1897, a combined expedition of Kamba warriors drawn from nearby Mukaa *ivalo* and a Maasai contingent of the East African Rifles defeated the *Akilungu* on their home ground (Munro, 1975, p. 46). Kamba "primary resistance" to British authority in Kilungu and elsewhere in Ukambani was then, practically speaking, broken.

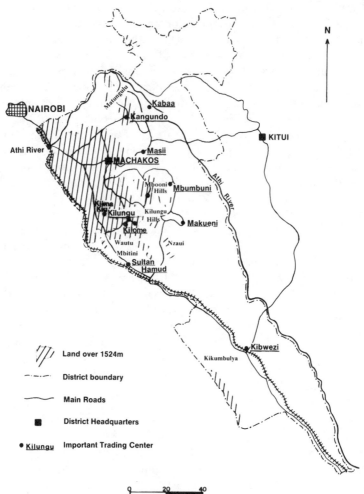

Figure 3-2. Colonial Machakos District.

As a people the Kamba came face to face at the close of the 19th century with a succession of relentless, overlapping natural disasters that also struck many other parts of East Africa (e.g., Kjekshus, 1977). These included locust swarms in 1894–1895, smallpox and influenza epidemics in 1897, and cattle losses from rinderpest outbreaks in 1897–1898. Finally, having suffered severe droughts and the crushing *Mapunga* (rice) famine of 1898–1899, the Kamba seemingly had little alternative but to yield their independence to British overrule. In 1898–1899 it is estimated that at least a quarter of the Kamba population had died from the famine and smallpox, and many others were left destitute and dependent on food procured for them at the coast by the colonial administration. These emergency rations were brought up to Ukambani on the new Uganda Railway and distributed through relief camps (Munro, 1975). The severely weakened population eventually recovered, only to find itself within a greatly altered and increasingly pluralistic social order which they did not control.

The emergent colonial system soon thrust the Kamba into a compromised position of inextricable political and economic dependency. Implementation of the Crown Lands Ordinance (1902) led to their restriction to a land "reserve" by 1906 and to the alienation of a portion of southwest Ulu District (Machakos from 1920) to European settlers. Although the Kamba effectively resisted mounting pressures to undertake employment in the cash economy until after World War I, soon thereafter men began to enter the police force and the King's African Rifles in increasing numbers. Those from Kilungu and other southern locations near the line of rail tended to sign on with the Uganda Railway.

Changes that permanently altered the Kamba way of life were not limited to the forced contraction of their land base and their incorporation into a wage economy. In addition, the imposition of British versions of civil and criminal law effectively undermined the democratic and egalitarian principles of traditional Kamba judicial institutions, as represented in the *ad hoc* local *nzama* councils of elders (*motui* level) and *king'ole* councils (*ivalo* level). Novel sources of authority, privilege, and prestige also emerged in the persons of colonial chiefs and their followers. Although these appointed chiefs assumed the traditional leadership roles formerly exercised by the *atumia* (elders) and *athiani* (war leaders), they did not supersede the Kamba medical-religious specialists.

Medicine men and women (*mundu mue*, s.; *andu awe*, pl.), were thus the third source of leadership in Kamba society (Ogot, 1974), and they remained key actors in the indigenous religious-cultural-health sphere during the colonial period. The *andu awe* occupied achieved-status positions, and according to Dundas (1913, p. 532) medicine women were more common than medicine men in Kitui. Kamba believed that their special insights and healing powers derived from a God-given ability to communicate with the ancestral spirits (*aimu*). They were expected to work for the welfare and harmony of the entire community. Aided by supernatural powers (*uwe*), they practiced beneficial skills such as divination, prophecy and fortune-telling, curing the sick, discovery of witches, and provision of countersorcery measures. *Andu awe* were custodians of sacred oathing instruments (*kithitu*) and the sacred groves (*Mathembo*) where sacrifices were performed. Communities looked to them for ritual protection of livestock, fields, and produce (Munro, 1975; Ndeti, 1972). The authority of the *andu awe* constituted the ultimate challenge to Christian missions when the latter initiated their assault on Kamba traditions in 1895.

Andu awe also emerged as leaders of indigenous political resistance to
the newly imposed system of alien administration that deprived Kamba of
control over many of their own resources and institutions for social control.
Between 1896 and 1911, and again very briefly in 1922, numerous *andu
awe*—in the roles of prophets and visionaries—were instrumental in the rapid
spread of essentially anticolonial millenarian movements. These movements
found dynamic emotional expression as new spirit-possession and spirit-
exorcism cults. The cults were known successively as *Kitombo* (1896–?);
Kyesu, or "Jesus" (1906–1908), which European administrators believed was
particularly attractive to young people, who were "seized by a peculiar form
of infectious mania" (Machakos District Annual Report, 1910–1911); and
Kathambi, named after the Kamba female water-spirit (Machakos District
Annual Report, 1910–1911).[7] The new types of possessing spirits and frenzied
dances (*ngoma*), drumming, and promise of deliverance associated with these
cults served to focus Kamba frustrations and their opposition to colonial rule
(Kitui District Annual Report, 1916). In 1911, thousands engaged in various
acts of passive resistance and civil disobedience that reportedly brought the
colonial administration's authority to the brink of collapse. Finally awakened
to the seriousness and emotional power of these developments, the government moved to restore its control through a show of military force, the arrest
and deportation of four *andu awe* it considered most responsible for the
movement, and the detention of 11 other cult leaders at Machakos for a year.
Although the movement was broken, by the start of World War I the Kamba
people had established a reputation within the colonial government as difficult to administer and as "adept passive resisters" who "cannot be driven
without coercion" (Ulu Province, Annual Report 1914–1915). This clash
between Kamba cultural values on the one hand, and the British sense of their
own indisputable cultural superiority on the other, is captured in dramatic
"imperial" prose written in 1910 by an utterly exasperated district commissioner (DC) in Machakos. "Effective administration," he wrote,

> is still much hampered by the difficulty experienced in inducing people to come
> in promptly when sent for. I do not think this is to be attributed to willful
> disobedience so much as to an innate contempt and disregard for our authority,
> that characterizes the whole attitude of the A-Kamba toward the government
> and the white man in general; nor is this to be wondered at. . . . We have bred
> up in the A-Kamba a spirit of independence and an idea of personal liberty that
> is altogether out of place in a subject people, low in the scale of civilization as
> they are. . . . Natives met with and called to instead of coming either remain
> standing where they were or subsequently make off as soon as a move is made in
> their direction (Machakos District, Annual Report, 1910–1911, pp. 16–17).

Another revolutionary movement marked by strong anticolonial sentiments arose briefly in rural Kilungu in 1922 under the leadership of Ndonye
wa Kauti, a young man "who blended the characteristics of a traditional

medicine man and the new Christian evangelist-teacher" (Tignor, 1976, p. 334). Ndonye appeared on the scene at a time of increasing strain in African-European relations. Disaffection with the colonial government's policies was widespread, and many if not most Kamba still seemed ambivalent toward the Christian missions and schools springing up in their midst. Ndonye, who claimed that he was instructed by the Christian God, prophesied the impending death of colonialism in Machakos and the emergence of a utopian social order in which people would be freed from the increasing tax burdens, conscript road work, inequitable access to wealth and political influence, and other grievances associated with the colonial administration's command over local resources.[8] As in earlier attempts to mobilize and sustain social action based on millenarian-style visions, Ndonye's considerable influence proved ephemeral and the protest movement he organized ended suddenly with his permanent deportation to Lamu Island at the end of 1922.

Ndonye's demise effectively ended public participation by individuals of *andu awe* status in organizing and channeling Kamba antipathy to alien rule and its involuntary social changes. Between 1896, when the Kitombo cult arose, and the early 1920s, only a handful of medicine men and women were actually involved as cult leaders.

After World War I most *andu awe* maintained their traditional roles as diviners and ritual healers even as mission stations, schools, and dispensaries were becoming more visible and influential. For the most part, the Kamba people "continued to look to the *andu awe* for help in securing their needs and treating their ills—especially in the eastern and southern plains where mission agencies had barely penetrated" (Munro, 1975, p. 149).

Kenya's first Christian missionary station in the interior was established in 1891 by Scottish Presbyterians at Kibwezi (Figure 3-2), on the Uganda Railway in southern Ukambani (Oliver & Mathew, 1963). Shortly thereafter, the American fundamentalist (and soon to be dominant) Africa Inland Mission (AIM) built its first station at Nzaui in 1895, on the threshold of the Machakos hill country. A second station was opened about 45 miles to the north at Kangundo in 1896. New AIM missionaries arrived shortly after the disastrous famine of 1898–1899. They opened stations at Machakos (Mumbuni in 1902), Mbooni (1908), Mukaa [taken over in 1909 from the Church Missionary Society (CMS) which had operated it for 6 years], and Okia in Lower Kilungu (1917). The latter was abandoned in 1923. An informal agreement among Protestant missions in 1909 enabled the AIM to regard Machakos District as its own "sphere of influence." In time the AIM extended its chain of missions into Kitui District, where it took over stations operated by the evangelical Lutheran Mission (Leipzig) following the deportation of Germans from the East Africa Protectorate to India during World War I; and into Kikuyuland and the northeast Belgian Congo (Tignor, 1976).

The Holy Ghost Fathers (HGF), or Spiritans, a mainly French Roman Catholic mission, entered Ukambani from western Kenya in 1912. Following a bitter dispute with the Protestant AIM over station sites and territorial interests, the HGF established its first station at Kabaa in the extreme north of Machakos District. They abandoned this in 1920 in favor of building a new station at Kikoko in Kilungu, where they were invited to establish a Catechist school and received direct assistance from Chief Malu of Kilungu (1911–1926). Both land and labor were given to the Spiritans, at least in part to counter the influence of the Kilungu peoples' old adversaries who lived downslope in neighboring Mukaa location where the AIM was established (Machakos District Annual and Quarterly Reports, 1918–1919). Despite "fierce opposition" by the AIM, the Spiritans succeeded in establishing their station and school at Kikoko. This move created a second, Catholic, sphere of influence in the southern hill country (Tignor, 1976, p. 114). Both the AIM and the Spiritans, however, neglected the southern and eastern plains of Ukambani, choosing instead to focus their efforts on the most densely populated northern and southwestern hill zones (Munro, 1975).

For years after the Christian missions gained a physical foothold in Ukambani, most Kamba remained strongly disinclined to embrace their programs of spiritual conversion and education. In both Machakos and Kitui, African attitudes toward the Christian doctrines put before them ranged from indifference to militant aversion. The missionaries insisted that the Kamba renounce many of their most important social and cultural institutions, forsake practices that helped to maintain their lineage system, and support schools that they perceived would subvert their children and withdraw them from the family economic system (Central Province Annual Report, 1916–1917; Kitui District Annual Report, 1912–1913; Machakos District Annual Reports, 1910–1911 and 1912; Tignor, 1976). Specifically, the missions demanded an end to Kamba rituals at the sacred *Mathembo* and required them to stop consultations with *andu awe*. They were also pressured to abandon use of the *kithitu* (fetish) oaths and the wearing of traditional clothing.[9] Intoxicating beverages and tobacco use were prohibited along with dancing, polygyny, clitoridectomy (the Kamba performed the minor operation only), bridewealth, and Sunday work. The AIM also placed a "fornication chair" in their churches as method of shaming people into conformity with missionary values on sexual conduct (Tignor, 1976).

As was true of most Catholic missions, the Spiritans were less rigid and more tolerant than the AIM regarding such practices as bridewealth, dancing, drinking and smoking. Nevertheless, both groups apparently "shared the same approach to polygyny and when it came to a discussion of the Christian ideas of marriage the Holy Ghost Fathers found the Kamba as uninterested as the A.I.M. did" (Munro, 1975, p. 106).

Not surprisingly, the Kamba perceived the American and European missionaries as extensions of the colonial administration. They saw the missions as a direct threat to their social order in general and to their traditional religious-medical system in particular. Before the 1920s, the missions attracted few adherents apart from small numbers of boys whose families lived close to the stations; displaced children, notably runaway girls; and a few men and women in adverse economic or domestic circumstances. Those Kamba who depended on the missionaries lived in small colonies on, or adjacent to, the station lands, isolated from and often scorned by the majority living beyond the station (Munro, 1975).

Attitudes to education and, hence, toward the mission schools underwent dramatic change during the 1920s. The first generation of Kamba to have contact with the missions had by now produced a small but highly influential class of men who were not only educated and Christian but, because of their achievement, held relatively well paid and more prestigious jobs. For these men, increasing status accompanied their higher living standard. Education was a vehicle to advancement, particularly in the agriculturally productive and prosperous northern and western locations of Machakos. Demand for additional church and government-run schools grew. Many younger Kamba now seemed more inclined to work within the colonial system rather than challenge it as their fathers had often done. They turned to the colonial administration, and particularly to the local native councils (LNC), to supply the schools, roads, and medical services that could provide the means to a better life (Gilks, 1933; Munro, 1975).

DISEASE, HEALTH CARE, AND THE COLONIAL MEDICAL SERVICES

A few widely scattered Christian medical mission stations were the only source of Western medical treatment available to Kenya's African population between the establishment of the East Africa Protectorate in 1895[10] and World War I. Although these mission medical services undoubtedly offered some amelioration of sickness in their immediate locale, they could scarcely have had beneficial value for the remaining tens of thousands of Africans in Kenya and elsewhere in East Africa who collectively suffered an enormous burden of acute and chronic disorders.

The successive waves of epidemic diseases and harsh famines of the 1890s are well known, but records of endemic diseases among the Kamba are scarce and fragmentary until the 1930s. Venereal diseases reportedly had been "rife" and spreading before and during the war, particularly in the vicinity of government administrative stations situated along the new railway line (Ukamba Province Annual Report, 1908–1909; Ukamba Province Annual Report, 1917–1918). Cerebrospinal meningitis burst upon Machakos District (Ulu) in 1913–1914. Africans claimed not to have seen this disease before, and it is said to have caused "considerable confusion" among a people already wracked by the "great prevalence of sickness" (Ulu Province Annual Report, 1913–1914). A few years later, a severe epidemic of meningitis claimed approximately 500 lives in Kitui and another 236 in Nairobi (Ukamba Province Annual Report, 1917–1918). Sporadic outbreaks of smallpox continued to be a serious problem as well. Some 25,984 Kitui Kamba were vaccinated in 1915–1916, but this did little to stop the disease from breaking out again in Machakos in 1919 (Machakos District Annual and Quarterly Reports, 1918–1919).

There were still no hospitals or Government medical personnel stationed in Machakos and Kitui Districts when the "terrible scourge" of pandemic "Spanish" influenza arrived in 1918. The disease was allegedly introduced into Machakos Township by *askaris* (local police) who had just returned from an official visit to Nairobi (Figure 3-2). Spreading quickly, the influenza caused 40 deaths among 895 confirmed cases in the town. It was readily transported into the adjacent Machakos Native Reserve, where 7,591 influenza deaths occurred between October 1918 and February 1919. Depending on the location, the case-fatality rate ranged from 2.8 to 15%. Another 5,187 people died in Kitui. The colonial administration blamed most of the high mortality on "the native practice of lying outside the hut when taken ill" (Machakos District Annual and Quarterly Reports, 1918–1919). However, mortality was "conspicuously small" in the neighborhood of mission stations. Youths and middle-aged people reportedly experienced the highest mortality, while young children and old people were the least affected. During this epidemic the Nzama elders, a traditional Kamba governing council coopted by the Administration, reportedly showed "an eager demand for medicine of any kind" (Machakos District Annual and Quarterly Reports, 1918–1919).

At least one British official blamed the spread of influenza and other diseases on the living pattern and mobility of town-based Indians. Sanitation on the plots they occupied together with their African employees was, he wrote,

> of the worst description simply and solely due to overcrowding. In the case of an epidemic such as Influenza or plague, the Indians at once try to leave the Township for a more open locality and thereby spread the epidemic. Their native employees follow their example (Machakos District Annual and Quarterly Reports, 1918–1919).

Until World War I the Medical Department of colonial Kenya catered almost exclusively to the tiny European community. Only secondary consideration was given to the African and Asian paid labor force upon whom the administration depended so heavily. The focus was on curative work and the enforcement of sanitary regulations, with particular concern for the prevention and control of epidemic diseases in the colony's emerging administrative centers (Beck, 1974). Apart from a reliance upon the few medical missions, no specific provision existed for the medical and public health needs of the millions of Africans in the reserves at risk to endemic diseases, malnutrition, and imported epidemic infections. Before 1920 only one European medical officer (MO) was assigned to all of the African areas of Kenya, and there was apparently "not a single Government institute that could properly be termed a hospital" (Adjala, 1962, p. 108).

Government consideration of African needs for medical attention was finally prompted by the appalling physical condition and subsequent mortality of African men recruited for military service during World War I. Most had malaria, yaws, and worms, and were malnourished and unable to sustain hard labor (Adjala, 1962). For example, 34% of the men recruited in Kikuyu district, adjacent to Nairobi, were disqualified as unfit for duty with the Carrier Corps. Another third were classed as unfit to work as laborers on estates. Approximately 70% of the 4,300 African troops who died during the war did so not from battle but from diseases. Similarly, 1 of every 8 of the 350,000 carriers died of disease (Beck, 1974).

Following World War I some European medical administrators began voicing their concern to the Colonial Office about the need for a progressive health care policy—one that would provide basic curative and preventive services for rural Africans. The notion that Africans might be entitled to health care on humanitarian grounds, as well as to enhance their efficiency as workers in the European sector of the economy, gained momentum in 1920. Two government dispensaries were opened that year in rural North and South Kavirondo in western Kenya. There were also discussions about establishing dispensaries in other parts of the country and about training more Africans as dressers and dispensers. With the implementation of new legislation in the form of a comprehensive Public Health Ordinance (1921), the Medical Department assumed responsibility for the organization and expansion of medical services, as well as for the maintenance and promotion of public health through local authorities. Administrative officers in municipalities such as Machakos Township, DCs, and MOs thereby received a considerably expanded charge for public health in the African reserves and townships throughout Kenya. Subsequent passage of the Native Authority Ordinance of 1924 was a major achievement for colonial administrative and fiscal policy. It opened the way for LNCs—the new instrument of local government in colonial Kenya—to identify specific needs in their own areas and to set budget priorities for public health and other social expenditures (Harlow, Chilver, & Smith, 1965).

The first MO for Machakos District did not arrive there until 1921. He soon opened dispensaries in Kilungu, Matungulu, Mbooni, Kiteta, and Ngoleni Locations in the Reserve, and constructed a "native" hospital of permanent materials in 1922 (Machakos District Annual and Quarterly Reports, 1922). However, most Kamba lived at a considerable distance from Machakos. Since they were also still unfamiliar with European medicine and its methods, it is not surprising that many chose to stay in their villages rather than attempt a long trek to the hospital in a weakened and perhaps painful condition (Machakos District Annual Report, 1924).

During the 1920s and 1930s a change in the basic strategy of colonial administration came into play in the shape of the Dual Policy, formal adoption of which amounted to a tacit recognition by the Kenya government of its neglect of African needs for economic and social development. Accordingly, the government would, at least in theory, henceforth commit itself to "the parallel and complementary development of the African reserves and the European 'settled areas'" (Munro, 1975, p. 165). In practice, not least because of the militant political power of the country's white settlers, this amounted to a policy of separate and unbalanced development in which only European production was promoted. Social services such as medicine and education were also organized along racial lines. Africans saw themselves getting whatever resources remained after Europeans and Indians were accommodated (Ogot, 1974).

There is evidence in the Machakos reserve that the Medical Department was more committed to advancing the original aims of the Dual Policy than was the case with the other agencies of the central government. Cooperation between the Medical Department and the Machakos LNC, which was formed in 1925 following authorization by the Native Authority Ordinance (1924), provided funds to build a new system of medical services that was to link a slowly expanding number of outlying dispensaries with the hospital in Machakos Township. Local native council staffs comprised headmen appointed and supervised by the DC and paid from central government funds (Beck, 1981). The Machakos LNC formulated development initiatives oriented toward small social welfare investments including roads, afforestation, support for agriculture, schools, and staff housing as well as medical dispensaries. It acquired most of its revenues from a local poll tax on adult males. A fund was thereby created that was not dependent on transfers from the central government treasury. It thus conformed to the colonial fiscal policy of minimal subventions for local African administration and development projects. Beginning with revenues of £2,300 in 1925, the Machakos LNC treasury funds increased to £5,200 in 1935, making it the wealthiest council in the eastern highlands. Approximately 50% of its funds went into medicine and education, the other half going to public works (Munro, 1975). By 1933 the District had 11 dispensaries, and the number of people seeking treatment then was considerably greater than in any previous year (Machakos District Annual Report, 1933).

Endemic, intercurrent diseases; sporadic epidemics; poor sanitation; and inaccessible biomedical services were the norm for Machakos Kamba in the decades following World War I. Abdominal disorders, bronchitis and other chest complaints including tuberculosis (especially in the higher elevations), eye diseases, malaria and tropical ulcer (notably in the lower, warmer regions), whooping cough, amebiasis, intestinal parasites ("chiefly tapeworm"), and dental caries with associated severe pyorrhea[11] were among the local health problems most frequently reported by European officers (Machakos District Annual Reports, 1921, 1927; Machakos District, Annual Report, 1928. Appendix No. 17. Medical Officer's Report). Poor nutrition aggravated the debilitating effect of infectious diseases. Administrators also believed malnutrition was exacerbated by the Kamba practice of selling local "protein-foods" such as beef, mutton, eggs, and honey outside Machakos District (Machakos District, Annual Report, 1928. Appendix No. 17. Medical Officer's Report).

Surgical conditions such as appendicitis, hernia, and cancer, and cases requiring amputation, rarely came to the notice of medical officers. From a European perspective, the Kamba were "very shy of the knife" and dreaded operations to the extent that they preferred to die in their villages than submit to surgery (Machakos District Annual Report, 1928. Appendix No. 17. Medical Officer's Report). Many Kamba apparently overcame such fears during the next several years. In 1936 "over 400 operations under general anesthesia" were performed at Machakos District Hospital, and it was "quite difficult to find accommodation" for the large numbers of Kamba seeking hospital treatment (Machakos District Annual Report, 1936).

In essence, the development of medical services in colonial Ukambani was heavily dependent upon local African initiative and local revenues. The London-based Colonial Office and its policy of "pay as you go" for development activities offered little scope for expansion of health resources by the colonial administration (Beck, 1974). Although a second MO was appointed to Machakos District in mid-1926, thereby facilitating several "medical safaris" into the reserve that year (Machakos District Annual Report, 1926. Appendix No. 17. Medical Officer's Report), the LNC shouldered an increasing share of the expenses for medical services, including buildings, drugs, and staff wages (Gilks, 1933). In the aftermath of the world depression the LNC found itself faced with a greatly increased demand for medical treatment (attendances at Machakos's ten dispensaries in 1932 numbered 49,712) and a colonial administration unprepared, due to financial stringency, to adopt a policy of expansion (Machakos District Annual Reports, 1932, 1933). In response, the Machakos LNC continued to push for new services and facilities, reflecting the increasing demands of the now rapidly growing African population for a better standard of living and their growing confidence in the efficacy of Western medicine. In 1934, for example, LNC funds supported the erection of a new dispensary in Upper Kilungu and a new ward for the hospital in Machakos Township. A child welfare and prenatal clinic followed the next year (Machakos District Annual Reports, 1934, 1935).

The cooperative spirit of partnership between the LNC and British officers in the Machakos reserve began in 1925 but lasted only about 8 years. Central government agents, recognizing their shallow control over an African society that increasingly expected more than just a nominal degree of local autonomy and participation in development activities, began to reassert their authority during the early 1930s. Local-level cooperation was no longer encouraged. In its place came

> a phase of ever-increasing regulation, compulsion, and monopoly, in which initiatives for promoting change passed from local hands and local interests became subordinated to the plans of central government (Munro, 1975, p. 178).

Local native council initiatives in health care were accordingly dampened by the government's increasing dominance in the 1930s. Nevertheless, the short era of cooperation between Kamba leaders and the DOs who supervised the LNCs' activities did produce important advances on several fronts. In addition to the construction of medical facilities, notable accomplishments included the training of African dressers, the introduction of sanitation and public health measures, and almost certainly the lowering of death rates (Munro, 1975). Indeed, nearly two decades later the enthusiasm of the Kamba LNCs for expanding health care (and social services generally) was still evident, as is vividly portrayed in the condescending remarks of the DC for Machakos. He observed that

> Hospitals come a close second to schools in the demand of the African for social services, particular stress being laid on the maternity side. . . . L.N.C.'s will gladly vote more money than they can afford for Health purposes and would like to raise more and larger buildings regardless of whether staff or funds are available to maintain them (Machakos District Annual Report, 1950).

In Machakos, as indeed in the other African districts in colonial Kenya, outspoken advocates for the development of more effective rural health services were often found among well-intentioned British medical directors and medical officers. That their ideals and the aspirations of the expanding African population were never realized attests to the inherent contradiction between the requirements for sustainable local development on the one hand and the imperial ideology of "indirect rule" on the other. Budgetary constraints imposed by the Colonial Office called for health and social services on a self-supporting basis. Local governments [African District Councils (ADCs) after World War II] were responsible for raising taxes for dispensaries, health centers (introduced after World War II), maternity services, environmental sanitation, and ambulances. Locally, the central government operated the district hospital and exercised its control over LNC and, later, district council health initiatives by retaining authority to approve annual estimates and by assigning MOs to supervise district medical services from a base at the district or provincial hospital. While this may have ensured "economy of personnel" (Fendall, 1960, p. 176) for the colonial administration, local district councils certainly were not positioned to finance and sustain even a minimal level of basic health care for the expanding African population.

By 1956, just 7 years before independence, the number of dispensaries in Machakos District had grown to 16, an increase of only 6 in 38 years. These dispensaries accounted for 92,146 annual patient-visits (Machakos District Annual Reports, 1927, 1956). This aggregate attendance level may appear substantial. However, it indicates that average attendance at each dispensary throughout 1956 was only about 24 persons per weekday. In the meantime, the district's population had expanded nearly fourfold from an estimated 138,400 (9.8 persons per square kilometer) in 1918 to 548,862 (39 persons per square kilometer) in 1962 (Morgan & Shaffer, 1966).

But the evidence of attendance levels at dispensaries is conflicting. By the 1950s, there are also reports of crowded facilities and steadily increasing demands by the Kamba for more medical services. These pressures had prompted the local British authorities to recommend to the ADC that people be charged a fee for dispensary and maternity visits. The ADC was also urged to solicit voluntary contributions for hospital construction. Despite strong initial resistance, the ADC introduced a Shs. 5/-(U.S. $0.70) fee for maternity services in 1952 and in 1954 a fee of 20 cents (East African) for the first visit to a dispensary in any 1 month. This last measure led immediately to a drop in attendance, but by the end of the year a recovery was said to have been underway (Machakos District Annual Reports, 1950, 1951, 1954). Fee-paying by outpatients and inpatients was introduced at Machakos Provincial Hospital in 1957, reducing the inpatient load by 13% and outpatients by 30%. In the words of the DC, "this fall in attendance eliminated the trivially sick and gave the staff a better opportunity to attend to those who were really in need of attention" (Machakos District Annual Report, 1957). This apparently cynical rationalization of bureaucratic expediency by the DC undoubtedly discounted many who were ill and could not afford the Shs. 5/- entry fee for health care.

Access by the majority of rural Africans to even a few curative medical services was neither intended nor feasible under the financial terms dictated by indirect rule. Even though local African authorities—the LNC at first and its successor the ADC—contributed one-third of the total expenditures for health by 1959–1960, the central government's two-thirds share amounted to only £2 million (Fendall, 1960), or roughly U.S. $0.70 per year for every man, woman, and child in Kenya. Furthermore the effectiveness of even this miniscule support was greatly diminished because of the strong urban bias of central government allocations for health. Most of the funds supported a comparatively elaborate and dense cencentration of primary, secondary, and tertiary medical services and specialist personnel in Nairobi and a few other towns, including Mombasa, Nakuru, Nyeri, Kisumu, and Machakos. Consequently, local authorities had to absorb the burden of financing rural health services. With resources of about $0.35 per person in 1959–1960, they were expected to build, equip, staff, and maintain dispensaries and health centers (introduced in some areas in the 1950s), and to be responsible for environmental sanitation (Machakos District, letter to Chairman, Machakos African District Council, 1963). As late as 1963, the year Kenya achieved independence from Britain, the Machakos District administration was quite unwilling even to support some of the struggling dispensaries already in operation. Given this political climate requests such as that which came from the people of Kivani for a health center in their area of Kilungu Location clearly had no chance of support (Machakos District, 1963).

However much the colonial administration failed to support a strong infrastructure of health care after the brief era of close cooperation between the LNCs and local British officers in the late 1920s and early 1930s, some additional positive developments for health can be noted. Apart from the expansion of the district hospital and limited support of the primitive dispensary system, two cottage hospitals (at Kangundo and Makindu), several tuberculosis (TB) clinics (Machakos District Annual Reports, 1948, 1956), and mobile clinics were established. General standards of sanitation and hygiene improved (Machakos District Annual Reports, 1935, 1954), and the threat of major epidemics was also reduced. Declining infant mortality (Central Province Annual Report of Provincial Commissioner, 1936) reflected achievements in reducing the frequency and severity of the classical epidemic diseases such as smallpox, influenza, and cerebrospinal meningitis. While the reduction of mortality rates was considered a high priority by the colonial administration, it also induced a quickening of population growth among the Kamba.

Despite these improvements in health, there is also evidence that biomedical services for the rural population actually declined during the last years of colonial rule. For example, in a letter to the chairman of the Machakos ADC on February 18, 1963, D. F. Hodman, the district MO of health, points out that little progress had been made in expanding health services in recent years and that

> indeed, we have moved backwards in closing a number of dispensaries. Council agreed in 1960 . . . and recently reaffirmed, that medical units would be built according to a list of priorities. So far . . . funds have not been available to make even a start.

> The people are, I believe, feeling frustrated at this stand despite the fact that it has been caused by their failure to pay rates (Machakos District, 1963).

TRADITIONAL MEDICINE UNDER COLONIALISM

It is arguable whether the colonial administration intentionally encouraged rural Kamba to rely on indigenous healers and midwives by minimizing its allocation of resources to health.[12] Certainly, biomedicine did not enter a vacuum upon its arrival in Ukambani. As elsewhere in Africa, the Kamba traditional medical system was part of a unified whole, interconnected with virtually every other aspect of social life and with ideas and practices that reflected a system of cosmological and earthly order. Traditional life and thought were not segregated into separate compartments labeled livelihood, religion, philosophy, medicine, legal affairs, and so on. Religious observance was just as essential to crop success as was soil quality. It was also an inseparable element of any procedure for healing the sick or determining guilt in a legal dispute. In this traditional Kamba society the *mundu mue* played a highly respected role as diviner and healer. He or she was the one called upon to interpret and mediate "the powerful influence of meanings derived from the interplay of the individual with his family and his culture and his bodily states" (Kiev, 1964, p. iv).[13]

It is not surprising that Africans instinctively maintained much of their own system of medico-religious beliefs and practices in the face of the massive challenge from Western culture. To do otherwise would have invited widespread social disorientation and cultural disintegration. At the same time they gradually accepted more and more of the European community's imported biomedical therapies.

Coincidentally with African accommodation of Western medicine, European minds were readily mystified by the psychological and behavioral manifestations of some of the indigenous societies' cultural attitudes and practices. This was particularly true in the case of supernatural explanations of sickness, and the interrelated uses of witchcraft as a mechanism of social control. Why, for example, would an outwardly healthy, Catholic Kamba shopkeeper with 10 years of formal education die of undetermined physical causes three weeks after submitting to the *kithitu* ordeal, through a court of law, in connection with a land dispute with his neighbor? Or, why would a trained, middle-class nurse employed at Kenya's largest hospital in Nairobi visit a diviner (*mundu mue*) in the shantytown of Mathare Valley for assistance concerning the sudden illness of her 2-year old child? Moreover, given the pervasiveness of witchcraft accusations in health and legal affairs during the colonial era, why did the witchcraft statutes of the East African territories fail to define the meaning and identity of a witch (Mutungi, 1977)? These examples suggest the disconformity and estrangement (at least in European eyes) of the traditional and biomedical explanations of human pathology that evitably emerged during their forced coexistence in the colonial era.

Even if traditional medicine and colonial biomedicine had been highly compatible in a cultural sense, the physical scarcity of the latter alone would have insured the predominance of the former system. In Ukambani, particularly outside the towns, traditional medical practitioners were consulted more often because of their reputations as specialists than as generalists. In the dispersed villages of Machakos and Kitui Districts it was (and is) customary to find, in addition to women who served as midwives (*mwisikya*, sing.), at least one man or woman with status as a diviner (*mundu mue kwausya*), an herbalist (*mukimi wa miti*), or one who "opens" a bewitched person (*mundu wa muvingui*).

Virtually every aspect of Kamba daily life, especially individual and group comportment, had medico-religious implications. For example, barrenness was and is an affliction of great significance for a woman's stature in life. Its presence required consultations with an infertility specialist (e.g., *mundu wa nthambya*), the taking of herbal preparations, and various rituals and sacrifices that required the husband's participation. After giving birth a mother would undergo purification and additional sacrificial rites, followed by *kwikia mwana mathaa*—the hanging of a special chain ornament around the child's neck on the sixth day after purification (Beresford-Stooke, 1928; Machakos District, Political Record Book, Up to 1910, (Pt. 2); Ndeti, 1973).

Some specialists operated exclusively as diviners, prophets, or healers of illnesses. In the Kilungu Hills, a distinction is made between one who practices divination only (*mue wa kwausaya*), and one who divines and also administers treatment to a client (*kwausya na kuiita*). Diviners were expected to demonstrate skill in discovering witches and in fortune telling. Some had a special ability to specify the identity of disaffected *aiimu* (spirits of ancestors) deemed responsible for someone's chronic sickness or other problem. The *mundu mue* was also the authority who designated the appropriate sacred place, such as a certain wild fig tree, or hills and rocks, where the offender would present food or other sacrifices to restore harmony with the offended *aiimu*. Herbalists, expected to have empirical knowledge of medicinal plants, treated symptoms and diseases such as malaria and other fevers, edema, gastroenteritis, burns, worms, pneumonia, headaches, conjunctivitis, and diarrhea. Other specialists in psychotherapy treated conditions such as epilepsy and mental disorders.

The art of healing also extended to the homestead, including gardens and fields, crops and livestock. Some specialists could predict the coming of the rains, while others would be called on to (ritually) dispel severe epidemics affecting people or cattle, such as smallpox or rinderpest (Lindblom, 1969; Machakos District, Political Record Book, Vol. 2, Up to 1910). Whether the Kamba people lost faith in their medico-religious healers and prophets during the deadly epidemics and famines of the 1890s will never be known for certain. In all likelihood the Kamba perceived these disasters as acts of God (*Mulungu*, or *Ngai*) which were therefore unfathomable to any mortal being.

Men and women known as *mwaiki* performed circumcisions on boys and girls (minor clitoridectomy only) during the age cycle designated for initiation rites (Machakos District Annual Reports, 1929, Appendix 29, Native Tribal Customs; Annual Report, 1930; & Annual Medical Department Report, 1931).[14] Other specialists greatly influenced individual and corporate health by administering and interpreting the results of various oaths and ordeals such as the *ndundu* (used to control deviance such as alcoholism and the practice of sorcery) and *muma*, which was sanctioned by the colonial courts and can still be used today as a last resort in the settlement of certain disputes (Mutungi, 1977). The *mundu wa muma* gives oaths over the *ki-thitu*,[15] a fetish object such as an earthenware pot, a woven basket, or a gazelle horn that he owns (and probably inherited) and which is filled with a hideous assortment of secret ingredients. This instrument embodies supernatural powers so terrifying that severe illness or death could come to any man[16] (and his relatives) who swore falsely over it (Murphy, 1926; Mutungi, 1977; Thomas, 1974; Ukamba Province Quarterly Report, October–December, 1909).

In colonial Kenya, few Europeans had sufficient cause to investigate or appreciate the systems of belief and behavior that reflected traditional medicine's underlying rationale, its functional organization, and its encompassing religious frame of reference. While the Kamba (and other African societies) did make substantial allowances for a naturally caused biological basis of human sickness, indigenous concepts of pathology were most firmly rooted in the belief that diseases are visited upon the living by human actions and by the offended spirits of one's ancestors who inexplicably govern the general behavior of individuals in society. Accordingly, "those who deviate from the normal activities in the culture, such as refusing to offer sacrifices to ancestors, disobeying cultural ethics, doing injustice to others, refusing to cooperate with others for the general good or ignoring one's responsibilities to himself and others must pay the price individually" (Ndeti, 1976, p. 18).

WITCHCRAFT, TRADITIONAL MEDICINE, AND COLONIAL ORDER

Belief in witchcraft and sorcery was and is an inseparable element of virtually all African systems of disease etiology, although not all African societies (including the Kamba) make a functional linguistic distinction between the two concepts. The late Mary Kingsley once ventured an opinion that "the African is almost as liable to die from a poisonous idea put into his mind as a poisonous herb put into his food" (Nottingham, 1959, p. 7). This syndrome is characterized by the perception of purposeful, malevolent behavior on the part of one's kin or associate(s) that is directed to oneself. Acting from jealousy, greed, or a desire for vengeance, the perpetrators are believed to call on evil powers to bring misfortune and harmful agents that induce mental distress and physical harm in the victim. In Kenya and other parts of Africa witchcraft was a pervasive phenomenon during the colonial era and remains so today. Because these beliefs remain so influential in human affairs I choose to discuss the system here in the present tense.

From an "objective" vantage point outside African society, witchcraft is a culturally conditioned, omnipresent fear system that encourages a type of individual and group hysteria which can grow into a continuing "fear process" (Miller, 1980). Several factors keep the fear process alive, including a folk history that confirms many supernatural beliefs; the persistence of myths, stories, and rituals that reinforce fears; and, especially today, a tendency to fall back on elements of past beliefs when they are useful to revive frightening images or ideas (Miller, 1980). In essence, witchcraft is the art of secrecy and imagination.

While no Kamba have ever seen witchcraft or can explain its nature, everyone discusses it and all know of its terrifying power (Ndeti, 1972). An Mkamba himself, Ndeti (1972, p. 123) contends that its power and destructive potential make it a "negative integrating force" that accounts for the presence of evil (*uoi*) in Kamba life. Witchcraft is one of the first principles of Kamba social organization and "a necessary condition for the development of the system of *mundu mue*" (Ndeti, 1972, p. 122).

A primary function of a *mundu mue* is to ritually neutralize sorcery power (*uoi*), a perceived cause of much illness and disruption within individuals and communities. Ideally a person of unassailable social and moral character who is dedicated to the community's welfare and peace, the *mundu mue* is traditionally held in the highest respect because he or she is the person nearest to God and is thus depended on to interpret the visible and invisible signs of spiritual and supernatural phenomena. Paradoxically, a *mundu mue* also embodies attributes that encourage awe and wonder, and it is in everyone's interest not to offend him or her for it is widely believed that "the person who knows the cure, knows the poison and the curse too" (Mburu, 1977, p. 172). While the *mundu mue* uses protective magic and acts to contain and mollify the fears that ebb and flow with cycles of sickness and other sources of insecurity in a village setting, the class of works known as *uoi* (witchcraft) is perpetrated by evil individuals (*mundu muoi*, s.) bent on causing injury. Their role is the reverse of that of the *mundu mue* (Mutungi, 1977).

Kamba witchcraft is connected with "old, wrinkled, ugly people, cripples, squint eyes, ugly facial features, or any crudeness in a person" (Ndeti, 1972, p. 122). When accidents occur, a child dies suddenly, a love affair disintegrates, an illness persists, when people or livestock behave strangely, or when any other events happen that could possibly be construed as unusual, or are otherwise unaccountable, bewitchment is normally seen as the underlying cause. Consultation with a *mundu mue* is crucial for confirmation of the cause and for guidance on appropriate countermeasures.

Uoi can be "sent" from a long distance. In Ukambani many of the notorious and feared practitioners of *uoi* have always been among the regionally more peripheral Kamba and Tharaka peoples in northern Kitui District (Kitui District Provincial Commissioner's Inspection Report, 1910; Lindblom, 1969; Nottingham, 1959). Explanations for this regional notoriety are vague. Usually, they reflect the belief that the region's semiarid environments are responsible for the growth of numerous drug plants that have extraordinary toxicity and potency. They are collected, prepared, and secretly available from individuals who traffic in poisonous plants and their concentrated extracts, as well as other fearful substances (such as a crocodile's spleen) that are prepared for the purpose of causing bodily harm.

The syndrome of belief, action, and response that nurtures and perpetuates the witchcraft phenomenon was a great concern in successive administrations throughout the colonial era in Ukambani. In the view of many colonial administrators, and quite likely among Africans too, no other ideology exerted a more deeply rooted and pervasive influence on the conduct and pattern of individual and community life. J. C. Nottingham (1959, p. 7), a British DO who took a particular interest in the problem during the 1950s, has written that "in Ukambani sorcery would seem on our evidence to be the very marrow of the structure of society." It is true that archival sources written and compiled by European colonial administrators can be interpreted as "class products," and overreliance upon them could conceivably create a distorted view of witchcraft's magnitude in the society if they are not crosschecked through oral fieldwork (Watts, 1983, p. 34). However, at least in Ukambani, the extent and significance of witchcraft is probably underestimated in the colonial administrators' reports. After all, these officials lived on government "stations" physically isolated from African communities, gathered intelligence primarily from Africans who worked for the colonial administration, and were transferred with great frequency.[17]

By 1910, colonial administrators had grown increasingly concerned about the widespread pattern of "magic, witchcraft, and the casting of spells," and among those accused of witchcraft there was a preponderance of women (Kitui District Provincial Commissioner's Inspection Report, 1910; Kitui District Annual Report, 1914–1915). Within a few months of his arrival the new DC for Kitui had been so impressed by the Kamba witchcraft phenomenon that he wrote: "I cannot help believing that there is some supernatural power we know nothing of" (Kitui District Annual Report, 1914–1915). He issued instructions to the "intensely superstitious" Kamba to report any case of alleged witchcraft, whereupon in the absence of satisfactory evidence the accused would be "removed to another Location and kept under observation" (Kitui District, Annual Report, 1914–1915).

Within a year this same DC was refusing to hear new witchcraft cases because the only evidence forthcoming "is that of a medicine man who, by divination, has named the accused" (Kitui District Annual Report, 1915–1916). With Kamba headmen pressuring him for a return to the traditional method of punishing those believed guilty (the *kingole* system provided for execution by bow and arrow on consent of the *nzama*), the DC declared that witchcraft had become "the most serious handicap to Administration people." Headmen and elders, he reported, "fear to do their work because someone might make *Mchawi*[18] against them; any death that is not absolutely accounted for is put down to witchcraft and it is impossible to convince them of the contrary even if one is oneself convinced there is nothing in it—and who can say that?" (Kitui District Annual Report, 1915–1916). This same DC also insisted, however, that witches "must not be confused with a medicine man . . . there is no connection" (Kitui District Annual Report, 1915–1916).

To their credit, some of the later DCs in Ukambani also took pains to distinguish "the harmless medicine man" (Machakos District Annual Report, 1933) from the evil-doers. In fact, to the very end of the colonial era, a few DCs showed considerable empathy and interest in understanding the traditional system. The Machakos DC, for example, insisted in 1955 that "care must be taken not to destroy the undoubted good which underlies white magic as a means of healing certain types of nervous disorders" (Machakos District Annual Report, 1955).

In Ukambani, and apparently elsewhere in Kenya, the forced superimposition of Western culture and the social change that necessarily accompanied it contributed in both direct and subtle ways to expanded and innovative forms of witchcraft. Although they are difficult to document, witchcraft and sorcery found new life when the old forms of social relations and bases of security started to come under attack. In a society that traditionally placed great value on egalitarianism, the new quest for an improvement in standard of living promoted visible inequalities among kin, friends, villages, and classes. The spread of materialistic values, and the advancement in status and privilege that was possible for some people through their participation in the cash economy and their access to formal education, all fostered a competitiveness that encouraged real and imagined advantages of one individual or family over another. Envy, greed, discontent, and fear were heightened in this new environment of increasing social and economic differentiation. These negative forces found a natural outlet for interpretation and expression in the institutions of witchcraft and sorcery.

For the colonial administration, the objective in dealing with the "problem" of witchcraft was to minimize its effects on public order. Different strategies were employed in an attempt to control it, including prosecution, oathing, and public confession, and burning of the paraphernalia of witchcraft. In 1936, for example, the DC in Kitui reported that despite court convictions the "evil" of witchcraft remained widespread. At the request of elders from Kitui's southern locations and supported by most LNC members, he took under advisement the possibility of bringing in a renowned traditional healer—Mwakatyengu of coastal Kwale District—whose power was reputed to be so great that no one would dare to practice witchcraft if he was physically present in the District (Kitui District Annual Report, 1936). It is not reported whether this measure was actually used.

While on safari in Mumoni Location that same year the DC wrote that he "was summoned to a *baraza* [meeting] of women who were really anxious to stop the practice amongst their sex, and I saw no objection to them taking an oath amongst themselves to lessen the evil" (Kitui District Annual Report,

1936). This comment is of considerable theoretical interest because it establishes a context of cross-cultural communication but leaves no clues to verify how the Kamba women actually interpreted *their own* situation, nor what outcomes resulted from their actions.

In 1954, during the height of the "Emergency"—the Kikuyu-led, nationalist, "Mau Mau" uprising against inequality and white domination in the Kenya Highlands—the Mau Mau-inspired practice of oathing spread into the Lower Mukaa and Mbitini Locations in Machakos. The geographical pattern of this movement took the colonial administration by surprise. It had been assumed that if Mau Mau sentiments did enter Machakos District the contact points would be "the more sophisticated" and politicized locations such as Kilungu, Kangundo, and Matungulu—that had "large numbers of educated young men" (Machakos District Annual Report, 1954). Instead, the movement began in the least advanced areas of Lower Mukaa and Mbitini. There, some 2,000 Kamba allegedly took the Mau Mau oath, which the DC characterized as having "crept in not along modern political-nationalist channels but through the dark sewers of sorcery and magic" (Machakos District Annual Report, 1954).

Apparently satisfied with the results of a campaign to "clean up" the Mau Mau activities in Machakos, the DC next proposed "right gently but firmly to tackle the witchcraft problem" in the southern locations. Chief Simeon Musyoki of Mbitini reportedly took the lead for the administration in this antiwitchcraft drive, and his example was followed in Nzaui and Kikumbulyu Locations. The actions taken in Mbitini, and their. perceived meaning, were related by the DC in his Annual Report for 1954. Recalling the imagery of the witches in Macbeth, he notes that

> the campaign culminated in a ceremony on November 5th . . . when a crowd of some five hundred Mbitini witches brought in and burned their paraphernalia ("eye of newt and toe of frog"; or more exactly, cuttings of hair and finger nails from the body of a person who has died from sorcery; human bones and excrement; sand and earth gathered from a crossroads; and so on). The witches gave their consent in unison to the burning (though two had committed suicide at their discovery) and the husbands watched, most of them having had no idea of what their wives had been doing, or keeping in their little bags. As the flames mounted, a few randy sorceresses came running up to add their contributions to the fires. It is satisfactory that only trade implements and not the witches themselves were burnt. And though one may tend to joke about these things, they are a very close and terrifying threat to these people, and a real manifestation of evil. This is another reason for helping in the work of the Women's clubs, and letting new thoughts and ideas into the old world, to aid the work of the Missions.

The preceding narrative succeeds in communicating the DC's sense of detached amusement concerning the so-called witchcraft practices, as well as his feeling that the "campaign" had achieved positive results. Once again, a colonial administrator's perceptions and definition of the scene are the only testimony available (Nottingham, 1959). One may reasonably expect that African interpretations of the causes and meanings of these public spectacles would be different. Certainly the events were exceptional. As they continued into 1955, Dr. Wilson, a government anthropologist, was brought in as a consultant to investigate "witchcraft in the area" and to recommend other approaches to curb it (Machakos District Annual Report, 1955).

During 1955 large ceremonial "witchcraft" burnings spread to many locations in eastern and southern Machakos District, including Nzaui in May. At Makueni, "a large cleansing ceremony . . . was attended by every woman in the settlement." In October, at Mbumbuni in Kisau Location, "700 witches, 50 warlocks, and 20 witchdoctors burnt their paraphernalia;" while in December "204 witches and 50 witchdoctors in the Wautu section of Kilungu openly admitted their arts and crafts and gave up a formidable number of implements" (Machakos District Annual Report, 1955). According to a later account by the DO, during the attempted purges carried out in three locations in 1954 and 1955 "some eighteen hundred women and one hundred and fifty men of all ages handed in witchcraft objects" (Nottingham, 1959, p. 5). These mass gatherings apparently ceased after 1955, although official reports for 1956–1958 continued to refer to the widespread prevalence of witchcraft among the Kamba in Kitui District as well as in eastern and southern Machakos.

Witchcraft accusations also remained a common charge brought before the African courts, whose jurisdiction included the Witchcraft Ordinance of 1925 (Government of Kenya, 1962; Mutungi, 1977). Although these courts could and did punish practitioners of "witchcraft," they provided no relief for the victims. According to the DC for Machakos, in one "recent case a novice witch found herself unable to lift the spell from her victim and finally, in desperation, she hanged herself." (Machakos District Annual Report, 1957).

What did the large public gatherings and burnings of "witchcraft paraphernalia" signify to the Kamba in 1954–1955? Colonial administrators' reports are colorfully descriptive and show concerned curiosity, but their reliance on undefined and stereotyped English terms such as "witch," "witchdoctor," "sorceress," and "witchcraft" created a semantic and conceptual tangle. The identity, variety, interrelationships, and significance of actors and events is either assumed or overlooked, and one is left with less than half of the story. Reliable interpretations of this particular episode of witchcraft "hysteria"—for that is apparently the source of energy which powered the contagious spread of these epidemic-like spasms of "catharsis"—must await the results of local fieldwork in oral history. Conditions in the mid-1950s seem to have paralleled those that gave birth to the messianic and cult movements that convulsed Ukambani during the 1890s and the first two decades of the 20th century. The fact that the "witchcraft" aberrations occurred during the tumultuous Mau Mau years, and were quite possibly an outgrowth of the political turmoil following the oathings that took place in Mbitini, seems a plausible link and a fruitful line of inquiry. Furthermore, the colonial administration itself engaged in direct manipulation of the unsettled conditions by conducting "small experiments" in the northern and southern areas of Machakos. This bit of social engineering reportedly uncovered 29 types of sorcery in use among the Kamba (Nottingham, 1959), and "revealed still further the extent to which both black 'Uoi' and white 'Uwe' witchcraft have a grip on the lives of the people" (Machakos District Annual Report, 1955). The extraordinary overrepresentation of women implicated in these sorcery inquisitions in Machakos during the mid-1950s remains unexplained. Perhaps the roots of this phenomenon were, as Nottingham (1959) suggests, embedded in women's inferior social status and power in traditionally male-dominated societies.

During the colonial period in Ukambami, as elsewhere in Kenya and tropical Africa, the positive association of traditional medicine and traditional healers with community health was eroded through contact with a European language and culture and as a consequence of unprecedented social change. Seventeenth-century English occult words such as "witch," "wizard," "witchcraft," "sorcerer," and "witchdoctor" (Miller, 1980, p. 3, calls the latter term a "logical impossibility that mixes good and evil practices") entered the popular idiom and were applied indiscriminately so that "witchcraft" became a virtual synonym for traditional medicine and "witchdoctor" became the common (and always pejorative) term for an African traditional healer.

The corrosive effect that word connotations in one language can have upon the integrity of another culture's institutions is vividly illustrated in a statement by the Machakos DC when he wrote that "witchcraft has a very considerable hold among the Kamba, and it is desirable to do all that can be safely done to reduce its practice; *particularly in its black* magic forms" (Machakos District Annual Report, 1955; emphasis added). The inference one easily draws from this remark is that beneficial traditional healing, or "white magic," is also "witchcraft." Consequently, even though the DC subsequently observes that "white magic" is of value in the treatment of some mental disorders, his words are evidence of the semantic blind alleys and inertia that can develop and undermine the communication of meaning between and within cultures.

A related problem of interpretation also emerged early in the colonial era in Kenya in connection with the imported British legal system. Legislation intended to suppress witchcraft by means of prosecuting and punishing practitioners was enacted in the form of Witchcraft Ordinances in 1909, 1918, and 1925. However, evidence sufficient to produce convictions proved elusive and inconclusive in the context of statutory and English Common Law. This transplanted code for colonial law and order never adequately comprehended the identifying characteristics of a witch, confused witchcraft with "the occupation of a medicine man" (Machakos District Annual Report, 1933, p. 33), and even denied the existence of witchcraft (so as not to encourage it) while simultaneously conceding its power to kill (Mutungi, 1977).

Use of conventional English terms such as "witch," "witchcraft" and, especially, "witchdoctor," date from the earliest years of contact between the European and African cultures. They have since degenerated to the level of mere caricatures and stereotypes that are incapable of precise definition and have little conceptual or comparative usefulness. It is now fashionable in many academic circles, for example, to use such terms as "traditional healer" or "medico-religious specialist" to describe a *bona fide* practitioner whose work is grounded in a traditionally based system. Nevertheless, the old and sensational, but grossly misleading, labels seem destined to retain their popularity over the presumably more precise and informative terms invented by social scientists. As Miller (1980, p. 6) observes, "even those who are acquainted with traditional medicine often mix terms, confusing practitioners, with malpractitioners, the trained with the untrained, established healers with quacks and charlatans." Herein lies a subtle but significant impediment to improving communication and thus fostering appropriate kinds of cooperation between biomedical workers and traditional medical practitioners in Kenya and other African countries.

NOTES

1. Republic of Kenya, *Population Census, 1979*; Morgan and Shaffer, 1966.

2. Sound land management findings in this zone suggest that there should be over 4 hectares per stock unit. Survey of Kenya, *National Atlas National Atlas of Kenya*. 3rd. ed. Nairobi, 1970, p. 28.

3. This legendary account of Kamba origins and dispersals is not well supported by linguistic or archeological evidence. Similar folk seem to have occupied the central highlands of Kenya for a much longer time than the present accounts of these peoples' traditions claim. The homogeneity and present distribution of the Thagicu subgroup of Bantu languages in the central highlands (including Kikuyu, Embu/ Mbeere, Cuka, Tharaka, Meru, and Kamba) indicates a highlands dispersal center for all of them. This apparently casts doubt on the Kamba claims of an origin in the Mt. Kilimanjaro region. Early Iron Age pottery, iron workings, and dometic livestock discovered at Gathun'ang'a and in Chuylu and dated to the 12th-13th and 15th-16th centuries show that agriculturalists, possibly Bantu, were living in the Kenya highlands well before the migrations recalled in the traditions (Spear, 1981, p. 59).

4. Eight famines have been documented for the 19th century, beginning with the *Yua ya Ilendu*, "the staggering famine" of the late 18th or early 19th century, and ending with the *Mapunga* (rice) famine of 1898–1899 when as much as half of the population is thought to have died. See Munro (1975), p. 21; and Owako (1971). For the nine famines in Kitui documented between 1908 and 1962, see O'Leary (1980).

5. Oliver (1965, p. 426).

6. Owako, 1971, p. 181; Lindblom (1969, p. 16) asserted that *kilungu* means "'part, portion,' a name which, whether intentionally or not, is very appropriate."

7. This account of the cults is heavily indebted to Munro (1975) and Tignor (1976). Munro (1975, p. 114) asserts that the cult leaders "normally came from emergent or aspirant *andu awe*, men and women on the margin of the achieved-status religious authority system, rather than from established practitioners."

8. Beginning in 1919, Africans in Kenya were compelled to supply European settlers with labor and were required to carry the hated identification card, or *kipande*, which the colonial government employed as an administrative tool to control Africans' activities (A. J. Temu, 1972).

9. The early Christian converts were readily identifiable since they were expected to wear European clothing (Tignor, 1976, p. 127).

10. In 1920 the East Africa Protectorate was renamed Kenya Colony and Protectorate. The "protectorate" was a narrow strip of land along the Indian Ocean coast that was leased from the Sultan of Zanzibar.

11. Dental disease was stimulated by the Kamba custom (both sexes) of filing the front teeth of the upper jaw to points after reaching the age of puberty. This was done to enhance one's physical attractiveness to the opposite sex (Lindblom, 1969, pp. 70, 392–397; Machakos District Annual and Quarterly Report, 1922; Machakos District Annual Report, 1927).

12. Thomas (1975) contends that the colonial administration purposefully fostered reliance on traditional healers.

13. Quoted by Ndeti (1972), p. 117.

14. Traditionally among the Kamba, "circumcision came rather late because theoretically nobody would participate in sexual intercourse with an uncircumcized person. An uncircumcized person participating in sexual affairs was unheard of, and anyone found doing so was thoroughly punished" (Ndeti, 1972, p. 87).

15. *Kithitu* is also a general term used to denote several categories of oaths and ordeals that vary according to the gravity of the dispute as well as with the seriousness of the oath's consequences once it has been administered (Mutungi, 1977).

16. "Women are never allowed to swear on the *kithitu*" (Mutungi, 1977, p. 93).

17. For example, in the 12 years from 1933 to 1945, nine DOs were posted to Machakos, including one who served a nonconsecutive term of 3 months. A few remained for as little as 5–8 months. In the 8 years from 1945 to 1952, nine different men served as DO in Machakos. In Kitui District, there were 32 DOs between 1920 and Uhuru in 1963. The longest term was that of R. J. Hickson-Mahony, who served from November 1956 to October 1960. The shortest terms in Kitui were 3 days in 1933, 17 days in 1942, and 50 days in 1932. In 1933 and 1942 Kitui had four different DOs in each year (Kenya National Archives list.)

18. *Mchawi* is the Swahili term for witchcraft.

4

Land, Livelihood, and Biomedicine in the Kilungu Hills

Kilungu Location—an administrative unit that forms the core of a Kamba region popularly known as the Kilungu Hills—is a rugged area of some 243 square kilometers (Figure 4-1). With more than 61,000 inhabitants (1979 estimate), it is also one of East Africa's most intensely humanized landscapes. As previously noted, I visited this area during the early months of 1978 to learn about the nature and relative importance of health care resources and options, and health problems, in a rural region. Historically, the people of Kilungu have been perceived as culturally conservative and "different" from other Kamba (Chapter 3). Also, although the Kilungu Hills are only an hour and a half away by car from Nairobi, the area remains "off the beaten track." Fieldwork began with the assumption that local biomedical services had received little attention from the government since Independence, and that traditional medical specialists still handled a large share of the population's acute and chronic illnesses.

This chapter presents the findings from fieldwork concerning the position of biomedicine in the ethnomedical system of Kilungu in the late 1970s. I begin by sketching the region's human geography in order to establish the context of disease and health care. I then consider the range of therapeutic options in Kilungu, analyze the biomedical evidence of disease, and assess the utilization and effectiveness of local biomedical services.

ENVIRONMENT AND POPULATION PATTERNS IN KILUNGU: A VIGNETTE

A half-hour out of Nairobi, near the Machakos turnoff, the main Mombasa Road heads southeast in long, straight stretches across the lava-based Kaputei Plains. Toward the east, on the far horizon, a gathering of clouds envelops the highest ridges of the densely populated Kilungu Hills. The plains landscape here is one overwhelmed by sky and a vast, undulating expanse of thinly wooded grassland and thornbush on "black cotton" soils. Bordering Masaailand, this is sparsely populated ranch country. During Kenya's colo-

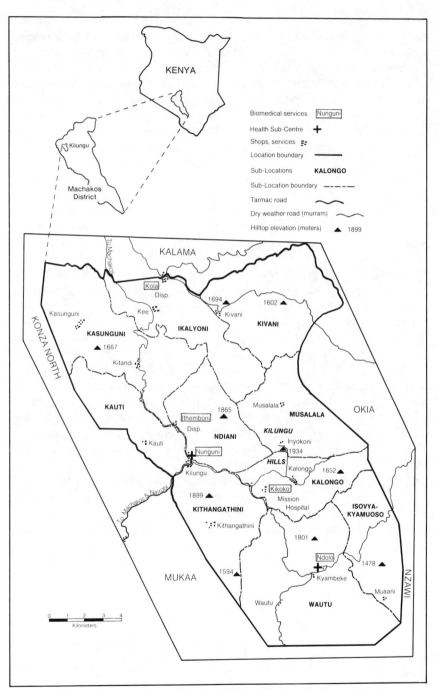

Figure 4-1. Kilungu location map.

nial days it was reserved in great tracts exclusively for European management. Widely dispersed herds of grade and cross-bred cattle graze the ranges, each supervised by an African herdsman. From the road, one can often see clusters of gazelle and, occasionally, a giraffe. These plains are situated at an average of 1,700 meters above sea level. They are punctuated by a few isolated hills that jut spectacularly some 200–300 meters above their surroundings. The roadbed's long grades are deceptively steep, and the drivers of tractor-trailers, lorries, and buses inevitably gear down several times while crossing these plains on the 500-kilometer run-up from the coast.

About 70 kilometers from Nairobi and some 12 kilometers before the turnoff to Kilungu, the land rises as the tarmac approaches the base of Kilima Kiu (Black Mountain) (2,018 meters). The signboard that comes into view at a road junction on the left reveals the stenciled name "John Muia Kalii, P. O. Ulu." Below the name an arrow points in the direction of a murram road that angles off into the bush and leads to the unique residential health care center at Kilima Kiu built and operated by Kalii. Over the years this middle-aged Mukamba man, a self-styled "traditional" practitioner (*mganga*) who wears a white physician's jacket over a shirt and tie, has created a powerful mystique about himself and his capabilities. He is undoubtedly the area's best-known citizen and a personality whose name is widely recognized in Kenya among medical personnel as well as the general population. Mercedes-Benz automobiles and lorries from the Mombasa Road are frequently seen on the dusty road leading to and from Kalii's place. Referring to himself as an herbalist, Kalii maintains a large stock of dried plants and powders stored in bottles in his "pharmacy"—located in 1 of the more than 15 buildings that make up his compound. Other facilities include inpatient wards, some ramshackle but the most recent a new, large concrete block structure with room for at least 25 beds which Kalii allegedly subsidized with "government" funds; staff housing; and a small modern-style restaurant–bar–motel complex, gaily painted with African "highlife" and wildlife scenes, for the use of patients and families who accompany them to the center. Extensive gardens are also maintained to provide food for patients and staff.

Kalii appears to have had great success in resisting efforts by outsiders to learn details of his therapeutic methods and herbal compounds. While he does show expected courtesies to visitors, including occasional groups of medical students who tour the site as a part of their third-year "community health" course at the University of Nairobi's Medical School, Kalii's technical procedures and results seem inaccessible to would-be investigators. This aura of secrecy surely strengthens Kalii's charisma among clients, while it undoubtedly insures his reputation as a curiosity—a "fringe" practitioner of dubious merit and uncertain virtue—among Kenya's rank and file biomedical community. Kalii is, nevertheless, a person of great psychological power and influence. Dr. O. K. Mutungi, a Mukamba who has served as Dean of the Faculty of Law at the University of Nairobi writes that Kalii told him that "not less than 175 potential candidates for the 1974 General Elections in Kenya visited him prior, and subsequent to, the August 1974 nominations" (Mutungi, 1977, p. 101).

Paradoxically, Kalii is said to work closely with a laboratory technician who is sent from the Machakos District hospital to collect sputum specimens from those patients of Kalii who present suspected active tuberculosis (TB). Following positive TB diagnosis at Kilima Kiu, patients reportedly are referred to Machakos hospital for treatment. In theory, at least, such cooperation is commendable and to be encouraged if it in fact brings positive health benefits. In Kalii's case, since evidence of the latter is unavailable, no verdict seems justified. Kalii's apparent recognition by the government has added to his reputation as a *mundu mue munene*—a very famous traditional healer.

Salama Trading Center is a gateway of sorts to the Kilungu Hills. Located on the Mombasa Road about 12 kilometers south of Kilima Kiu at the junction of the paved road to Nunguni in Kilungu, Salama is a regionally important service center consisting of a busy petrol station and a long, low-slung row of some 20 bars, *hoteli*, and shops that cater mainly to the needs and wants of travelers. On a typical afternoon a dozen vehicles—local and express buses, two petrol tankers, a Peugeot sedan, passenger vans (*matatus*), a Land Rover, and a Kenya Army lorry—line the highway shoulders and the access road that runs alongside the yellow, green, pink, and white fronts of the buildings. Groups of three and four men stand talking outside the Salama Yoani Kwenu Night Club, the Ngamyone Transporters Lodge and Bar, and the Hill Top Country Club Day and Night. Others sit inside at tables drinking bottled beer or Fanta served by local barmaids. Several patrons in the Hill Top are eating hot meals prepared on the premises—plates of starchy, white *posho* and beans served with goat meat, chicken, or beef, and soup. Outside is an old bus operated by a Machakos company. Light blue with four badly worn tires, it is making a slow, noisy, and smoky departure from the trading center with a full load of passengers. Its roof rack is loaded 4 feet high with old cardboard boxes and suitcases secured with ropes, bags of vegetables, chickens in a makeshift cage, two bicycles, and a small table and chairs. Upon reaching the road at the southeast end of the shops, the driver turns left to make the tortuous hour-long run up through Engai, Mukaa, and Kilome along the mountain ridges to Nunguni, 16 kilometers from Salama.

Located at the end of the only paved road into the Kilungu Hills, Nunguni is the principal central place, market center, and transportation hub of Kilungu Location (Figure 4-1). It is sited on a high ridge, and in clear weather the view south offers a magnificent vista over the hills and dry plains to snow-capped Mt. Kilimanjaro, 150 kilometers distant on the Tanzania border. Several dozen brightly colored buildings that serve as general merchandise shops, *hoteli*, bars, and butcheries line both sides of the road in Nunguni. Just beyond the central area of shops the road forks and forms the location's two main murram roads. One continues north past the location chief's office and court at Ithemboni, toward Machakos; the other road skirts Nunguni's Tuesday–Thursday periodic market and runs eastward to Kikoko, the site of a mission hospital and outpatient dispensary operated by the Precious Blood Sisters (Machakos Diocese); and thence to Kalongo Market. During the seasonal rains of March–May and October–December these murram roads and the ungraded tracks that feed into them are often muddy, slick, and extremely hazardous. They often become impassable for hours or days at a time.

Kilungu Secondary School is adjacent to Nunguni, while the mission-operated schools of Kikoko and Thomeandu are 5 kilometers to the east in Kalongo Sublocation. Some two dozen primary schools with widely varied facilities and resources also operate in Kilungu.

Nearby the Nunguni shops is the government-run Nunguni Health Sub-center, situated on the western edge of the location. A large proportion of Kilungu's inhabitants live beyond a 4-kilometer radius of Nunguni and are thus probably beyond the effective range of this rural health facility. Unfortunately for the sick and injured who come on foot, the health center was built on a hill-top site accessible only by a steep track leading up slope from the main road. However, once they have arrived at the health center's sky-top compound, patients can enjoy extraordinary panoramic views of the Kilungu Hills!

Kilungu's ten administrative sublocations are characterized by numerous ridges that frequently rise above 2,000 meters, and by deep, steep-sided valleys that occasionally open out to wide basins. Annual rainfall exceeds 1,200 millimeters on the higher elevations, and at this latitude of 1° 50′ south it has a distinctly bimodal distribution. Potential evapotranspiration ranges from 1,500 millimeters to 2,000 millimeters annually and is generally surpassed by rainfall only in April and November (Moore, 1979). Although these highlands do experience some precipitation in all months of the year, most streams have a distinctly seasonal flow. In the lower and drier northern extremity of Kilungu many small streams have been dammed to store runoff for cattle and domestic use.

Over 98% of Kilungu's 61,000 inhabitants are Kamba. With 56% of the population comprised of children under 16 years, Kilungu surely has one of the Third World's youngest populations. Assuming a 44.6% growth rate (identical to that for Machakos District as a whole) for the period 1969–1979, the mean population density of Kilungu now exceeds 250 persons per square kilometer (648 persons per square mile), which places it among tropical Africa's most thickly populated regions.

In Kalongo Sublocation, moreover, the estimated density exceeds 377 persons per square kilometer (976 persons per square mile). At this growth rate there would be more than 75,000 people in the location in the mid-1980s—approaching 310 persons per square kilometer (803 persons per square mile).[1]

The adult (16 years and over) sex ratio is also a significant demographic aspect of Kilungu. Within the location there are only 67 males for every 100 female adults, which is the lowest adult sex ratio among all 27 locations in Machakos District. This highly imbalanced ratio compares with 82 for the District overall and 72 in the neighboring Kitui District. Only five Machakos locations have ratios below 70. Comparison of ratios within Kilungu's ten sublocations reveals unmistakable similarity: Musalala (70), Wautu (68), Kauti (68), Kalongo (66), Ikalyoni (68), Kasunguni (60), Kithembe/Ndiani (65), Kithangathini (65), Isovya/Kyamuoso (71), and Kivani (67).[2] These figures suggest the extraordinary importance of male labor migration from Kilungu, a mobility stimulated in part by land shortage and deteriorating agricultural conditions in the highlands (Owako, 1971). A survey conducted by the Joint Machakos Project (JMP) in the northern part of the District suggested a similar situation. Approximately one-third of male household heads in the JMP had jobs outside the local area, while 50% of male household heads were absent at the time of inquiry (Slooff & Schulpen, 1978).

Out-migration of working-age Kilungu males distorts household composition and thus the structure of the location's resident labor force. Women, and to a lesser extent children and old men, have responsibility for the bulk of the agricultural work and day-to-day domestic responsibilities. Agricultural productivity is undoubtedly affected by this tradeoff. The low sex ratios, and thus the preponderance of women, children, and the elderly, also have implications for the kinds of health services the Kenya government supports and thus for health policy and planning.

Dispersed homesteads (*musyi*; pl. *misyi*) of nuclear and extended family groupings characterize the Kamba settlement pattern in Kilungu and other areas of Ukambani. Several *misyi* collectively define the threshold for a "village" unit (*utui*; pl. *motui*). Within an *utui* several of the 14 major and 11 minor patrilineal, exogamous clan (*mbai*) divisions of the Kamba may be represented (Ndeti, 1972). As elsewhere in Africa, house types and construction materials and the presence or absence of such features as pit latrines near a compound are useful indicators of a homestead's socioeconomic level. There is a considerable variety of house types in the Kilungu Hills. The round house (*musonge*) of mud daub and wattle, and thatched roof, is not uncommon, although rectangular structures—frequently of mud and wattle but increasingly of more permanent materials with corrugated iron roofing—predominate.

Latrines may actually interfere with hygiene and better health if improperly located and maintained. A study by the JMP in northern Machakos apparently found better sanitation among the more isolated, poorer, traditionally oriented households typical of the project's eastern sublocations—where nearly 80% of the compounds lacked latrines—than within the "richer," western areas where four of five households had latrines. This is because the dark, humid conditions of a covered structure with faecal matter concentrated where people place their bare feet, were assumed to form more ideal breeding grounds for parasitic helminths and other pathogens than the scattered, usually sunlit and well-ventilated sites utilized by nonlatrine owners (Slooff & Schulpen, 1978, p. 267).

At least a dozen small clusters of shops in Kilungu sell basic provisions such as sugar, salt, tea, soft drinks, beer, paraffin, bread, matches, and a few patent medicines. These centers link the rural population with the commercial economy and reflect a growing dependence on the exchange economy and extrafamily services. Some shops are open on an irregular basis only, and vacant or underused buildings are common. Cloth, tailoring services, shoes, bars, and barbers are also available in a few of the larger service centers such as Kalongo. Periodic markets operate at Kalongo (Monday and Friday), Muaani (Thursday), and Wautu (Wednesday), as well as at Nunguni and often draw several hundred participants at a time. Cultivation of food crops such as maize, beans, cassava, bananas, and some coffee, as a cash crop, constitute the focus of farming activities. Cattle and grazing land are scarce in the higher elevations. Most food crops are grown in a bush-fallow system developed on steep slopes with highly erodible soils. Bench terracing, in particular, and contour bunding are common management practices on the intensively cultivated slopes and lend a manicured appearance to the well-maintained gardens and tree crops; but many farmers still practice little systematic soil conservation. Visible evidence of the growing population pressure is widespread (Moore, 1979). The original mixed forests of the higher elevations have disappeared, replaced by crops, grasses, and small, widely spaced stands of imported and carefully guarded eucalyptus, black wattle, and conifers. Occasional fruit species such as mango and wild fig (the latter is a common marker of the remaining sacred groves, or *Mathembo*) are prominent among the few remaining trees in the Kilungu Hills. The extent of clearing is perhaps not exceptional considering that the region has been settled for at least two centuries (Chapter 3). Nevertheless, the current high rate of population growth coupled with population pressure on a shrinking per capita land base may present a land quality/productivity crisis with major implications for the nutrition and health of the latest generation (Moore, 1979).

THE CHANGING PATTERN OF HEALTH CARE OPTIONS IN AFRICA

Actions to alleviate an illness can range from a simple choice by the individual concerned to drink a medicinal tea prepared from an herb that grows near his or her compound, to complex negotiations that involve significant others and, eventually, visits to traditional medical practitioners (TMPs) and one or more of the local biomedical services (Table 4-1). Therapeutic choices are generally made one at a time. Decisions usually involve and may be determined by relatives, neighbors, or other authoritative persons acting on behalf of the sufferer, especially in more serious and threatening circumstances. As important studies by Janzen (1978) in rural Zaire and Chavunduka (1978) in urban Zimbabwe both confirm, lay authority continues to play a major role in the diagnosis and treatment of illness. Customarily, significant others within a sufferer's household and local community have the authority to act as "therapy managers" and thereby control the healing process. A diagnosis, as such, may be made by those in authority prior to consulting a TMP (or biomedical worker). A diviner, for example, may be expected to confirm a "causal hypothesis" suggested by therapy managers and then to recommend an appropriate course of treatment. This pattern of lay control illustrates a key feature of African therapeutics. As Feierman (1984, p. 31) observes in his superb critical review of the literature on African health and healing systems, "the people most qualified to make therapeutic decisions are the ones who have the least technical knowledge of medicine and the greatest personal knowledge of the patient."

Because African therapeutics are integrated with a community's social and economic structure, the inevitable changes, both subtle and direct, that continue to shape the peasant economy predictably alter authority patterns and the actors involved in therapy management (Feierman, 1984). Examples of such changes include the effects of out-migration of male household heads, the growth of female-centered households in both rural and urban areas, and the diminution or demise of elders' control in the face of state-authorized political and jural institutions. Several examples of altered patterns of lay authority in therapy management occurred during my research in Kenya. The most dramatic and well documented case, presented in Chapter 8, involved a terminal cancer patient in Nairobi.

THERAPEUTIC CHOICE IN KILUNGU

Whether one looks to direct, "objective" measures, such as daily outpatient returns, or to indirect, "subjective" indicators such as the local density of traditional medical specialists, disease and misfortune are never far removed from the lives of rural Kamba. The heavy burden of illness in Kilungu, as in other Third World societies, results in what Worsley (1982, p. 332) aptly describes as "the desperate search for health" among available and, sometimes, quite remote medical resources. It is appropriate to think of the collective therapeutic options as forming an open, inclusive "ethnomedical system" (Chapter 1) since any one person may use some or all of these resources during the course of a particular illness—and possibly in a different order for a later illness. In this sense health *seeking* is more descriptive of the

Table 4-1. Therapeutic Options in Kilungu Location, Kenya: The Range of Choice

Category		Source	Availability/place
Self-treatment			
Take no action		—	
Home remedies, incl. preparations from medicinal plants and advice from relatives		Local/external	Widespread
Shop medicines (Aspro, Malaraquin, Cofta Syrup, and so on)		Local/external	Trading center and markets
Antibiotics and injectibles		Clandestine, private sources via biomedical or commercial sector	Varies
Traditional Medical Practitioner		Local	Crude ratio = 1:350 persons
		External (infrequent)	
12 types of specialties represented			
Biomedical Facilities/Practitioners			
Health Subcenter	(O)	Local: government	2 (Nunguni and Ndolo)
Dispensaries	(O)	Local: government	2 (Kola and Ithemboni)
Small hospital	(O, I)	Local: mission	1 (Kikoko—no doctor)
Provincial hospital	(O, I)	External: government	Machakos town (22–45 km)
Private physicians	(O)	External	Machakos town, Nairobi
Private clinician (*kitali*)	(O)	Local	1 Kalongo market
Former officer in charge, Nunguni Health Subcenter			

Note: From field survey, Kilungu, 1978.
Abbreviations: I, inpatients; O, outpatients.

actual process of health care than health *delivery* by so-called providers (Worsley, 1982). John Janzen's (1978) title for his recent ethnomedical survey, *The quest for therapy in lower Zaire,* is clearly intended to reflect this often peripatetic process.

Residents of Kilungu have three possible sources of assistance in connection with their health problems (Table 4-1). As is common throughout the world, people begin with self-help. Thereafter, depending on the perceived need and characteristics of the ailment, and often on direction from "significant others" such as parents, relatives, or neighbors, help will be sought from a TMP, dispensary, or other biomedical service point. If an illness persists after such treatment the anxiety and desperation associated with it also grow. This can and often does lead the patient and his or her significant others to try a therapy that is different from the initial choice. In this way biomedicine, traditional healing, and self-treatment can and often do become intertwined—serially or concurrently—during the course of a single illness episode (Figure 4-2). This sort of pragmatism is a well-known characteristic of people in societies where medical pluralism is the rule. Previous studies of Kamba health care preferences and behavior indicate a perception of powerful alternative therapies in both traditional and biomedicine (Thomas, 1971). Furthermore, resort to a different therapy is more dependent on what has been seen or heard to "work" than on educational level or internalized biomedical concepts (Mburu, 1973; Mburu, Smith, & Sharpe, 1978).

From a strictly biomedical, curative perspective this meshing of different therapies introduces "noise" into the health care process in the form of environmental and behavioral variables that are little known and largely

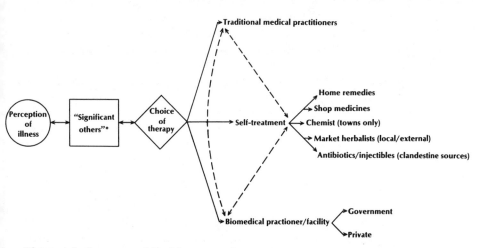

Figure 4-2. Structure of health care options and decision making: Rural Kenya. (*Therapy managing group [Janzen, 1978])

uncontrollable. Interactions of these factors promote a "black box" effect that is always potentially present when attempts are made to trace a patient through his or her illness episodes and to credit one or another therapy as the most efficacious.

HEALTH SERVICES AND DISEASE IN KILUNGU: BIOMEDICAL VERSION

Biomedical facilities (*hospitali* in Kikamba) operated by the Ministry of Health in Kilungu consist of two health subcenters (these are distinguished from health centers primarily by the absence of inpatient facilities and a smaller staff) located at Nunguni in Ndiani Sublocation and at Ndolo in Wautu Sublocation; and dispensaries at Ithemboni and Kola (Figure 4-1). They offer free outpatient and maternal and child health (MCH) services and a weekly family planning clinic. There is also a Catholic mission hospital at Kikoko which is operated by the Precious Blood Sisters (of Austrian origin) under the supervision of the Machakos Diocese. This facility, which functions on a fee-for-service basis, began as a dispensary and grew into a hospital shortly after World War II. The present Kikoko Mission Hospital was opened on the same site in March 1977. It is equipped with 40 regular and 12 maternity beds, and an outpatient clinic. However, there are no medical doctors and no surgical, x-ray, or laboratory services at Kikoko or elsewhere in or near the location. The provincial hospital is over 40 kilometers away in Machakos (pop. 20,000), the district's administrative and commercial center. Referrals from Nunguni Health Subcenter to the provincial hospital averaged 8.4 per month in 1977–1978 (Nunguni outpatient returns). In addition, two ex-clinical officers (COs) formerly in government service reportedly offer medical treatment and drugs on an occasional basis from rented rooms at two of Kilungu's markets.[3] Overall, a large proportion of the population requiring biomedical assistance in Kilungu and throughout Machakos District is poorly served. Moreover, once a patient arrives at an outpatient facility the wait is often many hours long but the diagnosis is quick.

In theory, the two government health subcenters and the mission clinic in Kilungu serve a location population of at least 61,000 people. In practice, outpatient registers indicate that substantial numbers of people from neighboring locations such as Mukaa, Konza, and Okia also depend on these services.

Nunguni, the larger of the two health subcenters, was staffed by a registered CO (this title signifies at least 3 years of medical training) who was not a Mukamba, three nurses, a mobile family planner field worker who was there once per week, plus two attendants and a watchman. Ideally, a health subcenter will have 14 staff.[4]

As previously noted, Nunguni's location near the western periphery of Kilungu creates poor access for many potential users. Ndolo, the other health subcenter, is situated some 12 kilometers southeast of Nunguni in a valley bounded by high ridges that serve to restrict access from other population clusters in the location, particularly to the north and west.

Nunguni is open daily for 7 hr, Monday through Saturday, and from 9:00 A.M. to noon on Sundays for patients who have been placed on a special antibiotic injection regime. Average monthly attendance was 2,216 in 1977–1978 (64% new cases, 36% reattendance).

The sublocations and villages of origin of people who came to Nunguni as outpatients during July 1977 are shown in Figure 4-3 and Table 4-2. As expected, females (59%) outnumbered males (41%), and children (52%) were seen with slightly greater frequency than adults.

Although a much longer period of observation is required to determine the pattern conclusively, a relationship between increasing distance from the health subcenter and decreasing attendance is evident. Table 4-2A reveals a distance gradient in terms of the proportion of visits relative to a subloca- tion's total population and its proximity to Nunguni, while Table 4-2B indi- cates that at least four of every ten outpatients lived within 2.3 kilometers of the clinic. Two-thirds of the people utilizing Nunguni lived within a radius of about 4 kilometers (Figure 4-3).

This pattern parallels studies of similar health facilities in other African countries where utilization is also characterized by a fairly regular decline in attendance with distance. In Uganda, for example, Gershenberg (1972) found that attendance rates of hospital and dispensary outpatients fell off by 50% every 3.2 kilometers. In Hausaland, northern Nigeria, utilization of govern- ment dispensaries declined by 25% per kilometer, meaning that the atten- dance rate at 4 kilometers was only a third of the rate at zero kms. (Stock, 1983). The regularity of the distance effect on utilization is greatest when *aggregate* (crude) rates are calculated for the total resident population found at specific distances from the health facility. This is also apparent for Nun- guni (Table 4-2). However, utilization rates for individual *villages* in Hausa- land had a much less regular pattern. The magnitude of difference among them was commonly 20:1. At the individual village level, behavioral factors were better predictors of utilization rates than distance, population, or access to modern transportation. Individual attitudes toward the efficacy of West- ern-type treatment relative to traditional therapies had the greatest effect on the utilization rates over distance. They varied significantly according to the type of illness. Fractures, for example, are most appropriately treated by a traditional Hausa bonesetter, in part because almost everyone believes that hospitals amputate fractured limbs. For treatment of TB, in contrast, hospi- tal medicine must finally be obtained because no effective traditional reme- dies are known. Other determinants of the "distance decay" effect included

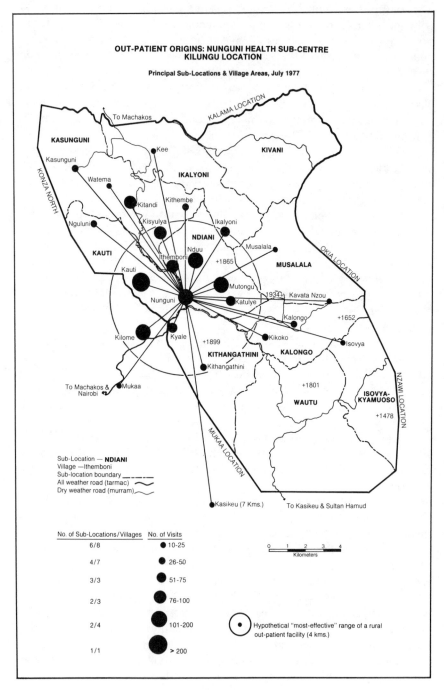

Figure 4-3

Table 4-2. Distance and Outpatient Attendance at Nunguni Health Subcenter, July 1977[a]

A		B		
Per capita visitation ratios by sublocation		Distance of village from clinic (km)		Total visits ($N = 1,808$) within 0–4 km (%)
Kauti	1:12	Nunguni	0–0.5	8.2
Ndiani	1:16	Ithemboni	1.6	4.5
Kasunguni	1:28	Kyale	1.8	3.7
Ikalyoni	1:55	Mutongu	1.8	5.6
Kithangathini	1:58	Nduu	2.2	5.9
Kalongo	1:70	Kauti	2.3	14.3 (42.2)
Musalala	1:121	Kilome	2.3	10.2
Muaani/Isovya	1:611	Katulye	2.7	4.1 (14.3)
Kivani	—	Kisyulya	3.2	4.2
Wautu	—	Kithangathini	3.7	2.6
		Ikalyoni	4.0	4.0 (10.8)
				67.2

Note: Nunguni Health Subcenter Outpatient Register, July 1–31, 1977.
[a]Linear distance.

quality of service (higher-order outpatient facilities such as hospitals or health centers attracted patients from longer distances than did dispensaries), sex (males in Muslim Hausa society travel farther for treatment than females), and age (children tend to travel shorter distances than adults). Distance, population, access to modern transportation, and seasonality did not prove to be reliable predictors of utilization rates at the village level (Stock, 1983).

The utilization rates for Nunguni Health Subcenter at the aggregate and village levels suggest much the same pattern as Hausaland (Table 4-2). Per capita visitation ratios decline roughly in accord with a sublocation's distance from the health facility. At the *village* level, however, places such as Kauti and Kilome sent proportionately more, and Kyale, fewer, patients to Nunguni than expected in view of their location (Figure 4-3). Kilome (just outside Kilungu Location) and Kyale are only a half-kilometer apart and both are accessible to Nunguni by an all-weather road. These patterns cannot be explained without a longer time series on outpatient utilization. Steep terrain is certainly a factor for some village areas such as Mutongu and Kithangathini, and even for Kyale when people come to Nunguni on foot. Unpaved (murram) roads are not affected by rains in July. The extent to which people attend the clinic when they have important business nearby—such as at the Nunguni shops or market, or at the chief's office or local court at Ithemboni—may also influence the monthly village patterns. Finally, as in the case of disease reporting, the extent to which the information on patient origins is accurate and reliable is open to question. For several reasons, including privacy, patients may report their sublocation name instead of their village. Each sublocation has a village with the same name. Also, there were 181 entries of village names (9% of the total) in the Nunguni outpatient register for July 1977, which could not be deciphered. Returns of patient origins may also be biased due to unsystematic recording by the facility's staff.

Kikoko Mission Hospital is located some 7 kilometers southeast of Nunguni in Kalongo Sublocation (Figure 4-1). It is staffed by a Mukamba CO, seconded by the Kenya Ministry of Health. He holds outpatient clinics 4½ days per week with the assistance of two nurses, a registered nurse-midwife (the European sister-in-charge at the hospital), two enrolled midwives, and three ungraded nurses. This hospital is a very important resource for the people of Kilungu because it is more central than Nunguni, provides inpatient as well as outpatient services, has a popular under 5's clinic, and offers the location's only biomedically supervised maternal services. Figure 4-4 shows the locations of the 24 principal village areas served by Kikoko (out- and inpatients) according to an analysis of hospital records for 1977 (Kikoko Mission Hospital, Annual Statistical returns; January 1–December 21, 1977). Only 5 of the 24 villages shown on the map are situated within 4 kilometers of this facility. Except for Kivani in the northeast, all sublocations are represented at the hospital, as are several areas outside Kilungu. Furthermore, Kikoko draws its patients from each of the sublocations served by Nunguni, and also serves all of Wautu Location to the south. The substantial overlap of the Kikoko and Nunguni service areas, particularly in the northern and western areas of Kilungu, underscores the functional specialization—most notably Kikoko's maternity and inpatient capabilities—and complementarity of the two facilities.

The fact that patients who attend Kikoko are charged a fee for services rendered, in contrast to the free services at the government facilities, may prevent some sick individuals and pregnant women from utilizing the hospital. Adult patients paid Shs. 10/ = (U.S. $1.25) if the usual injection (*kutonya* = to inject) was included with an outpatient visit, while the fee for delivery was Shs. 60/ = ($8). Charges for children were Shs. 5/ = ($.67), and for infants Shs. 3/50 = ($.47). In-patients were charged according to their type of treatment and length of stay, with a minimum basic fee of Shs. 22/ = ($3).[5] In actual practice, in the minds of Kikoko's outpatients a fee for service was possibly perceived to have less bearing on their utilization of the clinic than the fact that, in contrast to Nunguni, the CO who diagnosed and treated them was a Mukamba.

During the past decade preventive and family health services, including immunization and antenatal, child welfare, and family planning clinics, have received increasing emphasis from policymakers and planners in Kenya's Ministry of Health. For the period 1979–1984, an ambitious national pro-

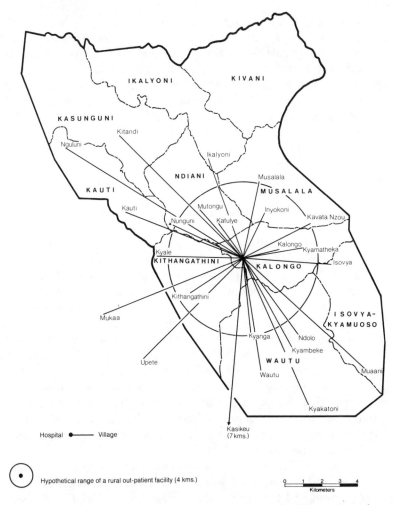

Figure 4-4. Patient origins: Kikoko Mission Hospital, Kilungu Location, 1977, principal sub-locations and village areas. (Source: Annual Statistical Returns, January 1 to December, 31, 1977.)

gram was proposed by the government to reduce various rural health problems such as infant mortality, maternal problems, and nutritional diseases by 50% or more. Goals set for increased attendance at clinics by 1984 are as follows: school children screened (from 0% to 90%); toddlers seen and immunized against diseases such as diphtheria, whooping cough, tetanus, polio, measles, and TB (5% to 50%); pregnant women (15% to 33%); deliveries supervised (10% to 25%); and infants followed up (15% to 80%).[6]

The extent to which these goals have been realized in Kilungu is so far unknown. In view of the Location's estimated population growth, the large national shortage of trained health workers in the existing system, and the continuing urban bias of biomedical services, the prospects for significant progress in health care coverage in Kilungu seemed slight in 1978. In the case of childbirth, for example, Kikoko Mission Hospital is the only facility available for an area much larger than Kilungu. In 1977, maternity cases ($N = 356$) accounted for 31% of all admissions to the hospital. The 347 deliveries performed under biomedical supervision represented only 11.2% of the expected births ($N = 3,106$) in Kilungu for 1977 (Kikoko Mission Hospital, Annual statistical returns, 1977). This compared with a national average of 15%, and illustrates the high proportion of home deliveries (89%) and the crucial role of traditional midwifery in Kilungu and the rest of Kenya.[7] As one *mwisikya* (traditional midwife) said, "most women give birth in their homes. They go to the hospital only when they get a major problem that the *mwisikya* cannot solve."

In northern Machakos District a study of traditional and modern antenatal and delivery care included 2,223 women who gave birth to 2,246 children during 1975 and 1976 (Voorhoeve, Kars, & Van Ginneken, 1982). Although the rate is more than twice as high as that estimated for Kilungu, only 26% ($N = 582$) of the women delivered in the hospital. Of the remaining 74% ($N = 1,641$) who delivered at home, 64% ($N = 1,043$) were attended by a traditional birth attendant (TBA), or *mwisikya*; 27% (445) were assisted by a friend or relative; and 9% ($N = 153$) gave birth alone. Distance from the hospital, lack of public transport, previous hospital delivery, and age were significant factors affecting a woman's decision to have a hospital delivery. Observations of births managed by TBAs uncovered "few harmful practices."

Curative services are the predominant activity and reach the largest populations at both of Kilungu's government health subcenters, the dispensaries at Ithemboni and Kola, and at the mission hospital. Total average monthly attendance (first visits + reattendance and referrals) at Nunguni from May 1977 to February 1978[8] was 2,216. Persons reattending during the same month accounted for 36% of the average monthly visitation. Attendance at Nunguni's child welfare ($N = 3,833$) and antenatal ($N = 2,924$) clinics added another 6,757 visits (or 676 per month) for this period. Over the same period only 93 attended Nunguni's family planning clinics for the first time (9.3 per month), while only 207 made return visits. Although the period to which these statistics refer is very brief, they are quite suggestive of the very low level of family planning acceptors throughout Kenya and the correspondingly high user dropout ratio of about 80% (Republic of Kenya, 1978, p. 130).

Comparable monthly outpatient statistics for Kikoko Mission Hospital (1977) are 1,007 average first attendance plus 26% for reattendances. Visits of women and children to Kikoko's antenatal and Under Fives clinics added another 12,181 attendances, or an average of 1,015 per month (cf. Spring, 1980). The much higher (50%) average monthly attendance at Kikoko's

MCH clinics may help to explain the apparent lower incidence of diarrheal disease recorded there. Attendance may also have been spurred by free dietary supplements. Once a month, after their child has been examined and weighed at the under-fives clinic, a partially cooked preparation consisting of bulgar wheat, soybean flour, milk powder, and oil prepared in the United States was distributed to mothers.

Table 4-3 shows the 12 most common outpatient diagnoses at Nunguni (May 1977–February 1978) and Kikoko (January–April 1977). The CO determined these diagnoses during a rapid appraisal of the patient's signs and symptoms.[9] Diagnoses for Nunguni are labeled according to the standard classification used by the Ministry of Health, whereas the CO at Kikoko followed an independent system which, among other features, was possibly more accurate in that it recognized simultaneously occurring diseases in a patient. Other representative combinations of disease in the Kikoko entries included malaria and bronchitis, measles and pneumonia, and diarrhea and vomiting. Inconsistent and incomplete outpatient records at Kikoko Mission Hospital presented serious difficulties for analysis. Outpatient *complaints* were recorded on accessible forms only for the months of January through part of May 1977. Thereafter, the clinic staff recorded the drugs and treatment they had administered but not the nature of the illness![10]

Communicable, infectious, and parasitic diseases formed the predominant diagnoses in both clinics, which again underscores their persistent nature in rural Africa (Table 4-3). However, striking differences between the clinics could not be easily explained. Some of the contrasts may have been artifacts of the classification and terminology selected by or imposed upon the diagnostician. For example, the CO at Nunguni used only 20 different disease labels in registering diagnoses for official records. Any condition that did not fit was placed in an "all other diseases" category. In other cases a true proportional difference in the maladies patients presented at the two clinics may have existed. In the absence of comparable terminology and a longer time series, the validity of these incomplete and short-term data cannot be confirmed. Half of the diagnostic categories used at Nunguni were missing from the list of Kikoko's 12 leading maladies. They included acute respiratory infection, accidents, intestinal worms, acute eye infections, ear infections, and measles. These diseases appeared at lower levels of incidence in the Kikoko returns, often with more specific labels such as "burned hand," hookworm, otitis media, and septic eye. Categories in the Kikoko ranks that did not appear in Nunguni's disease list included bronchitis, septic wound, abdominal pain, and wound. Bronchitis was undoubtedly subsumed under "acute respiratory infection." However, the lack of a separate category for abdominal distress at the government clinic is puzzling because it ranks so high among the conditions that TMPs treat, and it was the second most common complaint (18.6%) in a recent study of 6,828 illness episodes among Kamba in northern Machakos District (Schulpen & Swinkels, 1980).

Table 4-3. Twelve Leading Outpatient Diagnoses at Two Kilungu Clinics, 1977–1978

	Nunguni Health Subcenter (May 1977–February 1978)			Kikoko Mission Hospital (January 1977–April 1977)	
Rank	Diagnosis	First visits (%)	Rank	Diagnosis	First visits (%)
1	Acute respiratory infection	32.5	1	Malaria	29.1
2	Disease of the skin	11.5	2	Pneumonia	17.2
3	Malaria	9.4	3	Bronchitis	8.7
4	Diarrheal disease	5.6	4	Malaria and pneumonia	7.6
5	Accidents (incl. fractures, burns, and so forth)	3.5	5	Venereal disease	5.1
6	Intestinal worms	3.3	6	Skin disease	3.6
7	Pneumonia	2.2	7	Diarrhea and malaria	2.5
8	Acute eye infection	1.8	8	Measles	2.5
9	Rheumatism, incl. joint pains	1.3	9	Septic wound	2.2
10	Ear infection	1.0	10	Rheumatism	1.5
11	Gonorrhea	1.0	11	Abdominal pain (incl. lower)	1.0
12	Measles	.8	12	Wound	0.8

Note: Nunguni outpatient returns, May 1977 to February, 1978. Kikoko Mission Hospital, annual statistical return, January 1 to December 3, 1977.

Seasonal factors such as wet and dry and hot and cold months often have a significant impact on the transmission and incidence patterns of many diseases in the Third World (Chambers, Longhurst, & Pacey, 1981). Seasonality also influences nutritional health when it is associated with chronic preharvest food and cash shortages and, as discovered by the Joint Machakos Project, the lower breast-milk yields of lactating Kamba mothers (Van Steenbergen, Kusin, & Rens, 1980). Outpatient records at Nunguni also suggested the effects of seasonal influences on acute respiratory conditions (ARC). While the monthly average for ARCs was 31% of all new visits at Nunguni, the relative incidence was greatest in July (48%) and August (42%). These are the coldest, driest, and most overcast months (Moore, 1979).[11]

Malaria transmission, mainly by *Anopheles gambiae* or *Anopheles funestus*, is also markedly seasonal in most of Kenya. In highland areas such as Kilungu which rise above 1,700 meters elevation, malaria is unstable, occurs sporadically or in local epidemics, and has a short transmission season of 1–3 months (Roberts, 1974). The Nunguni disease returns suggested, on the one hand, that the greatest local incidence of malaria may occur in July and January, following the end of the long (March–May) and short (October–December) rains by 4 to 6 weeks. In July 1977, the staff of Nunguni Sub-Health Center reported 144 cases of malaria which moved it from third to second place in the rankings (9% of all first visits). In January 1978, malaria also accounted for 9% of first visits, compared with 7% for the ten-month period. On the other hand, the most critical aspect of the malaria reports for Nunguni and, especially, Kikoko is the accuracy of diagnosis, particularly since neither clinic has laboratory facilities. For example, it is very unlikely that malaria could have a confirmed proportional morbidity rate of 29% over the 4-month period covered by the Kikoko returns. The 53% rate recorded for January 1978 (351 of 664 first visits) is even more remote. This extraordinarily high attribution level surely reflects overdiagnosis of "clinical malaria"—a common practice in Kenya and elsewhere in Africa (Diesfeld & Hecklau, 1978). In fact, "malaria" is often used interchangeably with fever (Kikamba = *ndetema*) and a host of other clinical conditions. Roberts (1974, pp. 306–307) notes that in many of Kenya's district hospitals "more than 80% of cases diagnosed as malaria among outpatients are not confirmed by laboratory methods."

As the preceding discussion reveals, questions about comparability, reliability, and validity of recorded data greatly complicate the interpretation of reports from rural areas of Africa. Such concerns as accuracy of diagnosis, disease labeling, and incomplete or short time series data for longitudinal analysis must be confronted and carefully weighed to properly assess local health and epidemiological conditions. In addition, the profiles of patients and proportional levels of diseases recorded or otherwise noticed at biomedical facilities in areas such as Kilungu may also be influenced by the ethnicity and popularity of the clinical staff. Kenya's Ministry of Health commonly posts medical staff to areas with a different ethnic background from their own in order to promote a "Kenyan" consciousness. This may affect the practitioners' acceptance locally, and limit their ability to identify important factors influencing the condition and behavior of their patients.

Other factors affecting the reliability and comprehensiveness of disease profiles for rural areas may include whether services are free or paid; physical access; local ecological relationships of environment, behavior, and disease; whether an illness is deemed more suitable for treatment by a traditional specialist at some point in the therapy search; and the perceived availability or efficacy of medicines and drugs at a source location. On Pate Island in the Lamu Archipelago astride Kenya's north coast, for example, there are six dispensaries. Attendance can shift dramatically from one dispensary to another in accord with rumors concerning the availability of drug stocks, or the time at which medicine is believed to be particularly effective.[12] Undoubtedly, some of the people who attended the Nunguni clinic also used Kikoko, and vice versa, for the same or different courses of therapy. This practice would lead to overestimates of disease incidence.

Hospital records provide an alternative and supplementary guide to the biomedical version of disease recorded at small rural clinics such as Nunguni and Kikoko (Table 4-4). Machakos Provincial Hospital serves all of Ukambani and is the major point of referral for difficult cases to Kenyatta National Hospital in Nairobi. Between 1967 and 1974 a total of 12,475, mainly Kamba patients admitted to the medical wards of Machakos Provincial Hospital were interviewed, examined, diagnosed, and registered by highly trained medical staff. Analysis of these records, which presumably reflect greater accuracy of diagnosis, yielded an approximation of the biomedically defined disease pattern that is probably fairly representative of rural Machakos District.

As expected, communicable and infectious diseases—those easiest to prevent—were the most common disorders seen at the Provincial Hospital. This pattern was broadly similar to the conditions found in the Kilungu clinics. Bronchopneumonia (13%), measles (9%), and gastroenteritis (7%) were the leading diagnoses overall by a large margin, accounting for 29% of all hospital admissions. Nearly 60% of measles cases were complicated by other conditions such as bronchopneumonia, laryngotracheobronchitis, and encephalitis. Intestinal diseases, particularly gastroenteritis, dominated in the hospital's pediatric and medical wards. Salmonellosis, shigellosis, typhoid, and amebiasis were roughly equal as causal agents of enteritis. Hospital records also showed that liver disease posed an "extremely common" problem in Ukambani. It is causally linked to childhood malnutrition, chronic alcoholism, infectious hepatitis, and irritant agents such as parasites and eggs of *Schistosoma mansoni* (Oomen, 1976). The high prevalence of *S. mansoni* in biopsies at Machakos confirmed the Ukambani pattern indicated in other surveys. For the Mbooni Hills, an area adjacent and ecologically similar to

Table 4-4. Disease Pattern of Ukambani Based on Records of Machakos Provincial Hospital, 1967–1974

Rank	Disease group	Cases	Death rate (%)	Most frequent diagnoses (%)	
1	Infectious diseases	5,279	15.3	Measles	26
				Whooping cough	6
				Malaria (clinical)	6
2	Lung diseases	3,581	13.5	Bronchopneumonia	59
				Lobar pneumonia	9
				Pulmonary TB	9
3	Digestive tract diseases	1,613	11.8	Gastroenteritis	68
				Typhoid	5
				Duodenal ulcer	4
4	Various	1,095	13.4	No diagnosis	25
				Diabetes mellitus	18
				Surgical conditions	14
5	Liver diseases	878	15.0	Hepatosplenomegaly	28
				Jaundice	23
6	Malnutrition	873	28.1	Kwashiorkor	37
				Marasmus	22
				P-C malnutrition	18
7	Blood diseases	774	4.7	Hypochronic anemia	42
				Anemia (unspec.)	26
				Normochromic anemia	11
8	Heart diseases	760	13.9	Heart failure	31
				Hypertension	23
				Rheumatic valvular disease	22
9	Neuropsychiatric disorders	702	16.8	Encephalitis	14
				Cerebrovascular accidents	21
				No proper diagnosis	12
10	Urogenital diseases	335	10.4	UTI	29
				Glomerulonephritis	20
				Uremia	15
		15,902	14.5		

Note: Adapted from "Disease pattern in Ukambani, Kenya" by L. Oomen, 1976. *East African Medical Journal*, 53, pp. 341–349. (Copyright 1976 by Copyright Holder. Reprinted by permission.

Abbreviations: P-C, protein-calorie; TB, tuberculosis; UTI, urinary tract infection.

Kilungu (Figure 3-1), analyses revealed a 30% frequency of *S. mansoni* eggs in stool examination during the period 1958–1972 (Diesfeld & Hecklau, 1978). Several other helminths also remained endemic in Ukambani. Collectively they form a "Kamba type" (Diesfeld & Hecklau, 1978). Their distribution by type and area are shown in Table 4-5.

Highest death rates at Machakos Provincial Hospital occurred in cases of malnutrition, particularly kwashiorkor (30%), and among premature babies (49%). A large proportion of these deaths have been attributed to

Table 4-5. Helminthological Distribution Pattern, Ukambani: Relative Frequency of
Egg Findings from 1.1 million Stool Specimens, 1958–1973

	Area	
Type	Ukambani (%)	Mbooni Hills (%)
Schistosoma mansoni (schistosomiasis)	41	30
Ascaris lumbricoides (roundworm)	35	53
Ancyclostomatidae (hookworm)	12	11
Schistosoma haematobium (schistosomiasis)	11	—
Taenia spp. (tapeworm)	11	6

Note: Adapted from 54 "Annual laboratory returns" of Kenya hospital laboratories as
presented in Map 7A in _Kenya—a geomedical monograph_ by H. J. Diesfeld and H. K.
Hecklau, 1978. Berlin: Springer-Verlag. Used by permission of authors.

inadequate follow-up by nursing staff due to lack of knowledge, interest,
training, and equipment (such as proper scales for weighing). Malnutrition
was also correlated with the large number of deaths linked to infectious
diseases such as bronchopneumonia and gastroenteritis (Oomen, 1976).

Anemias remained common in the hospital wards, but accurate, specific
diagnosis was, once again, hampered by inadequate laboratory analysis.
Heart diseases accounted for under 5% of admissions, as expected at this
stage of Kenya's epidemiological transition. Neurological diseases included
cerebrovascular accidents (21%) and encephalitis (14%). Patients with mental
disorders are treated in a separate ward and were not included in the data.
The hospital records failed to note that mental patients are frequently taken
to traditional specialists either before or after a hospital stay.

Reported cases of urogenital diseases ($N = 335$) fell well below what was
believed to be their real incidence. Oomen (1976) asserted that urogenital
infections accounted for the majority of complaints heard in an outpatient
department. Such a claim is not strongly supported by data from Nunguni
Health Subcenter, where "gonorrhea" is the only urogenital disease reported
separately in the outpatient returns. At Kikoko Mission Hospital, however,
the generalized label "venereal disease" was applied to over 5% of all first
visits. The extent to which shame and the use of Kikamba euphemisms for
venereal diseases influenced the accuracy and validity of reporting of urogenital disease is not known with much confidence (cf. Van Luijk, 1982). Oomen's
(1976, p. 345) observation that "the importance of urogenital infections in a
population with a high birth rate and high incidence of gonorrhea may be
greatly underestimated" seems appropriate.

The provincial hospital at Machakos maintains laboratory facilities and
trained personnel sufficient to detect many disease agents and symptomatic
clues, some of which are endemic to the region but pass virtually unrecog-

nized in the Kilungu clinics. Examples include nutritional diseases, schistoso-miasis and its complications, diabetes, meningitis, typhoid, various anemias, urogenital diseases, and liver disease. These conditions were rarely reported, if at all, by COs at Nunguni and Kikoko, and very few patients get referred to the Provincial Hospital (a total of only 84 patients in 10 months from Nunguni).

Symptomatic treatment of many preventable diseases, based on a "scatter-gun" approach that relies on heavy use of available antibiotics and injec-tions, is the archetypical pattern of biomedical practice in Kilungu. This includes the use of vitamin B_{12} placebos where no other vaccine is indicated in order to satisfy the powerful therapeutic effects Kamba attribute to *kutonywa singano*—the injection—which is greatly preferred over tablets and liquid medicines. Other observers have also underscored this phenomenon. Mburu (1973, p. 94), for example, concluded from his field study of Kamba health care practices in an upland area of northern Machakos that the "cognitive orientation towards the use of the injection is so intense . . . that many people regardless of their educational level think it is the only valid type of therapy." Indeed, it seems that a large share of the success and prestige enjoyed by biomedicine is attributable to the demonstrated therapeutic effect of the injection on the symptoms of bacterial infections and acute ailments. Swantz (1979), who studied traditional and modern conceptions of illness and healing in Tanzania, believes that the injection needle was readily incor-porated into local ethnomedical systems because the general population perceived it as an essentially magical process with a proven record of "suc-cess." Hence, the paradox, now widespread in Africa, whereby certain bio-medical procedures are compatible with traditional healing when their effi-cacy is empirically demonstrated according to the local culture's standard of rationality (Dingwall, 1976).

The extent to which the perceived benefits of biomedicine extend to *preventive* medicine due to positive experiences with immunization against childhood diseases may still be slight. Mburu (1973) found that Kamba mothers do not reject preventive medicine for measles, pertussis, and acute diarrhea—diseases to which their children are highly vulnerable. However, he also asserted that the mothers did not understand how vaccinations and other preventive measures worked against childhood diseases or why they should be obtained. Consequently, most young children brought to a biomedical facility had already experienced the three diseases. Undoubtedly, many other children in this population needlessly succumbed to one of these diseases, but the fatality rates are not recorded.

This chapter has examined the nature, functions, and extent of biomedi-cine as a therapeutic option in the ethnomedical system of the Kilungu Kamba. It is evident that local biomedical services are mainly curative. They are also minimal in relation to the magnitude of ill-health and to the region's rapidly expanding population. The few service points are readily accessible only to a small proportion of the population. Some form of village-based care incorporating a strong emphasis on preventive and promotive services is imperative if the people of Kilungu are to experience demonstrable progress toward the World Health Organization (W.H.O.) goal of "health for all by the year 2000."

As in times past, TMPs remain—after self-treatment—the primary sources of health care in Kilungu's villages. The cultural, spatial, and functional characteristics of this traditionally based sector of the local ethnomedical system are considered next.

NOTES

1. These figures are based on calculations from the 1969 and 1979 Kenya Population Census and do not consider possible out-migration. Kenya's overall rate of natural increase is estimated at 4% (Population Reference Bureau, 1984), whereas the doubling time is less than 18 years. With total fertility between 8 and 9 children per woman, this may be the world's fastest growing population.

2. Based on the ratio in the Kenya Population Census, 1969.

3. Personal communication, clinical officer, Nunguni, March 3, 1978.

4. If staffed according to plan, a health subcenter consists of a clincal officer, two enrolled community nurses, a public health technician, two family health educators, a statistical clerk, a laboratory technician, a driver, and five support staff.

5. These monetary equivalents reflect 1979 exchange rates. Mimeographed paper on rural health units and comprehensive health care supplied by Dr. Roy Shaffer, African Medical and Research Foundation (AMREF), Nairobi.

6. Republic of Kenya (1978).

7. The estimated crude birth rate (CBR) for Kenya's Eastern Province as 54.1 per 1,000 in habitants in 1969. Kilungu's estimated total population for 1977 was 57,412. Assuming that the location's CBR is similar to the province as a whole, there would be 3,106 expected births ($57,412 \times 0.0541$).

8. Outpatient and special clinic records for Nunguni prior to May 1977 were not available.

9. Clinical officers at two (unidentified) rural health centers in Kenya reported to a visiting analyst that they "averaged 0.8 of a minute per patient" (Family Health Institute, 1978, p. 81).

10. Outpatient registers indicated standard information that *should* be recorded, including serial and reattendance numbers, patient's name, age, sex, address, complaint/treatment/remarks, paid, not paid, and drug/medicine supplied.

11. See Survey of Kenya, *National Atlas of Kenya*, 3rd ed., 1970, pp. 16–21.

12. Personal communication, chief, Faza Location, Pate Island, at Lamu, July 7, 1979.

5

Kamba Culture and Contemporary Traditional Medicine in Kilungu

Similar to other Africans (Koumare, 1983), people in Kilungu depend on a broad range of traditional medical practitioners (TMPs) to help them cope with the heavy burden of physical illness and mental stress. Virtually all TMPs in Kilungu are small-scale peasant cultivators. Only a small minority realize substantial material benefits from their role in health care. They are generally easily accessible and available to the local populace. My visits with many TMPs in the Kilungu Hills provided abundant evidence that some of them, particularly herbalists, continue to provide a significant share of the health care in local communities. Some practitioners in the biomedical sector confirm this pattern. For example, it was the opinion of the clinical officer at the Kikoko Mission Hospital outpatient facility that, after self-treatment, TMPs are usually the local population's first recourse. Hospital medicine is typically their second line of choice. The clinical officer believed that few people are inclined to openly admit their involvement with TMPs because they fear the possible negative consequences such revelation might invite from potentially hostile biomedical staff.[1] Comparatively, our study of the knowledge, attitudes, and practices of outpatients at the Kenyatta National Hospital in Nairobi (Chapter 2) found that Kamba are more likely to admit to the use of traditional medical resources than most other Kenyan ethnic groups represented.

Most people visiting a TMP in Kilungu are accompanied by at least one relative, neighbor, or friend who has some authority in therapeutic decisions and who provides moral and possibly financial support. Collectively, TMPs in Kilungu are called on for advice and treatment in connection with many physical, psychosomatic, and psychosocial complaints. Patients[2] almost always travel to the TMPs residence or other accustomed workplace to receive therapy. Exceptions include child birth and the "decontamination" and preventive "protection" rituals such as *kuvinga musyi*, which require the physical presence of the practitioner at the client's homestead. A cure or resolution of a problem may involve return visits spread over a period of a few days to several weeks.

The contemporary Kamba ethnomedical system is pluralistic in ways that could not have been envisaged prior to the introduction of biomedicine and its codependent pharmaceutical link (Table 4-1). Today, traditional Kikamba medical resources and practices persist as a highly elaborated subuniverse of this wider system. Geographically, the traditional medical system still serves the widest areas and covers the largest population "on the ground." Education, science, Christianity, Islam, and unprecedented population mobility and cultural mixing have fundamentally changed many of the ways people learn to think and act today. Nevertheless, for the "average" person in Kilungu, belief in the omnipresence of supernatural forces—of *aimu* and sorcery power—is a significant epidemiological fact of life that also influences the management of natural illnesses and helps to define the character and position of traditional medicine in the 1980s.

COSMOS AND CULTURE: THE HUMAN CONTEXT OF DISEASE AND HEALTH

Current Kamba ideas and behavior concerning sickness and its cure have little meaning if divorced from the worldview and religious referents of their traditional culture. Indeed, traditional medicine and religion are inseparable elements of a coherent system that also incorporates various features of biomedicine. Regardless of the emphasis, procedure, or materials that characterize traditional medical practices in this culture, as in the rest of Africa "religion is the predominant factor" (Koumare, 1983, p. 31). Worsley (1982) refers to this holistic conceptualization as the "metamedical," ideological frameworks in which medical systems are integrated. It applies to the biomechanistic model of Western medicine—which functions poorly or not at all if deprived of its "science" and logical positivism—as well as to astrology, Chinese medicine, and the many medical systems of African tradition.

Today, of course, the Kamba are counted among the "cultures of medical pluralism" that predominate in Africa and most Asian and Latin American societies (Worsley, 1982). Their metamedical universe continues to change as individuals and the larger society decide that certain traditional patterns of thought and behavior are outmoded, and as they selectively accommodate the new concepts and forms of imported biomedicine and religion. To understand their contemporary approach to illness it is essential to recognize the nature and strength of their traditional cultural fabric and world view, which are logically interwoven with ideas about disease and misfortune.

The Kamba are a monotheistic people who recognize *Mulungu*, or *Ngai*,[3] as the creator and preserver of all things in the world. Although he provides children, rain, and food, and certain diseases are attributed directly to him, *Mulungu* is a distant, impersonal force above and beyond the daily affairs of individual Kamba (Mbiti, 1971; Ndeti, 1972).

In contrast, the *aimu*, or spirits of the departed, are the heart of Kamba religious beliefs and practices and figure prominently in the everyday life and well-being of all members of society. While Kamba do not "worship" their ancestors they do venerate their experience (Ndeti, 1972). Mbiti (1971) distinguishes two types of *aimu*: the spirits of recently deceased persons (up to three to four generations removed), or the "living-dead"; and the spirits of ancestors who died many generations ago and whose names have been forgotten. Both kinds of spirits are called *aimu* (*iimu*, s.) in Kikamba. This world of spirits is an integral, inseparable part of being, and traditionally the Kamba resolutely believe that *aimu* are actual "people" who live much like the living Kamba in a land that is a physical copy of Ukambani. Thus when death occurs the spirit is not destroyed but travels to the spirit world where it becomes a living-dead in community with household members who died earlier. Although details vary from one ethnic group to another, this concept of the living-dead is evidently nearly universal in African belief systems. Senegalese poet C. A. Diop's testimony to this theme is emphatic: "Those who are dead are never gone" (Ndeti, 1972).

Aimu represent one of the five modes of existence in Kamba ontology (Mbiti, 1971). These are (1) God, the originator and sustainer of everything; (2) *Aimu* and the living-dead; (3) the human being, who physically represents the dead, the living, and the unborn and embodies the principles of life, spirit, and immortality (Ndeti, 1972); (4) animals and plants; and (5) inanimate objects. Mbiti (1971) underscores the "intensely anthropocentric" character of this ontology—a unity centered on the human being such that death disintegrates the body but cannot annihilate the soul. The exception is when someone dies without leaving a child or survivor to carry on his or her name and lineage (Ndeti, 1972). Destruction of the spirit would effectively destroy the coherence of being and also the Creator. Immortality is not a goal, however, since Kamba view death as man's inevitable fate. Their primary concern is the manner in which *aimu* affect life in the world of the living. Consequently, "the world of the departed presses hard on ordinary people. It pervades their whole life, their vocabulary, folklore, social relations, acts of sacrifice, etc." (Mbiti, 1971, p. 138).

Kamba believe that *aimu* play a major, inseparable role in their physical and mental health. They can cause diseases (*mauwau wa maimu*), especially chronic conditions such as infertility and madness, and misfortunes of all kinds. Such problems materialize in response to various human acts of omission or commission, and ritual error. Traditionally, these problems are diagnosed by a medicoreligious practitioner (*mundu mue*), who contacts the spirits by means of divination (*kuwausya*) to learn the "real" cause of the illness. Thereafter, the *mundu mue*, with the spirits speaking through him or her, instructs the sufferer about the appropriate measures that must be taken in order to "please" the spirits and gain relief. Many of the human errors identified by the *mundu mue* in cases involving *aimu* are sufficiently general so that particular persons or incidents need not be directly identified (Thomas, 1971). There are numerous culturally appropriate alternatives to which the *mundu mue* can refer. Other kinds of transgressions that may provoke *aimu* include breaches of taboo and improprieties such as proscribed or prohibited (e.g., intergenerational) sexual relations. The consequences of such acts by the parent(s) of a young child may include *thavu*, a syndrome manifested in the child which shows up biomedically as the wasting and edema associated with malnutrition.

Ceremonies to restore the equilibrium between humans and *aimu* which has been disrupted by ritual error or omission often customarily demand kin, age, or cult group participation. Typically, beer is brewed (some is poured as a libation for the spirits), a goat is sacrificially slaughtered and roasted (some of its blood and small pieces from each part of its body are placed in a calabash which is then emptied on the ground for *aimu* to eat), and food is cooked and eaten. These kinds of culturally prescribed actions, together with confession, requests for forgiveness, other sacrifices (*kuthemba*), *kilumi* dances to exorcise demonic *maimu* implicated in diseases that cause delirium (Ndeti, 1972), and various other purification rites are relied on to help restore a sense of balance and harmony between the individual, his or her social and physical environment, and the spirits.

Earlier (Chapter 3) I drew attention to the pervasive influence that sorcery and witchcraft (*uoi*) and magic have in Kamba life, past and present. They are unequivocally implicated by Kamba as the causal instruments of much sickness and misfortune. This perception is also evident almost everywhere else in Africa during the current turbulent transition to a new social order. Today, the depth of these persistent supernatural beliefs, which British social anthropologist Monica Wilson once defined as "the standardized nightmares of a group" (cited in Hautvast & Hautvast-Mertens, 1972, p. 407) is barely comprehensible to most Western minds. Nevertheless, Professor John Mbiti, the well-known Mukamba theologian, leaves no doubt about the present influence of the occult in contemporary Kamba society:

> The people are saturated with beliefs, fears and superstitions connected with these practices. Every Mukamba, whether Christian or otherwise, has a dormant or active share of these beliefs. The people have not been sufficiently armed to fight against witchcraft and sorcery, in spite of many years of contact with Christian teaching and Western education. It will take many more years to eradicate beliefs in witchcraft, sorcery and magic from the lives of the Kamba (Mbiti, 1971, p. 9).

KAMBA DISEASE CLASSIFICATION

Given the relative position and prominence of God, *aimu*, witchcraft and sorcery in the thought and life of the Kamba people, it is essential to examine the ways in which they systematically relate these concepts to disease etiology and classification. The various ordering systems in use have evolved over time in response to knowledge gained through adaptation to different environments, the experience of trial and error, and religious concepts that channel individual responses and permeate ethnic consciousness. Added to these motivations are the legendary mobility and individualism that have shaped, and been shaped by, Kamba culture over time. These traits produced a comparatively open system which stimulated economic and cultural exchange with diverse peoples such as the other Central Bantu groups, the Mijikenda peoples (e.g., Giriama), the Islamized Swahili and Arab cultures on the coast, the Maasai, and the Europeans. All of these groups have contributed in varying degree to the Kamba characterization and treatment of sickness. Biomedicine, still cloaked in its own distinctive Western metamedical garb, is at once the most recent, systemic, and potent of these external influences.

With allowance for minor subregional variations, including local orthographies, there is a coherent and widely recognized (if not universally accepted) "Kikamba" system of disease etiology and classification. The most prominent system among the rank and file of the population in the Kilungu Hills and elsewhere in Ukambani is classification according to cause—specifically *ultimate* causation. Dr. J. N. van Luijk (1983, 1984), who for several years conducted research on Kamba health behavior in northern Machakos District in association with the Dutch-sponsored Joint Machakos Project, recently produced a detailed analysis which distinguishes four categories of disease classification: according to cause, treatment mode, disease characteristics, and affected person attributes (Table 5-1).

Van Luijk's classification for northern Machakos broadly conforms to the pattern I found some 70 kilometers south of the Joint Machakos Project in the Kilungu Hills. With allowances for local distinctions, the Kamba etiologies are similar to those of many other African ethnic groups ranging from the neighboring Kikuyu people to the Kongo (Janzen, 1978) of Zaire and the peoples of the Ivory Coast (Lasker, 1981). My discussions with 35 TMPs in Kilungu yielded information concerning more than 85 diseases and syndromes, including signs, symptoms, etiology, and therapies. Classification by *cause* (Table 5-2), by *body part* affected (Table 5-3), and by *signs and symptoms* (Table 5-4) are the most commonly used and heard categories among Kilungu TMPs.

Illnesses attributed to natural causes (A, E, F in Table 5-2) include colds, various allergies, malaria, short-term abdominal pains, constipation, measles, chickenpox, gonorrhea, and syphilis. The four latter diseases are also understood as contagious, and the sexually transmitted diseases are commonly perceived as having originated from outside Ukambani during the early decades of this century. In Kilungu, as among the Luvale of Zambia (Spring, 1980) and many other African peoples (Warren, 1979), naturally caused conditions have a prominent place alongside "illnesses of man" (Janzen, 1978,

Table 5-1. Akamba Systems of Disease Classification

System Related to (Ultimate) Cause (5 Classes)
  Mauwau wa Ngai Mwene (diseases of God)
  Mauwau wa Aimu (diseases caused by spirits of the departed)
  Mauwau wa Kuoa (diseases caused by witchcraft)
  Mauwau wa mundu kwiyika newe mwene ayende (diseases caused by people who harm themselves by their own wish)
  Mauwau wa mundu kwikwa ni ungi (diseases caused by someone else—but not witchcraft)

Systems Related to Treatment (Two Types)
  By specialists who cure/heal the disease (6 classes)
  By type of treatment (25 classes)

Systems Related to Disease Characteristics (Four Types). According to
  Severity (3 classes)
  Curability (2 classes)
  Communicability (2 classes)
  Hereditary characteristics (2 classes)

Systems Related to Characteristics of Affected Person (Three Types). According to
  Sex (3 classes)
  Age class (8 classes)
  Body part affected (±19 classes)

Note: From utilization of modern and traditional medical care by the Kamba of Machakos, Kenya. Part II. Survey on perceived morbidity and medical care: Static and dynamic analysis. Draft report, Department of Tropical Hygiene, Royal Tropical Institute, Amsterdam, January 1983, by J. N. van Luijk. Copyright 1983 by Copyright Holder. Reprinted by permission. And from van Luijk, J. N. The utilization of modern and traditional medical care. In J. K. van Ginneken & A. S. Muller (Eds.). Maternal and child health in rural Kenya. London: Croon Helm, 1984. Reprinted by permission.

Table 5-2. Kilungu Akamba: Etiology of Diseases

A. Diseases of God, or God-given illnesses/natural diseases (*Mauwau wa Ngai*, or *Ukuno wa Ngai*)

B. Diseases caused by ancestral spirits (*Mauwau wa aimu*)

C. Diseases caused by witchcraft/sorcery (*uoi*)

D. Evil eye (*iithuko ithuku*). Inborn trait of carrier.

E. Contagion, or "close contact"

F. Allergy/seasonal symptoms associated with flowering plants (*kuthuwa ni kindu*)

Note: From field interviews, Kilungu, February–April 1978.

Table 5-3. Twelve Disease Labels According to Anatomy:
Kilungu Akamba

Disease (Kikamba)	Description
Uwau wa metho	Eye infection; usually purulent conjunctivitis
Uwau wa ngoo	"Heart attack"; heart problems
Uwau wa kithii	Disease of chest (many possible symptoms)
Kwalwa ni kithui	Pain in chest (many possible symptoms)
Kwalwa ni ieo (maeo)	Toothache
Kwalwa ni maleenyu	Pain in joints
Kwalwa ni ivu	Abdominal pain (nonspecific cause)
Uwau wa nondo	Breast problems (lactating woman)
Uwau wa matu	Infection of the ears
Kwalwa ni mutwe	Headache
Kitau kya mutwe	Intense, chronic headache
Kwalwa ni muongo	Backache

Note: From field study, 1977–1979.

Table 5-4. Selected Diseases Labeled by Signs and Symptoms: Kilungu Akamba

Disease (Kikamba)	Description
Ngome	"Tight ring around throat," difficulty swallowing or spitting
Ndetema	Fever; malaria
Muluo	Painful urination
Kyambo	"Pneumonia"; sharp pains in chest
Kuua	Continuous menstruation
Ngungu	A barren woman
Nyamu	Hard swelling on head or other body parts
Kutavika	Vomiting
Kwituwa	Diarrhea
Kwituua nthakame	Bloody diarrhea
Ikua	Common cold
Muuku	"Jaundice"
Ndetema ila nene yaani	"Yellow fever"; hepatitis (?)
Kitau	Wounds
Kiveti kite yiia	Mother lacking breast milk
Kukooa	Cough
Kwalwa ni mwii	Pain all over the body
Nduuka	Mental illness, madness
Kisonono	Gonorrhea
Kinguu	Scalp ringworm

Note: From field study, 1977–1979.

p. 8). The latter are brought on by ancestral spirits, interpersonal conflicts, or breach of taboo. They are symbolized by accusations and acts of witchcraft, sorcery, and evil eye.

Kilungu Kamba also characterize some God-given or natural illnesses as incurable, including sleeping sickness (*uwau wa kwombosya*) and a form of madness (*nduuka*). In such cases, say TMPs, "the person has to die." Incurable assignations at the outset of an illness seem rare, however. Time is required for the signs of an incurable illness to materialize. Moreover, few would dare to foretell an event that only God (*Ngai*) decides.

In common with a practice that is widespread elsewhere in Africa (Whyte, 1982), the Kamba also distinguish between symptomatic and etiological treatment. Disease and misfortune are philosophically accepted as normal events in life. As in the United States, Taiwan, and Nigeria (Kleinman, 1980; Maclean, 1971), self-treatment of many symptoms such as those associated with the common cold (*ikua*), toothache (*kwalwa ni maeo*), fever (*ndetema*), or headache (*kwalwa ni mutwe*) is the rule among the Kamba.

The recent Machakos Project uncovered a "striking . . . frequency of self-medication" (Schulpen & Swinkels, 1980, p. 348). In a sample of 800 households, over 35% of persons who felt unwell said they used self-medication exclusively (Schulpen & Swinkels believe self-medication is underreported by at least 60%). This phenomenon of self-treatment is evidently associated with a shift to shop (patent) medicines, use of which has been stimulated by the introduction of biomedicine, advertisements by pharmaceutical companies, and the general desire to adopt a "modern" life-style (Schulpen & Swinkels, 1980; van Luijk, 1984).

In general, self-treated complaints are natural, mundane, and frequently self-limiting. Various patent medicines, including analgesics (Aspro, Algon), antimalarials (Nivaquin, Malaraquin), antiworm and antidiarrheal medicines (Santonin, Philip's), medicines for constipation and upset stomach (Actal, Eno), and preparations for coughs and sore throats (Cofta, Strepsils) can be purchased in many of the local shops throughout Ukambani. Some individuals also engage in potentially dangerous self-medication with various controlled drugs of dubious authenticity and appropriateness which they may obtain illegally from individuals supplied from government or private sources (cf. Van der Geest, 1982). In addition, many people retain at least a limited knowledge of medicinal plants from which herbal teas and other home remedies can be prepared. If the home cures and self-medication do not succeed and the symptoms persist, help may be sought from a local herbalist or the health center. If the symptoms still do not yield to the therapies tried, often the next strategy is to consult a *mundu mue*, who is expected to uncover the underlying or ultimate cause of the illness and to recommend appropriate measures for relief. There is thus a redefinition of the illness in terms of etiology. This step does not mean that symptomatic treatment is abandoned.

Instead, it is perceived that this method cannot succeed on its own because it does not attack the real cause. (A possible analog in modern psychiatry concerns the treatment of "depression." Thus cognitive therapy, which is directed toward long-term symptomatic improvement and prevention through self-manipulation of negative thought patterns, may have limited effect on the patient if the disorder is inherited or caused by a chemical imbalance.) In Kilungu, as Whyte (1982, p. 2057) found among the Nyole of Uganda, "it is important to understand that the distinction between symptom and etiology is a difference in . . . analytical level, not an either-or dualism. There is no contradiction between the two levels of treatment."

Time is thus a crucial enabling factor in the emergence, classification, and treatment of disease among the Kamba and other African societies. Most so-called "Kikamba" illnesses, such as *nyamu* (a mysterious, hard swelling on the body), *thavu* (a special form of malnutrition), *kavoo/kavaso* (pain and swelling in the sternum and lower ribs), and *ngungu* (a barren woman), which are or were believed to be immune to biomedical remedies, manifest themselves after the symptoms become chronic.

Kamba taxonomies present practical difficulties when they are used as a basis for ordering descriptive data about the etiology, symptoms, and treatment of diseases (van Luijk, 1982). For example, there is considerable overlap between and within the classifications. Thus *kwalwa ni ivu* (*kuiwa ni ivu*), or abdominal pain, ranks among the common maladies most frequently attributed to witchcraft (*uoi*); yet in Kilungu it is also known to be caused by eating unusual or spoiled foods, thereby locating the cause as natural or self-inflicted disease (Table 5-1 and 5-2). If the condition becomes chronic and a *mundu mue* is consulted, a strong psychosomatic element is usually apparent. Through divination the problem will probably be attributed to *uoi* or *aimu*.

Nduuka, or madness, provides another illustration of the many afflictions that are attributable to several possible ultimate causes. It is a serious, chronic psychotic disturbance that can lead the sufferer to withdraw socially, run around naked, adorn himself or herself with rags and trash, talk incoherently, inflict self-injury, and engage in violent antisocial behavior. Individual circumstances allow for several possible etiological interpretations of *nduuka*. In the Kilungu Hills it can be diagnosed as spirit-caused (*nduuka ya maimu*), a consequence of witchcraft (*uoi*), or a natural affliction brought on by God himself (*Ngai Muene*). In the Machakos Project study area ritual cursing was also mentioned as a cause of *nduuka*. It is accomplished through a petition to *aimu* or *Ngai* for revenge by the parents of a child who has shown gross disrespect for their status and authority. Special forms of *nduuka*, including madness caused by malaria (*nduuka yumanitena ndetema nene*), by addiction to marijuana (*nduuka ya mbangi*),[4] and by alcohol dependence (*nduuka ya uki*) have also been reported from northern Machakos (van Luijk, 1982).

Throughout Ukambani, indeed anywhere Kamba have settled, *nduuka* in certain persons has also been one of several possible traditional signs that they are "called" by *aimu* to become a *mundu mue*—to exercise their God-given beneficial supernatural powers (*ue*) of divination and healing. These powers could never be simply inherited or learned in apprentice fashion. If the call was ignored, it was believed (and still is by some of the population) that the *aimu* would never allow the disease to leave the person.

A second practical problem connected with Kamba disease classification and labeling, or for that matter any other taxonomy, is how to achieve a concordance or workable conversion system between traditionally based classifications and the conventions of biomedicine. The latter are represented by the World Health Organization's International Classification of Diseases (ICD). This issue is vexing because the ICD is itself permeated by cultural assumptions. Janzen (1978, p. 192), for example, reported that "only 55 percent of the entries in the . . . ICD are 'scientifically diagnosable' entities, that is, reducible to single, universal and duplicable, sign-symptom complexes. Remaining entries of the ICD are independently varying signs and symptoms classified somewhat arbitrarily according to body parts and problem focuses."

However inconsistent, fragmented, and imperfectly applied, both the traditional Kikamba and versions of the ICD approaches to disease classification are widely used today. They form an integral part of the Kamba *ethnomedical system.* Van Luijk (1982) has made an interesting attempt to reconcile the Kikamba descriptive system of disease signs and symptoms (his code list contains 213 entries) with the 70 "cause groups" in W.H.O.'s ICD[5] and the 30 diagnostic codes for outpatients used by the Kenya Ministry of Health. These cross-tabulations could lead to better-informed practitioner-patient dialogue in settings where biomedical services are offered. Nevertheless, a more comprehensive "ethnomedical" approach—one that includes concepts of ultimate causation as well as the cultural idiom for expressing the experience of illness—is required if biomedical and traditional practitioners are to understand each other's technical language and effectively relate this to their own and their patients' explanatory models (Chapter 1).

Whether one's vantage point is the Kilungu Hills, Nairobi, or Dakar, there are no shortcuts across the contemporary landscape of African medical pluralism. It is unlikely that anyone can devise a unified, coherent "metamedical" classification of disease signs, symptoms, causes, and labels that is simple enough to be practical. Consequently, ways must be found to provide biomedical workers with an informed understanding of the traditionally based medical cultures of the ethnic groups represented in the localities where they are posted. It is equally important that this knowledge be applied to improve the outcome of patients' visits to health centers and dispensaries. An ethnomedical-clinical approach that goes beyond treatment of a patient's apparent disease symptoms to a consideration of etiology and social nexus will lead to more effective diagnoses and health care of higher quality in Kilungu and elsewhere.

DIVERSITY, IDENTITY, AND LOCATIONAL CHARACTERISTICS OF KILUNGU'S TRADITIONAL MEDICAL PRACTITIONERS

As in many other cultures (Worsley, 1982), numerous specialists are found among the TMPs of Kilungu. The skills and esoteric knowledge used in the treatment of various diseases and illnesses are not widely distributed among either the TMPs or the general population, although most households seem to have at least one member who is familiar with common plants of medicinal value. However, the term *mundu mue* (pl. *andu awe*) is a broad designation that all Kamba understand to mean a medicoreligious specialist of great wisdom (*mundu* = person; *ue* = wisdom)[6] who uses divination (not necessarily exclusively) as a tool in his or her practice of healing the sick and looking after the well-being of the community. The position of *mundu mue* is not hereditary in a technical sense, although such persons do cluster with greater frequency in particular families and lineages. In contrast, *uoi* (pl. *aoi*) is the Kamba term for witchcraft and sorcery. The *mundu muoi* is a practitioner of disruptive and evil activities and represents forces in the society which the *mundu mue* serves to combat and neutralize.

Andu awe commonly employ a special calabash (*ketiti*) that contains numerous seeds, pebbles, and other objects of similar form (*mboo*). The wire-stringed "magic bow" (*ota*), a musical instrument that is used to set the scene and tone of a divination session and, sometimes, to induce a trance, is also standard. These three elements are the primary material instruments of divination and remain in wide use. To reach the insights required for a proper decision (e.g., concerning the cause of a disease or a verification of sorcery and the sorcerer's identity), the *mundu mue* systematically shakes the calabash and produces, in hand, variable lots of *mboo* which are then cast onto a special animal skin or cloth. Numerous series of *mboo* lots are cast, during which time the *mundu mue* will develop insights by directing seemingly casual but shrewd questions to the client and those who accompany him. Ultimately, the *mundu mue* fashions a prediction or diagnosis based on his or her interpretation of the numerical or statistical results of the exercise. Lot-casting can take from 5 min to an hour or more depending on the circumstances of the case.

Very noticeable differences exist among Kamba diviners with regard to their specific instruments, procedures, and systems for interpreting the significance of the lots cast. Some, for example, gather information through such aids as mirrors, snuff, dreams, or Islamic-inspired recitations. Nevertheless, extended observation of many *andu awe* lend support for an important principle of divination proposed by Moore (1969), that is, each practitioner follows a rather consistent system which is independent of any particular occasion of divination. In other words, the procedures of individual TMPs are standardized but as a group their outward procedures and paraphernalia are highly idiosyncratic.

Much of the literature on traditional medicine in Africa conveys an impression that diviners, herbalists, and midwives represent essentially uniform, undifferentiated subgroups of traditional practitioners. In reality, as recent studies illustrate, there is considerable internal variation within such categories of TMPs in terms of their skill levels, specialization, knowledge and beliefs, the types and organization of therapy they provide, and personal characteristics (Good, 1980a; Stock, 1983). Such diversity needs to be more widely recognized because it has important policy implications in areas where formal cooperation between biomedicine and TMPs is under study (Stock, 1983).

Eleven distinct categories of traditional medical and medicoreligious skills can be identified in the Kilungu Hills (Table 5-5). Although divination, herbalism, and midwifery are the most common skills represented, the numbers in the table are more indicative of actual proportions than precise guides. Kilungu TMPs typically have at least two such skills and only about half practice divination (Table 5-6). Although they are generally able to provide their services on demand, traditional medicine is a part-time activity for all rural practitioners encountered. In addition to farming, some also perform roles such as sublocation Chief, lay preacher in the African Brotherhood Church, and local chairman of the Kenya African National Union (KANU), the national political party.

By definition, a *mundu mue* is set apart from other TMPs by virtue of having acquired supernatural power (*ue*). He or she has the power to divine (*kwausya*) and to transform certain medicinal herbs into a substance (*ng'ondu*) which has magical healing and cleansing properties. *Andu awe* are considered skilled at fortune-telling, diagnosing illness, and prescribing a course of treatment, although some do not provide the treatment themselves.

The principal "barrier" to recruitment to the role of *mundu mue* is experiential. A person must be "called" to practice divining and healing magic (*ue*) through an act of God. In some cases there may be signs from birth that a child is destined to become a *mundu mue*—for example, if it is born holding ritual objects (*masyawa*) such as seeds (used as divining counters) or a twig in its hand, or with "seeds" or other special markings on its umbilical cord (*mukauti*). Typically, the sign of the professional calling materializes in later childhood or adult life and is manifested in a lengthy illness syndrome whose symptoms may include dreams, hallucinations, socially aberrant behavior, vision problems, and inability to concentrate (e.g.,

Table 5-5. Diversity of Medicoreligious Skills in Kilungu, 1978[a]

Specialist	English interpretation	No. individuals with skill
Mukimi wa miti	Herbalist	38
Also		
Muiiti wa ukima miti	Herbalist	
Muiiti wa miti	Herbalist	
Muiiti wa kukaa miti	Herbalist	
Mundu mue kwausya	Diviner	37
Also		
Mue wa kwausya	Diviner	
Mue wa kwausya na kuiita	One who divines and treats	
Kwausya na kuiita	One who divines and treats	
Mwisikya	Midwife	14
Mwisikya na kuvindua syana	Midwife who can rotate a fetus	1
Mundu wa muvingui	One who "opens" a bewitched person (exorcism and catharsis)	6
Mwaiki	Circumciser	5
Mundu wa nthambya	Ritual specialist who uses *ng'ondu* in infertility therapy, provides protective medicine for newborns, and performs purification rites following sexual misconduct (a cause of *mwimbo*, or edema)	3
Kulikwa ni aimu	(Treatment of) one possessed by spirits of the dead	2
Mukuni wa kilumi	One who beats the drum in *kilumi* (spirit cult dance)	1
Ngui ya kilumi	Singer/leader in *kilumi* dance	1
Muusya wa kithitu	One who removes effects of another's oath against you, or who treats effects of an oath taken and broken	1
Mundu wa muma	Specialist who administers the *muma*, a particular type of *kithitu* (oath) designed to resolve a grievance against another person (has legal sanction in local Kenyan courts)	

Note: From field survey, 1978.
[a]Four sub-locations.

on schoolwork); or a series of misfortunes that effect the individual's entire family. Of particular importance among female *andu awe* is a personal history of perceived barrenness, or of their children dying. It is also believed that once the signs of *ue* in a person have been verified by other *andu awe*, no relief from the persistent difficulties is likely until he or she acquiesces to the call and agrees to undergo the ordination ceremony (*mboka ya ukususya*

Table 5-6. Combinations of Traditional Medical Practitioner
Specializations: Kilungu

Specialization	No. TMPs
Diviner/one who divines and treats	22
Diviner and ritual expert in infertility and protection of newborn	3
Diviner, herbalist, and midwife	2
Diviner and treatment of spirit possession	2
Diviner, midwife, and "opener"	1
Diviner, herbalist, and "opener"	1
Herbalist	10
Herbalist and "opener"	1
Herbalist and one who removes ill effects of a *kithitu*	1
Midwife	8
Midwife who can rotate fetus	1
Midwife and herbalist	1
Midwife and diviner	1
Circumciser	2
Circumciser and herbalist	2
Drum beater in *kilumi* dances	1
Singer/leader in *kilumi* dances	1
One who treats a person possessed during *kilumi*	1
Others not classified	4
	65

Note: From field survey, 1978.

uwe) required of a practitioner. Today, this demand is often obeyed with considerable reluctance. Acceptance of the call to become a *mundu mue* may limit one's vocational choices and life-style. At least, to young people with some formal education it symbolizes unprogressive values.

In contrast, a *mukimi wa miti*, or herbalist, is one who has a large repertoire of medicinal herbs and uses these singly and (usually) in combination to treat specific ailments such as joint pains, fever, or infant gastroenteritis. Herbalists themselves insist that they do not employ *ue*, or *aoi* (sorcery power), and are simply the vehicles through which *Ngai* (God) heals with *miti* (herbs). Unlike a *mundu mue* (who can also be an herbalist), practitioners of herbalism who do not perform divination are not religiously compelled to practice their *miti shamba*. However, talented herbalists do find a strong demand for their skills and treat a wide range of illnesses.

A group of 26 TMPs (including some who divine *and* treat) who "regularly use *miti shamba*" in their therapy perceived themselves as "best qualified to treat" the complaints shown in Table 5-7.[7] Of those seven TMPs in the census who said they do not use *miti shamba*, five were diviners (who also treat possession or perform *nthambya*) and two were birth attendants. It is no accident that all of the leading responses, and many others shown in Table 5-7, are illnesses which often become chronic among the Kamba and have strong psychosomatic associations. In Kilungu, attributions of witchcraft or sorcery, for example, are frequent in cases of persistent stomach disorders, headache, pain in the sternum area, and edema, and there is confirming evidence for this pattern from central and northern Machakos District (Thomas, 1971; van Luijk, 1982). Consequently, TMPs are sought out sooner or later, and sometimes exclusively (cf. Thomas, 1971), for treat-

Table 5-7. Illnesses Kilungu Herbalists[a] Perceive Themselves "Best Qualified to Treat"

Illness/complaint	Percentage of all conditions reported
Adominal pain/stomachache (*Kwalwa ni ivu*, and so forth)	14
Headache (*kwalwa ni mutwe*)	10
Pain in sternum area (*kavaso/kavoo*) (frequently translated as "peptic ulcer")	8
Edema (*mwimbo*)	7
Diarrhea (*kwituua*, and so on)	7
"Pneumonia" (*kyambo*)	5
Chest pains (*kwalwa ni kithui*)	5
Heart "attack," heart "failure" (*ngoo*)	5
"Syphilis" (*teko*)	4
Gonorrhea (*kisonono*)	3
Generalized pain in bones or rheumatism (*mutambuko*)	3
Common cold (*ikua*)	3
Boil (*muimu*)	3
Hard swelling on head, body (*nyamu*)	3
Infant gastroenteritis (*nyunyi*)	3
Vomiting and diarrhea (*kutavika na kwitua*)	3
Epilepsy (*kisala*, or *mungathuko*)	3
Swollen breasts of lactating woman (*uwau wa nondo*)	1
Infertility/barrenness (*ngungu*)	1
Whooping cough (*mutitino*)	1
Circumcision wounds (application of *mnooya*)	1
Pain in joints (*kwalwa ni maleenyu*)	1
Toothache (*kwalwa ni maeo*)	1
Conjunctivitis (*uwau wa metho*)	1
Ringworm (*kiea* and *kinguu*)	1
Fever (*ndetema*)	1
Jaundice (*muuku*)	1
Pancreatic disease, nonspecific (*wasyungu*)	1
	100

[a] $N = 26$.

ment of these and many other illnesses such as *mutambuko* (rheumatism), *nyamu* (hard swellings on head or body), *kisala* (epilepsy), *ngungu* (barrenness), and *wasyungu* (probably a pancreatic disorder). Many of these and other illnesses such as *ngome* (a tight constriction of the throat that interferes with swallowing) and *kindu nda* (foreign objects in the abdomen) are not seen in the records of the Nunguni and Kikoko out-patient clinics. This constitutes another measure of the inadequacy of the existing biomedical approach to health needs in rural Africa. Traditional medical practitioners commonly label such afflictions as "Kikamba illnesses."

Herbalists know the symptoms and have traditional remedies for the chronic and psychosomatic illnesses, but they usually leave interpretation to the *andu awe* and profess ignorance of the causes. The emic descriptions of herbalists underscore once again, in the context of diagnosis and therapy, the complexity of illness semantics, the importance of possessing relevant cultural information, and the therapeutic value of knowing the patient's social and physical environment (Chapter 1).

Three illness syndromes that are common in Kilungu illustrate the complexity of disease interpretation and management. Two are interdependent ailments and all three occur in the same general abdominal region.

1. *Iwethyu* describes "a swelling of the flesh around the base of the ribs (which protect the heart and liver). This swelling can lead to vomiting, and the person may become jaundiced if treatment is not obtained. After some time this produces 'yellow fever.' It always affects the right front side of the body. One cannot have *kavoo*, or *kavaso*, without first having *kwethyu*." (Herbalist in Kalongo Sub-location)

2. *Kavoo* or *kavoso* is commonly translated as "peptic ulcer." These anatomical terms appear to be used interchangeably to denote the sternum cartilage or sternum bone. The affected area is the upper abdomen at the base of the sternum. The discomfort of nausea and vomiting is perceived to be caused by a protrusion of the sternum into the abdominal cavity. To relieve the symptoms, the TMP (who first applies oil or soap to his or her hands and warms them over a fire) must skillfully "lift up" the sternum and return it to its "normal" position. The patient is also given a medicine prepared in powder form from the roots of *mulingula*, which is added to tea water or porridge (*uji*). Thomas (1971, p. 152) reported that *kavaso* is "a reaction to psychological stress" to which women are at greatest risk. Van Luijk (1982) found that *kavaso* is also related to *nyama ya kati* (pain in diaphragm). Old people clearly distinguish the two complaints, but young people may confuse them.

3. *Wasyungu*, which is biomedically a nonspecific disease of the pancreas,[8] was characterized by a prominent herbalist in Kilungu as a swelling under the left side of the rib cage accompanied by severe pain and fever. This swelling produces *ndumo*, a further swelling of the abdomen caused by the collection there of excess body fluids. As the illness progresses, the sufferer's

arms and legs become thin. People who live in drier areas (such as in Kilungu's Kithangathini Sub-location) are perceived to be at greatest risk because the disease is associated with those who drink water from sandy sources, such as holes dug in river beds. A cure may take from 3 days to 3 months, depending on whether the symptoms are old or new. To treat *wasyungu*, the herbalist lifts the *iwethyu* [see (1) above] and has the patient drink a tea prepared from the roots of *mulingula* and *muthekethe*, the leaves of *kauthilu*, and the roots and leaves of *kiema uvunyie*. If improvement does not occur, the herbalist will make cuts on the affected region and rub in a powder made from crushed and burnt roots of *muthekethe*.

Midwifery is the third most common health-related skill found in Kilungu, after herbalism and divination. Midwives (*isikya*, pl.) tend to be among the oldest women in their respective communities, averaging over 60 years of age. They started in this role after bearing children of their own and had thus been serving the women of their communities for several decades. Six of eight surveyed also possessed other skills, such as herbal treatment of certain illnesses, and a few had their reputation based primarily on their role as *andu awe*.

It is notable that projects designed to promote cooperation between biomedical and traditional medical practitioners in Africa (and other areas) have with few exceptions focused on the role of traditional midwives rather than traditional medicoreligious specialists (Warren et al., 1982). Feierman (1984, p. 113) argues that this pattern signifies that midwives (by implication females) form "the heart of community care-giving networks." He implies that other kinds of TMPs (by implication males) are not part of the same networks. But this ignores the fact that, as in Kilungu, female *andu awe* may be in the majority. Furthermore, in some areas of Africa such as southern Ghana, males provide a large share of the traditional midwifery services (Neumann & Lauro, 1982). Pillsbury (1982) seems closest to the mark in observing that TMPs' practices are often based on supernatural ideas and other beliefs that are difficult for biomedically trained practitioners to understand and accept. In contrast, midwives are mainly concerned with the "technical processes of childbirth" and their procedures are less alien to biomedical workers.

Most *isikya* in Kilungu had at some time worked as midwives or practiced other traditional medical skills in areas outside their present sub-location, ranging from neighboring Makueni and Mbooni to Nairobi, Kikuyu-land and, in one instance, Tanzania and Zambia. Limited information concerning the fee structure of *isikya* suggests that charges for delivery ranged (in 1978) from a minimum of 1 to a maximum of 10 shillings (U.S. $.125 to $1.25). Food is often accepted from close relatives or friends in lieu of cash payment.

In traditional Kamba society two rites of circumcision symbolize a person's passage into adulthood. *Nzaiko ila nini* (small circumcision) involved the surgical removal of the foreskin of boys and clitoridectomy for girls who were about 5–8 years old. The resulting scar symbolized adulthood and wholeness. This ceremony requires a skilled medicoreligious specialist (*mwaiki*, s.; *aiki*, pl.). A second stage, *nzaiko ila nene* (big circumcision), occurs some years later and centered on a week or more of ceremony and indoctrination. Mbiti (1971, p. 122) observes that these acts and the moral training that accompany them "represent the highlight of one's life, qualifying a person to become a complete Mukamba" and signifying acceptance of adult responsibilities.

Today, there is considerable local variation within Ukambani (and elsewhere in Kenya where circumcision is customary) in the extent to which the surgical circumcision procedures are still performed by traditional specialists (*mwaiki*, s.; *aiki*, pl.). Fifteen years ago, in an area 40 kilometers north of Kilungu called Masii, Thomas (1971) observed that the male and female operations (*nzaiko ila nini*) were still important prerequisites to marriage. However, the lengthy rituals that traditionally accompanied and followed them had been substantially reduced. More recently van Luijk (1982, 1983), who conducted fieldwork in an area of northern Machakos which has been deeply influenced by the presence of several Christian churches and relatively greater access to biomedical facilities, concluded that nowadays virtually all male children are circumcised without fanfare at a local dispensary or hospital. He was unable to discover any evidence that clitoridectomy of girls is still locally performed, unless some parents arrange clandestine operations.

In the Kilungu Hills the customary circumcision of both boys and girls is still widespread and commonly performed by traditional specialists. Nevertheless, here too the group aspect of the rites and training for adulthood that once provided the essential context for the surgery (Lindblom, 1920) has evidently been abandoned. Today the circumcisers carry out the operation in the privacy of a child's home.

During my TMP census in four sub-locations of Kilungu, five elderly persons, including a married couple, identified themselves as circumcisers (*aiki*). According to the wife of the couple, she and her husband still circumcise boys and perform clitoridectomy on girls during the "hot" months of August and December, a seasonal pattern confirmed by other adults in Kilungu. The circumcision wounds are said to heal more quickly at these times than in the cooler months of the year. Another local informant who is an herbalist and a midwife, but not a *mwaiki*, confided that in this part of Machakos District circumcision is still done to both boys and girls. Long ago candidates used to be gathered in one place but now the circumciser visits them in their houses. The parents make appointments with the circumciser.[9]

In traditional Kamba life the *mwaiki* was a religious functionary whose acts of circumcision and clitoridectomy had medical implications. A small proportion of Kamba youth, girls in particular, were certainly at risk to hemorrhage, shock, infection, and even death following the surgery. However, since their operation was restricted to excision of the clitoris and part of the labia minora, Kamba girls were undoubtedly less prone to hazardous complications than girls from some other ethnic groups in Kenya which practiced more extensive excision and/or infibulation. Today, there is heated public and private controversy over female "circumcision" in Kenya (where it is traditional for some 25 ethnic groups) and elsewhere in Africa. The debates are stimulated by an increasing awareness of the associated health hazards, which are frequently reported by the Kenyan press, by rapid social change which encourages the educated and/or Christianized elements of the Kenyan population to question the custom's social value, and by the global womens' rights movement which has produced documents such as the *Hosken's Report* (Owano, 1978). Such studies tend to take strong positions against "female genital mutilation." In view of these factors and the improbability that the few remaining *aiki*, who are very old, could find younger people motivated to learn circumcision skills from them, there is little reason to expect that the practice of clitoridectomy will continue in Kilungu into the next generation. Male children will continue to be circumcised. However, this will be done at Nunguni, Kikoko, or one of the other biomedical outpatient facilities rather than by a local *mwaiki*.

My census in four of Kilungu's ten sub-locations identified 65 individuals whom the local population defined as TMPs.[10] More than two-thirds of the TMPs are women, many of whom did not work as traditional healers until after they were married.[11] Economic forces such as "competition" and "market penetration" apparently have little bearing on the spatial distribution or relative location of Kilungu's TMPs. Instead, their physical locations in the rural settlement fabric are determined by broader considerations that reflect traditional Kamba principles of social and territorial organization. This is particularly evident in the case of women, whose homes in adulthood reflect clan (*mbai*) exogamy and patrilocal residence rules. Hence, when a married woman leaves her home to join her husband on his land the act incorporates her into a new clan but occurs independently of her role as a TMP.

Interviews with over half ($N = 33$) of the Kilungu TMPs (primarily those from Kalongo and Musalala) (Figure 4-1) provided a profile of their personal experience and socioeconomic characteristics. Except for two Kikuyu women who moved to Kilungu because they had married Kamba men, all are ethnic Kamba and all are farmers, although two are now incapacitated, respectively, by blindness and infirmity. Although they range from 35 to 75 years of age, eight of ten exceed 50 years and the average is about 60 for both women and men. The TMPs are thus an elderly population. On average, they had practiced their arts for about 28 years, the men reporting somewhat longer experience (29 years) than women (24 years). A strong attachment to place is indicated by the fact 88% of the TMPs were born in Kilungu Location, and over 60% were born in the sub-location where they currently reside.

Although on religious grounds one could not become a *mundu mue* without a calling from the spirits verified by other *andu awe*, and, quite often, some physical evidence (e.g., *masyawa*) of bestowed supernatural powers, there is little evidence that practitioners of traditional medicine are restricted by age, heredity, or clan affiliation. Gender is significant only in relation to midwifery (males are rare) and circumcision. Traditionally among the Kamba only men could circumcise boys; but this is not the case today in Kilungu. There is, however, evidence of a professional "predisposition" of families since about half of the TMPs identified at least one other person in their extended family—most often sisters, brothers, and husbands—as a practitioner. In one instance, a woman who practices as a diviner, herbalist, and midwife noted that her mother and father as well as two brothers and a sister are also *andu awe*.

Ties of kinship and descent may play an important role in the maintenance and transfer of traditional medical skills and knowledge. This is an important issue because it relates to how accumulated knowledge, improvements, and innovations spread throughout a population of TMPs, but it was not specifically addressed during fieldwork in Kilungu. Nevertheless, it was soon evident that the TMPs (by their own admission) generally practice their crafts in relative social isolation from one another. Consequently, their therapeutic activities, choice of medicaments (Table 5-8), and styles reflect self-perpetuating idiosyncrasies and divergence. This factor will undoubtedly loom importantly if and when a role for TMPs in primary health care receives serious consideration from Kenya's Ministry of Health. Again and again, individual TMPs expressed fear about the consequences of coming into contact or competition with other TMPs, and they described the strong jealousies among themselves that inhibit professional cooperation. Hence, it was not surprising to discover the absence of any formal (or informal) organization or guild structure among the Kilungu TMPs.

Overall, the ratio of TMP's ($N = 33$) to the total population in the census portion of Kilungu was 1:378, while the separate values were 1:254 for Kalongo, 1:429 for Ndiani, 1:439 for Musalala, and 1:458 for Wautu. The actual values are probably underreported by 10% to 20%. Van Luijk (1984), for example, recently calculated a ratio of one TMP per 92 residents in the Machakos Project. He qualified this by reporting that only a handful of the TMP's were "professionals"—a term he did not explain. These Kilungu values compare favorably with the limited evidence available for other rural and urban areas of Africa. At this macrolevel of analysis TMPs appear easily accessible to the people of Kilungu—especially in comparison with the national ratio (1:2,560) of paramedicals (clinical officers and nurses) to population in Kenya (*Weekly Review*, 1981). For physicians (none are based in Kilungu) the Kenya ratio is about 1:70,000 (Family Health Institute, 1978).

Precise village locations were mapped for 37 of the 65 TMPs in the census.[12] Information concerning the distribution of TMPs at the village level is most complete for Kalongo, where marked clustering of practitioners

Table 5-8. Herbalists' Remedies for Seven Illnesses

Illness	Herbalist	
	Nzeki	Kutenga
Headache (kwalwa ni mutwe)	For pain in temples, muti (Aspilia mossambicensis?) leaves are squeezed and added to water. Drops of this mixture are placed in both ears. For pain on top of head, leaves of mulavuta (Helichrysum odoratissimum?) and muemaaka are squeezed and added to water. Mixture dropped in each ear.	Leaves of kalaku (Fuerstia africana?) and kamwooya are squeezed and dried; crushed root of mukoi wa ivia are mixed with water.
"Peptic ulcer" (kavoo/kavaso)	Herbalist applies oil or soap on his or her hands, warms them over a fire and lifts up the kavoo (sternum cartilage). A powdered medicine prepared from mulingula roots is added to tea water or porridge for the patient to drink.	The burnt bark of mulingula and dried and crushed stem of mikoi wa ulenge are added to water. Patient drinks this.
Fever (ndetema)	A tea prepared from boiled muthaa and mukomoa leaves is drunk by the patient.	A tea prepared from the bark of mwaitha (Entada leptostachya? or Dalbergia lactea?) and muvindavindi (Fagaropsis hildebrandtii?) is given to the patient.
"Pneumonia" (kyambo)	Juices squeezed from fresh leaves and roots of mususu (Indigofera spicata?) and fresh leaves of muloo plant are mixed with water and drunk by patient.	A tea prepared from dried and crushed root of mwany'a nthenge and crushed back of mukolekya is given to the patient.
Edema (mwimbo)	Leaves of ng'ondu (Talina portulacifolium?), mwisya, and muvindavindi (F. hildebrandtii?) are boiled together with roots of muvou. Patient drinks this decoction.	Dried and crushed root of mwany'a nthenge is mixed with milk and drunk. Also, juices from leaves of mukoi wa ivia and ndulu (Solanum nigrum?) are squeezed into warm water. Patient drinks mixture.
"Syphilis" (teko)	Muvou roots are burnt, crushed into powder, and applied to sores. Liquid from the roots of musanzaikyaitu and muthaa is given to the patient to drink.	Root of muungua/mundua (Dichrostachys cinerea?) is dried, crushed and mixed with water.
Gonorrhea (kisonono)	A mixture of extracts from songe leaves and roots, watima (grass), mukweo bark (Bridelia micrantha?), and musanzaikyaitu root is given to patient in liquid form.	Barks of kiatine, muvuti (Eryhrina abyssinica), muvunu, mulingula, and mukenia (Fagara chalybea?) are crushed and steeped in water which the patient drinks.

147

(about three per square kilometer) is noticed in the southeastern portion of the sub-location (Figure 5-1). At this scale the crude ratio of TMPs to population in Kalongo suggests that the practitioners are readily accessible to people in the villages. This impression is reinforced by the TMPs' own reports: more than 8 of 10 indicated that they are available day and night, every day. In terms of these purely *quantitative* measures Kilungu TMPs appear to satisfy two conditions that attract patients: they are easily accessible and convenient. In practical terms, however, these advantages are diminished when other characteristics of the TMPs are considered, such as the

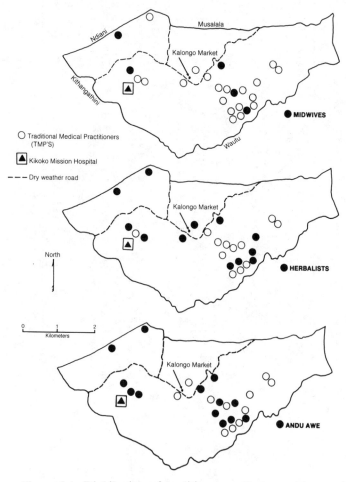

Figure 5-1. Distribution of traditional medical practitioners: Kalongo Sub-location, Machakos District, Kenya.

spatial and proportional distribution of their specializations, age, physical and mental abilities, and their individual reputations and charisma. The need to evaluate the actual proportional and spatial allocation of traditional medical *skills* in an area has been largely overlooked, although a few studies from Africa suggest an awareness of the implications of this issue (Feierman, 1979; International Development Research Center, 1980; Stock, 1983; Thomas, 1971).

Kalongo Sub-location, for example, has 23 TMPs. Most of them practice at least two of seven types of specialization. Table 5-9 shows the actual frequency (proportional allocation) with which these skills occur in Kalongo and the other sub-locations examined. Thus in Kalongo the specializations are represented 40 times among 23 TMPs. Although there is one (unspecified) skill for every 146 people in the sub-location, there is only one herbalist for every 532 people and only one midwife for every 975 people (1:259 adult females). Compared with the crude ratios, these *skill-specific* ratios represent a superior method for assessing the actual type and availability of traditional medical resources in an area. When mapped, the skill-specific patterns enhance understanding of the type, availability, and accessibility of TMPs in relation to communities and administrative areas (Figure 5-1). Such basic

Table 5-9. Allocations of Traditional Medical Practitioners' Specializations in Four Sublocations of Kilungu

	Sub-location				
	Kalongo	Musalala	Ndiani	Wautu	Total
Est. population (1978)	5,849	4,389	7,290	7,422	24,597
Skills/specialization					
Diviners (*andu awe*)	14	6	8	9	37
Ratio to population	1:418	1:732	1:911	1:825	1:665
Herbalists (*akimi wa miti*)	11	8	9	9	37
Ratio to population	1:532	1:549	1:810	1:825	1:665
Traditional midwives (*asikya*)	6	2	3	4	15
Ratio to population	1:975	1:2,195	1:2,430	1:1,856	1:1,640
Circumcisers (*aiki*)	1	2	1	1	5
Ratio to population	1:5,849	1:2,195	1:7,290	1:7,422	1:4,919
Mundu wa nthambya	2	—	1	—	3
Ratio to population	1:2,925	0	1:7,290	—	1:8,199
Muvingui	3	3	—	—	6
Ratio to population	1:1,950	1:1,463	—	—	1:4,100
Others (*kilumi*-related;	3	2	2	—	7
muusya wa kithitu;	1:1,950	1:2,195	1:3,645	—	1:3,513
mundu wa muma)					
Total no. specializations	40	23	24	23	110
Ratio (overall)	1:146	1:191	1:345	1:323	1:224

Note: From field survey, 1978.

information is essential to any planning that may envisage a role for TMPs in primary health care (cf. Stock, 1983).

Three other factors also influence the actual utilization of traditional medical skills in an area. First, all TMPs do not have the same reputation or networking potential. They are not restricted to clientele who reside in the local community or even their own sub-location, and it is conceivable that some draw most of their custom from farther afield. For instance, although nearly all practitioners in Kalongo draw many of their patients from Kalongo, they also identified 13 other sub-locations within and outside Kilungu as source areas. The evidence is similar for nine TMPs who live in neighboring Musalala Sub-location. All attract clientele from Musalala, but they also identified eight other sub-locations as sources of patients. Second, as a rule TMPs in Kilungu reflect the trait of mobility that has long been a characteristic of Kamba culture. Despite the strong bonds of TMPs with the local territory noted earlier, nearly three-quarters of those interviewed had at some time practiced traditional medicine outside of their present sub-location. Most cited experience in neighboring sub-locations and other areas of Machakos District such as Mukaa and Makueni. A few had worked in Nairobi or Mombasa and, in one case, coastal Tanzania and Zambia.

The two factors noted above are interrelated with a third important locational consideration: the preference of many clients and *andu awe* not to be treated by, or to treat, respectively, someone who is well known to them, whether personally or through others. Additional evidence of this pattern in the form of client/patient origin areas is presented below. As Thomas (1971) observed in an earlier study elsewhere in Machakos, many cases brought to a *mundu mue* involve strained interpersonal relations and sorcery and are thus highly sensitive. Such a client will often search for a healer who is known by reputation only in order to discourage gossip and preserve confidentiality through anonymity. For his or her part, to establish credibility and expectancy a *mundu mue* may prefer to treat strangers who are less familiar with his or her style and techniques. Moreover, since those accused of sorcery by the *mundu mue* are customarily among the client's kin, friends, or coworkers, the practitioner is not keen to draw the hostility of the accused and perhaps jeopardize himself or herself and the client. Elsewhere in Kenya the implication of other individuals in patients' illnesses reportedly has caused "feuds, fights, and even murders" (Acuda, 1983, p. 13).

Spatial separation and social distance are thus useful devices that enable a *mundu mue* to maintain a sense of awe and social control in many therapeutic situations, particularly those involving alleged sorcery, "home problems," ritual error, or some protracted illness. Such afflictions not only "infect" a specific individual but create wider imbalances which can disrupt family and social relations. They are thus far more serious and threatening than an acute ailment such as a fever, gastroenteritis, or wound for which a person would typically seek assistance at a biomedical facility. In the final analysis, the search for therapy and a resolution of the socially disruptive problem often takes the afflicted person and his or her "significant others" far afield from his or her home area. Once a problem is perceived as a threat to the social fabric the expense and travel required to locate an effective *mundu mue* become secondary concerns. Consequently, the usual "distance-decay" effect seen in utilization patterns of biomedical services (e.g., Kilungu, Chapter 4) is often absent. Many of the patients we observed in the "clinics" of urban TMPs in Nairobi, for example, began their quest from a rural location and had traveled considerable distances to obtain therapy at great financial cost. Seen in this light, the romanticized notion of the "friendly, accessible village TMP" may bear little resemblance to reality in the case of *andu awe*. On the other hand, those herbalists and midwives who are not specialists in ritual therapy and supernatural phenomena have a larger local practice. It is these latter TMPs who are probably in the best position to become community health workers in cooperation with the biomedical services.

DISEASE PATTERNS, HEALTH SERVICES, AND BEHAVIOR: THE TRADITIONAL MEDICAL VERSION

Traditional medical practitioners interviewed during the Kilungu census ($N = 33$) were asked to recall the number of patients they had seen during the preceding week and to identify the problems they had presented (Table 5-10). Nearly two-thirds of the TMPs had seen at least 1 patient, and the total was 112 patients. Three herbalists had treated nearly two-thirds ($N = 71$) of these patients. This disproportionate utilization of a few select TMPs reinforces the point made earlier concerning their differences in popularity, esteem and, apparently, effectiveness. Most of the remaining patients consulted with *andu awe*, whose qualifications enable them to divine and to provide diagnosis and therapy for someone who has provoked the spirits' displeasure or to assist those who have been targets of witchcraft and sorcery.

Considerable overlap exists between the conditions herbalists and diviners treat. As a rule, the *andu awe* are sought out when an illness persists for an unusual length of time. Abdominal complaints, for example, are frequently chronic conditions that may require the services of a *mundu mue* who is skilled at removing the foreign objects from the patient's body (*kumya kindu nda*) that are believed to cause the discomfort. *Andu awe* are always consulted for conditions such as madness (*nduuka*), epilepsy (*mungathuka*), spirit possession, unusual weight loss or wasting (*thavu*), bewitchment (*kuowa*), and sorcery protection or neutralization rituals such as *kuthambya* and *kuusya musyi*. The latter ceremony cures a family of the effects of sorcery. In 1978 the usual fee for this service in Kilungu was approximately Shs. 300/- (U.S. $40.) and a goat.

Kuthambya and *kuusya musyi* both involve the use of *ng'ondu*, which is

Table 5-10. Illnesses/Problems of Traditional Medical Practitioners' Patients
($N = 112$) in Week Before Interview: Kilungu

Illness/problem	Rank
Abdominal pains (*kwalwa ni ivu*, and so on)	1
"Spirit possession" (*aimu*)	2
"Peptic ulcer" (*kavoo/kavaso*)	3
Headache (*kwalwa ni mutwe*, and so on)	
Vomiting and diarrhea (*kutavika na kwituua*)	
Deliveries (birthing)	
Hard swelling (*nyamu*)	

Other Conditions or Services Reported at Least Once

Severe discomfort around trachea (*ngome*)	Madness (*nduuka*)
Worms (*syoka ya nda*)	General body pains (*kwalwa ni mwii*)
Continuous menstruation (*kuua*)	Vomiting blood (*kiathi*)
Fever/malaria (*ndetema*)	Diarrhea (*kwituua*)
Asthma (*asima*)	Pain in limbs (*kwalwa ni mamutha*)
General edema (*mwimbo*)	Weight loss (*thavu*)
Measles (*mukambi*)	"Syphilis" (*teko*)
Pancreatic disorder (*wasyungu*)	Pneumonia (*kyambo*)
Cough (*kukooa*)	Chest pains (*kithui*)
Conjunctivitis (*uwau wa metho*)	Whooping cough (*mutitino*)
Rotation of fetus (*kusosolwa*)	Sorcery protection (use of amulet)

Note: From field survey (recall method), 1978.

prepared by a *mundu mue* from certain plants and used for ritual cleansing. In Kilungu certain plants with thick leaves are called *ng'ondu*. These plants are burned to create a powder or squeezed to extract juices—processes that transform them from mere herbs to substances with magical power. *Ng'ondu* is used, for example, by a *kuthambya* specialist, or *mundu wa nthambya* (Table 5-9), to treat a woman who has suffered a miscarriage. Juice from the plant is swallowed by the woman and this "washes" away the evil responsible for her misfortune.

Comparison of the illnesses actually presented to TMPs, the conditions that they believe they are most qualified to treat (Table 5-7), and the leading outpatient diagnoses at Nunguni and Kikoko (Table 4-3) suggests possible overlap in cases involving diarrheal diseases, "pneumonia," venereal diseases, and a few others. More striking is the tendency for TMPs to treat numerous complaints that, except for gastrointestinal complaints, are rarely if ever recorded *as such* at the biomedical clinics or provincial hospital (Table 4-4). It is, of course, impossible to know with any certainty how much or how little overlap and complementarity actually does exist because of inconsistencies in the data. At least three systems of diagnosis and classification are represented. Moreover, the data were recorded under entirely different conditions and represent different time periods.

Despite these shortcomings in disease reporting, the comparisons are instructive because they highlight fundamental issues of value, meaning, and communication within and between the lay and specialist groups in medically pluralistic societies. For example, what is an illness? What assumptions and expectations surround it? In how many ways is it described and labeled? What are the conceptual or structural limitations of biomedicine that obscure the recognition of significant illness when it is presented? What illness syndromes recognized in the traditional medical sector have epidemiological importance? What stressors are associated with abdominal pains, spirit possession, "peptic ulcer" (*kavoo*), "heart attack" (*ngoo*), edema (*mwimbo*), *nyamu*, *ngome*, and *thavu*—illnesses that are widely reported by TMPs and their patients but are totally absent or obscured by different diagnoses or labels in reports from biomedical institutions? And what proportion of such complaints are fundamentally mental disorders, which may explain why they fail to show up in the biomedical records?

The problems of accurate diagnosis and, hence, appropriate therapy are illustrated in a recent psychiatric study by Ndetei and Muhangi (1980). Over a period of 30 days they examined 140 outpatients at a clinic in semirural Athi River, a small manufacturing and commercial center situated between Nairobi and Machakos. Most of Athi River's population consists of low-income immigrants from other rural areas of Kenya, although about a fifth are pastoral Maasai and peasant Kamba families. Ndetei and Muhangi assert that their analysis points to a high prevalence of psychiatric disorders in rural Kenya. Similar findings in Ethiopia (Giel & van Luijk, 1969) and Uganda (Muhangi, 1970) add credibility to this view. At Athi River, 20% of the outpatients were diagnosed as primarily psychiatrically disabled, an unexpected 70% of whom were married; 59.3% were primarily physically ill. Diagnosis was uncertain for the rest (20.7%). None of the patients was psychotic. Joint pains and backache, abdominal discomfort, constipation, headache, pain in the iliac fossa, tinnitus, and eye ache were the commonest symptoms. Other complaints included tiredness, weakness, disturbed sleep, palpitations, impotence, and diarrhea. Duration of the symptoms was less than 3 months in 16% of the patients, 3–12 months in 30%, and more than a year in 54% of the patients. Many of the patients had previously been diagnosed and treated, with little effect, as cases of malaria, stomach ulcer, heart disease, slipped disc, or chronic headache. None complained of subjective symptoms such as fearfulness in cases of anxiety or sadness, or of guilt or nihilism in cases of depression. Direct questioning of the patients about such feelings also failed to produce evidence of their presence. Ndetei and Muhangi assert that such complaints would be improbable anyway since those concepts do not have medical implications in the local culture. They conclude that more forms of anxiety and depression exist in Africa than is commonly recognized, in part because in African patients those conditions are "masked by somatic symptoms" (Ndetei & Muhangi, 1980, p. 244). The failure of biomedical practitioners to understand this pattern often creates needless delays in diagnosis and management.

Many of the most common symptoms Ndetei and Muhangi (1980) found among the clinic patients at Athi River, such as headache and abdominal complaints, are also among those most often reported by TMPs in Kilungu (Tables 5-7 and 5-10). A high proportion are psychiatrically originated pains whose localization is strongly influenced by what Muhangi (1971) calls "organ hierarchy." In a large number of his patients suffering with anxiety and depression, the abdomen was the most frequent focus of distress. Illiterate, rural folk, he reasoned, tend to perceive their abdomen as the most important body part because of its basic function in nutrition and reproduction. Many cases involving these two systems were observed in the clinics of urban TMPs in Nairobi from 1977 to 1979.

This interpretation of an "organ hierarchy" is supported by Van Luijk's case studies and reviews, which indicate that pain in the digestive system is a major signal for Kamba to begin a search for diagnosis and treatment. Much time and money may be spent in the search for a cure, even for minor ailments (van Luijk, 1982). Lindblom (1969, p. 316) observed 65 years ago that Kamba "remedies for stomach affections seem most numerous of all." Similarly, there is always high anxiety among both men and women about real or imagined symptoms localized in the reproductive tract. These may lead to such feared consequences as sterility, impotence, and, worst of all, childlessness. Jacobs (1961) observed that the "maintenance of sexual vigor" is a major culture theme of the Kamba (cited in van Luijk, 1982).

Since the majority of biomedical clinicians have neither the training (e.g., in psychiatry and cultural analysis) nor the time (given the structure of biomedical procedure) to accurately identify and interpret symptoms presented by a large share of their patients, the propensity to misjudge or overlook such cases is of considerable consequence in actual patterns of therapy selection. Patients whose illness is not effectively dealt with at a biomedical facility, for instance, are thus motivated to consult other clinics and TMPs, and vice versa. As Janzen (1978), Mburu (1973), and others have shown, it is not necessary for the patient to have a clear grasp of the internal rationale and logic of an alternative therapy in order to utilize it. Biomedical workers and TMPs *both* see large numbers of patients who have been misdiagnosed or unsuccessfully treated at other biomedical facilities or by other TMPs. Some of these are cases where individuals may have been "cured" in a biomedical sense, yet remain anxious and feel compelled to consult a *mundu mue* to learn *why* they, or perhaps their child, have been singled out for a particular illness.

Although Janzen's (1978) analysis of therapy choices in the Zairian context emphasizes the central role of the therapy managing group, his evidence also implicates ineffective treatment in both the biomedical and traditional sectors as a major source of motivation for the joint or serial utilization of multiple therapies. In other words, biomedicine and the traditionally based therapies each stimulate the use of the other because they both have therapeutic and structural deficiencies that limit their efficacy and exclusive adoption. A recent study of psychiatric morbidity among 200 randomly selected patients at two hospitals in western Kenya, for example, found that up to 50% had consulted a TMP for the same illness *prior* to coming to the hospital (Acuda, 1983).

How, then, is efficacy to be determined, given an increasingly complex ethnomedical system made up of diverse therapies? As Feierman (1984, p. 28) correctly observes, it is not that African medicine lacks therapeutic efficacy. The difficulty is that each therapeutic tradition has a "legitimate claim to efficacy," and one cannot exclude the use of the others during an illness.

For some who become ill the search for therapy may be conducted indefinitely over a period of weeks, months, and even years. Specific cases, described in the following chapter, illustrate how people often adopt a peripatetic search for a cure. In cases of chronic and terminal illness, the sufferers or "significant others" who have a decision-making role in choosing therapy may also be desperate enough to consult unscrupulous town-based astrologers, "herbalists," and other money gougers in the "popular" medical sector.

Ndetei and Muhangi's study (1980) uncovered health needs and utilization patterns that have important implications for the general planning of health services in Kilungu, in rural Kenya generally, and elsewhere in Africa. Their findings also support a key argument of this book. First, at least 20% of the people who attend biomedical clinics are psychiatrically disabled, but the existing screening process is incapable of identifying them. Second, there is a demonstrable need to change this unnecessary and wasteful pattern through the provision of psychiatric services in rural and urban clinics. This need is particularly critical in view of the fact that untreated mental disorders tend to become chronic and result in searches and ministrations for physical illness that consume much time but do not improve the patient's condition. Third, it follows that "there is a need to extend psychiatric training and methods not only to doctors but to community nurses, health visitors, clinical officers, and other members of the health team" (Ndetei & Muhangi, 1980, p. 249). Collaboration with specialists in the traditional sector may also be considered in this connection. Finally, in contrast with the outpatient returns from Nunguni and Kikoko clinics, the symptomatic findings in the Athi River study not only have greater comparability with the information provided by the TMPs in Kilungu; they also highlight the strong similarity and frequency of symptoms manifested in two quite different medical arenas. This last observation suggests that the choice of therapy in the *initial* stages of many illnesses is determined more through an evaluation of the availability of and prior experience with alternative clinics and other TMPs than by the type of illness.

KEEPERS OF TRADITION: PROFILES OF TRADITIONAL MEDICAL PRACTITIONERS

Thus far it can be seen that the traditionally based medical system in Kilungu continues to be widely relied upon for the treatment and management of many illnesses, the provision of midwifery services, and as an element of social regulation. How much longer this continuity of "tradition" will remain recognizable as such is problematic. Rapid social change brought about by greater popular access to formal education, competing religions, and economic forces; the age of practicing TMPs; and certain internal structural weaknesses of traditional medicine *as* a system, particularly its unreliable methods for recruiting new TMPs and their lack of cohesiveness, undermine its integrity and cast shadows on its capacity to reproduce and survive.

Other, recent studies document a decline in the numbers of indigenous practitioners in *rural* Third World settings (Foster & Anderson, 1978). The evidence already cited suggests a similar reduction of TMP numbers in Kilungu.

In his recent study in northern Machakos District—also a Kamba area—van Luijk (1984, p. 304) found a "low frequency of use of traditional medical care." However, this finding was not conclusive because it was not feasible to record actual visits to diviners (*andu awe*). In this area of strong Christian influence, reports van Luijk (1984):

> a visit to the *mundu mue* is a secret activity that hardly anyone . . . would openly admit or discuss. It is too secret, too private, to be reported in a survey. Christians of all denominations are forbidden by church dogma and by the church elders to go to a *mundu mue*. . . . People go secretly at night. (p. 304)

At the same time, Kamba living in the eastern margins of the study area had been less affected by Christianity and modernization and their behavior was much more traditionally based.

The following profiles offer a closer appreciation of individual TMPs and their knowledge base in traditional medicine, highlight their continuity with a body of Kamba traditions, and suggest some possible directions of change.

Nzeki and Kutenga—Two Herbalists of Kilungu[13]

Nzeki and Kutenga are both senior citizens, 60 and 70 years of age, respectively, whose homes are about 3 kilometers apart in different sub-locations in the central hills of Kilungu. They are two of the region's most respected and popular herbalists (*akimi wa miti*) and have practiced their art for a combined total of 85 years. Kutenga has practiced as a herbalist in several other areas of Ukambani including Kangundo, Makindu, Nzaui, and Mukaa; Nzeki has no outside experience. Both are small-scale farmers and active in many other areas of community life. Nzeki also operates a general merchandise shop in a local trading center, is the chief of a sub-location with over 4,500 inhabitants, is chairman (*muthiani*) of his clan, and a member of the governing committees of two primary schools. Kutenga, a devout Christian and lay pastor at a Salvation Army Church, is also a local official of KANU, the national political party, in his sub-location and chairs the committee for a local primary school. Between them, Nzeki and Kutenga have 5 wives and 29 living and 5 deceased children.

Nzeki and Kutenga both practice herbalism exclusively. Unlike a diviner, their decision to take up this profession in the 1930s was a personal choice that was not influenced by special signs from the spirits (*aimu*) and did not require formal ordination. Both men claim that they had no apprenticeship or formal tutoring in *miti shamba* (medicinal plants). Their knowledge was implanted in them by God, although they allowed that one can also be trained by observing or working with other herbalists. When he was a teenager Nzeki did live with an uncle who was an herbalist, although he insists that he did not receive specific instruction from him. At first he treated only *kavoo/kavaso* ("peptic ulcer"), *wasyungu* (pancreatic disease), and *mutwe* (headache), but with "God's help" his healing extended to a broad range of illnesses.

After his conversion to Christianity in 1936, Kutenga practiced as a faith healer for 20 years among his fellow Christian converts. During this time he used no medicinal herbs, but relied instead on treatment of Christian believers through prayer. His decision to complement the prayers with herbal therapy came about suddenly during the 1950s (after the Mau Mau emergency) when he experimented with herbs in treating a man's leg wound. Having seen the wound heal encouraged Kutenga to use both herbs and prayer at the next opportunity, which was a case of "epilepsy." With that case he started on his career as a "pure herbalist" and began to treat nonbelievers as well.

Nzeki explained that different herbalists may use different *miti shamba* to treat the same illness, although no one would openly say that one treatment is more effective than another. Nzeki and Kutenga both agreed independently on the criteria that people use to informally rank herbalists. These include the number of illnesses one can successfully treat relative to other practitioners, the length of time it takes to demonstrate a cure, and the numbers of people who patronize the healer. A truly famous practitioner (Kutenga referred to Muia Kalii of Kilima Kiu as an example, see Chapter 4) might be called a *mundu mue munene*. Because some practitioners are more effective than others, jealousy is common and, as Nzeki said, "they do not love one another." Consequently, the TMPs tend to avoid contact with each other. They fear that someone else will possibly take advantage of their knowledge and diminish whatever personal reputation they have established. Even in the ordination of a new *mundu mue*, only one practicing *mundu mue* is called to lead the ceremony with the leaders of the individual's extended family. To call more than one would invite jealousy and disagreement.

Both herbalists normally treat roughly equal numbers of adults and children, including members of their own families, and make themselves available daily (Nzeki excludes Sunday) in the mornings and afternoons. They rarely make housecalls. Nzeki treats most of his patients in the back room of his shop at the trading center (about 3 kilometers from his home), and the rest at his house. Kutenga sees patients at his home and is often called out of his gardens by a family member to treat someone who has arrived at the compound. Nzeki and Kutenga draw many patients from Kalongo and Ndiani Sub-locations (Figure 4-1), but each also attracts substantial numbers from two or more areas in Kilungu that are not adjacent. Each herbalist had treated more than 20 patients during the week before the interview. My observations before and after confirm the validity of these numbers. The volume of patients they see apparently does not vary much throughout the year, except during the long rains of March to June and during July, which is the coldest month. The extent, characteristics, and reasons for this seasonal pattern remain uncertain. Limited evidence of seasonality from Nunguni clinic, for example, suggests a higher than normal incidence of acute respiratory infections and fevers during the drier, colder months. The only certainty is that the epidemiological study of illness in the traditional medical sector is virtually nonexistent in Kilungu and elsewhere in Africa.

In keeping with the custom of practicing in isolation, both herbalists indicated that they never refer patients to other TMPs. Nzeki, however, said he sometimes advises a patient with chest pains to go to the hospital for x-ray examination if he has failed to cure the illness. Evidence of such flexibility and willingness to use what works is widespread in Kenya and elsewhere in Africa. Janzen (1984) seems correct in his conclusion that African medicine has always been pragmatic, eclectic, and open to innovation in terms of consultation patterns, practitioners' specialties and tools, and in the utilization of multiple therapy traditions. In this sense, the philosophical basis of African medicine is only typical of most medical traditions because it is universalistic and thus "unexceptional." What makes African medical thinking unique, Janzen (1984) asserts, "is the exceptional degree of its openness." Nevertheless, although the *systems* are open most African TMPs are not and cannot be innovators, particularly when they practice in isolation from one another. There is a general receptivity to change, but this is coupled with a desire on the part of individual TMPs to protect whatever "competitive edge" they perceive themselves to have.

Neither herbalist kept written records about his patients, which would be unlikely since together they had only 7 years of formal education. Relying instead on powers of recall, Nzeki and Kutenga had treated a total of 13 different illnesses among the 40-plus patients who came to them in the week before the interview. These cases included infant gastroenteritis, leg ulcers, "heart attack," chest pain, swollen knee, continuous menstruation, "peptic ulcer," worms, swollen rib cage, fever, toothache, joint pains, and headache. Headache was the only similar complaint that both men treated during the period.

Nzeki's and Kutenga's perceptions of their own capabilities and their attitudes toward various illnesses and therapies are shown in Table 5-11. Their responses represent independent, but not comprehensive, answers to the same questions. There is no apparent overlap in the three illnesses they are "best qualified to treat." Since both men have lived and practiced relatively near to each other for several decades and each has a large clientele, it is plausible that they have evolved an informal division of labor that is complementary and mutually advantageous. This explanation also gains support from a comparison of specific therapies for 46 illnesses that Nzeki and Kutenga independently provided (Appendix A). Only 7 of the 46 illnesses are common to both herbalists' lists (Table 5-8).

The herbalists also see the specific strengths of traditional medicine quite differently; yet they do agree on biomedicine's effectiveness in treating "pneumonia" and fever. Their apparent difference of opinion concerning epilepsy may simply reflect the fact that Nzeki evidently lacks a treatment for this condition, whereas Kutenga claims to know a specific cure for it (Appendix A).

With "peptic ulcer" (sternum pain) the only exception, Nzeki and Kutenga use entirely different herbal remedies (*miti shamba*) to treat the *same* illnesses (Table 5-8). Although greater overlap could be expected in a comparison of a larger group of herbalists, the range of plant species used in therapy would also expand. This pattern of divergence among TMPs in their choice of the drug plants used in therapy illustrates their individualism. However, it does not suggest that the choice of plants occurs randomly or without systematic knowledge of their properties. Different plants can have common active principles of therapeutic value. Unfortunately, to date little progress has been made in the chemical analysis of Kenya's local drug plants in order to determine their pharmaceutical properties (Kokwaro, 1976).

Traditional medical practitioners' perceptions and beliefs about disease patterns constitute information that can be important to the development of appropriate community and public health policies in the biomedical sector. Table 5-12 summarizes the responses of Nzeki and Kutenga concerning the incidence and prevalence of diseases, including recently imported diseases. These are augmented and compared with the responses of another local, female TMP 45 years of age, a *mundu mue* who combines the roles of

Table 5-11. Herbalists' Perceptions, Attitudes, and Beliefs About the Treatment of Illness

| Questions | Herbalist (mukimi wa miti) | |
	Nzeki	Kutenga
Three illnesses you are "best qualified to treat"	*Kavoo/kavaso* ("peptic ulcer") *Nzoka* (worms) *Wasyungu* (pancreatic disease)	*Mwimbo* (edema) *Kuua* (continuous menstratuation) *Kyambo* ("pneumonia")
Illnesses you cannot treat or cure	*Kisala/mungathuko* (epilepsy)	Surgical conditions Fractures, bone injuries Boils
Illnesses traditional medicine treats most successfully	*Kavoo/kavaso* ("Peptic ulcer") *Wasyungu* (pancreatic disease) *Kwalwa ni mutwe* (headache)	*Uwau wa nguu* ("heart attack") *Mwimbo* (edema) *Kuua* (continuous menstratuation) *Kisala* (epilepsy)
Illnesses modern medicine treats most successfully	*Kyambo* ("pneumonia") *Ndetema* (fever) Tuberculosis	*Kyambo* ("pneumonia") *Ndetema* (fever) *Mutwe* (headache) *Nduuka* (madness)
Illnesses that neither traditional nor modern medicine can cure	*Uwau wa kwombosya* ("sleeping sickness") *Maitia nthi* (edema of the legs) *Nduuka* (madness)	A disease that has spread to many organs in the body

Note: From field survey, March and April, 1978.

160

Table 5-12. Traditional Medical Practitioners' Perceptions and Beliefs About Disease Patterns in Kilungu

	Traditional medical practitioner		
	Nzeki	*Kutenga*	*Mundu mue*
Which diseases, if any, are *endemic*?	"Asthma" Kavoo Fever	"Pneumonia"	Edema
Which diseases, if any, are *epidemic*?	Measles Chicken pox Uvee[a]	"Plague" Smallpox	Measles Whooping cough
Are there any *recently introduced diseases* that were previously unknown?	Gonorrhea Syphilis	Gonorrhea Syphilis Continuous menstruation Edema "Heart attack" Liver infection Nyunyi[b]	Gonorrhea Syphilis Continuous menstruation Edema

Note: From field survey, March and April 1978.
[a]Whole body itches, but particularly the hands and area above tailbone.
[b]Specific form of infant diarrhea, cause of which is a problem affecting the quality of the mother's breast milk.

diviner, herbalist, and midwife. With the exception of measles, the TMPs' perceptions of epidemic diseases are dissimilar. I believe that a larger cross-section and a more comprehensive, intensive survey of Kilungu's TMPs would demonstrate a greater correspondence of responses.

In discussing traditional approaches used to control or limit epidemic diseases the three TMPs gave remarkably similar answers. Their agreement is not surprising because the ritual procedures were communal. They were conducted by certain elders (*atumia*) who were also priests (*athembi*) whose primary role was to conduct sacrificial ceremonies, particularly in association with prayers for rain, at the tree shrines (*Mathembo*). When an epidemic broke out everyone in a village would collect small portions of food in their homes and deposit it in a calabash. A goat or chicken might also be contributed. All the villagers then assembled and were led by the *athembi*, who carried the calabashes to the village border. There the food was thrown across the boundary and the people would say that they had "escorted the disease" (*kumaasya mauwau*) out of their village. People in the next village followed the same procedure. A variant form of this ritual was described by the *mundu mue* included in Table 5-12. In addition to calling on everyone to collect food, the *athembi* would take a nest of a bird called *Nguni* (which like the chicken is a common cultural metaphor) and have all the villagers spit into it (a traditional form of blessing) to ensure success in driving out the disease.

Accompanied by the villagers, the *athembi* then carried the bird's nest and food to the village border and threw it outside.

Agreement among the TMPs concerning "previously unknown and recently introduced diseases" is striking, particularly in the case of venereal diseases (Table 5-12). These are invariably understood as sexually transmitted diseases. They are commonly referred to as *kavithe* (literally, "that which is hidden")—a euphemism that usually refers to gonorrhea (Oendo, 1982)—or generally as *uwau wa ukani/uumeni*, which means disease of the female/male sexual organ (van Luijk, 1982). *Uwau wa ukani* can also refer to "continuous menstruation" (*kuua*), a condition that two of the three TMPs believe is also of recent origin (Table 5-12). Van Luijk (1982) concluded from his study in northern Machakos that only young and very old men feel free to talk openly about diseases such as gonorrhea (*kisonono*).

As to the local origin of gonorrhea, a very old female informant explained to van Luijk (1982, p. 8) that the Europeans who first visited Ukambani took some prostitutes away with them to foreign countries where they got *kisonono* through sexual relations with dogs. When the women returned to Ukambani they infected Kamba men. Nzeki told me that both gonorrhea and syphilis began in the towns of Kenya and then spread to the rural areas. These diseases started to be seen in Kilungu, he said, "around 1929." This perception is compatible with the documentary evidence of widespread venereal disease before and during World War I cited in the preceding chapter. The problem was greatest near government stations along the Uganda Railway, which today forms the southern boundary of Machakos District for more than 130 kilometers.

Edema (*mwimbo*), the swelling of the body due to excess fluid concentration, was also viewed as a recent import by two of the three TMPs whose responses appear in Table 5-12. It affects young children (it is a major sign of kwashiorkor) as well as old people and is one of the illnesses that TMPs in Kilungu think they are best qualified to treat (Table 5-7). *Mwimbo* is commonly attributed to *aimu* or sorcery, but Nzeki and Kutenga both described their own remedies for it without reference to possible supernatural aspects. Although data are at best scanty and circumstantial, it is unlikely that *mwimbo* was actually a new disease in Kilungu and Ukambani during the colonial era. The TMPs' perception of the problem possibly reflects an increase in cases of kwashiorkor they see among very young children, who because of changes in weaning and feeding customs are often at great risk of developing edema due to protein malnutrition.

Kethi and Mutheu: Religious-Medical Specialists of Kilungu

Kethi, 35, and Mutheu, 45, are among the youngest female TMPs in the Kilungu study area. Small-scale cultivators, they live in adjacent villages which are about 300 meters apart on opposite sides of a hill in Kalongo Sub-location. Kethi is a diviner who also treats with herbal medicines (*mue wa kwausya na kuiita*). She had been practicing for 6 years when we met her. Kethi has lived in the same village all her life and never attended school. Her husband was also born in this village. He is a TMP who specializes in the treatment of possession, including the mild possession (*kithio*) that affects some people during *kilumi* dances, and madness (*nduuka ya maimu*). Kethi has produced nine living and four deceased children, and has no cowives. She attends the local Catholic church on Sundays with her children.

Mutheu is a specialist in spirit-caused illnesses, uses medicinal plants in some of her therapies, and is also a midwife (*mwisikya*). In 1978 she had practiced as a TMP for 12 years. Mutheu was born in a neighboring sub-location and, like Kethi, has no formal education. She is a widow with five living and five deceased children (the total child mortality rate for Mutheu and Kethi is 39%) and is also a member of the Catholic Church. Two of her five living children had attended primary school for 4 years, two were still in primary school, and one had completed secondary school.

Although there are other *andu awe* in the extended families of both Kethi and Mutheu, neither of these women became a traditional religious-medical specialist by personal preference. Their cases thus stand in sharp contrast to Nzeki and Kutenga, who chose to become herbalists on their own accounts and could, in theory, retire without fear of retribution or offending the spirits (*aimu*). The stories of the "recruitment" of Kethi and Mutheu to the role of *mundu mue* reflect a continuity of traditional Kamba religious ideology that is validated through community participation. Although personalized in their details, these stories, told below, are remarkably similar in their content, patterning, and outcome to dozens of other accounts of the professional calling of Kamba *andu awe* recorded in Kilungu and Nairobi. They are, in effect, representative "cultural scenes."

Mutheu related her experience as follows:

> After I married, I gave birth and the child died. My second, third, fourth, and fifth children also died. My husband consulted a *mundu mue* to know what was wrong with me, and was told that his wife has a call as a *mundu mue*. After some time a *kilumi* dance was held in our village. A large group of women gathered for this and danced through the night to please the *aimu* (spirits).
>
> The day after the *kilumi* dance I treated my first patient, who was possessed (*kathambi*). I went to the sick woman's home and played the *kithembe* drum until the spirits finally left her. I started treating possessed people from then on. Two or three weeks later I was ordained as a *mundu mue* in a ceremony (*mboka ya mukono*) conducted at my house by other *andu awe*. There was a chief *mundu mue* who instructed the others in attendance about what they should do. Ordinary people also came to witness the ceremony. Maize and beans, and chicken, were cooked and there was bread, tea, and beer. A goat was killed and roasted. People gathered to eat the food. The old men were served the beer.

In contrast to most accounts of persons "called" to the role of *mundu mue*, Mutheu said she received two months of special instruction from a female TMP who is a family friend. Kethi also admitted that she had received some training from her mother, who was an active TMP. Kethi recalled that she became a *mundu mue* in 1972.

> I was made very sick by the spirits. The sickness was not serious at first but after some time it became worse. This forced my family to seek help from the nearby Mission Hospital. I was treated there but showed no improvement. I did not know what the sickness was except that it felt like *kyambo* ("pneumonia") underneath both my breasts and brought much pain. The pain increased and then I would be unconscious. The people from the Mission Hospital also came to treat me at home. When the hospital people were completely defeated my family thought of seeking help from somewhere else.

> When I was sick I was doing extraordinary deeds. I could drink a tank full of water within a very short time. I could ask for porridge (*uki*) and drink three or four very big gourds of it. With these funny behaviors I was bothering my husband and the neighbors because they were forced to draw much water and also to buy a lot of flour for preparing porridge. All of this required money and they had little.

> After the hospital treatment failed my husband and other family members decided to take me to a *mundu mue*. This person was my 'mother'. When she saw me she said that the spirits were troubling me because I had not responded to the call to become a *mundu mue*. If only I would agree the sickness would disappear. After I accepted the call, an ordination ceremony was arranged, after which I became well again.

The similar experiences related by Mutheu and Kethi are typical for female Kamba *andu awe*. Both were affected after they had married—Mutheu by the calamitous misfortune of successive deaths of her children, and Kethi by a chronic illness (possibly part of a trauma associated with her husband's death) against which hospital treatment had no apparent success. Both cases also highlight the role of "significant others" in decisions about therapeutic strategy (Janzen's [1978] "therapy managing group"). Validation of the calling by another *mundu mue* and legitimization by means of the public ordination ceremony are the primary ceremonial acts of recruitment which remain essential to the integrity of the office. In Mutheu's case, the holding of a *kilumi* dance illustrates another appropriate (but not obligatory) element of illness therapy. In one form *kilumi* is a seasonal dance for married people held to celebrate the ripening of crops. However, *kilumi* is also widely used as a form of positive psychotherapy for a person who manifests some form of delirium and, according to a *mundu mue*, is possessed by demon spirits (Ndeti, 1972). Relatives and close friends join the patient in an energetic, even violent dance form which is characterized by shoulder-spinning, singing, and tactile stimulation designed to exorcise and propitiate the offending spirits. It is conducted through the night by a leader (*ngui ya kilumi*) whose singing is accompanied by the rhythmic beats of a special drum played by a *mukuni wa kilumi* (Tables 5-5 and 5-6).

Kilumi is the Kamba expression of the *ngoma* possession (dance) therapies which are an ancient and widespread healing device among the Bantu-
speaking societies of central and southern Africa. A recent assessment of
contemporary *ngomas* in these regions suggests that this therapeutic tradition
is highly adaptable to changing epidemiological, social, and political circumstances, not least in the context of the personal and social disintegration that
are so much a part of life today in urban Africa (Janzen, 1984, pp. 26–27). Is
ngoma therapy effective psychotherapy? Janzen (1984, p. 26) argues that for
afflicted individuals who experience *ngoma* therapy, as in the *kilumi* dance
held in Kilungu for Mutheu, "there is an obvious build up and release of
emotional energy; whether through the taking of drugs, or the commence-
ment of cortical driving (brain wave alteration), and intensive autosuggestion,
an environment conducive to psychic relearning is set up."

To function as *andu awe*, Kethi and Mutheu required specific items of
equipment. Since Kethi is a diviner she customarily received a magic bow
(*ota*) and a half gourd from a male elder in her clan, in this instance her
father. Another magic bow and half gourd, and a gourd (*ketiti*) which holds
divining "marbles," or counters (*mboo*), were given by her husband. Spirits
brought the "marbles" at night. Sometimes Kethi found them on the bed or
elsewhere in the house; at other times the spirits placed them in her hands
while she was asleep. Because Mutheu does not do divination, she needed
only two medicine bags (*nthungi*), which she made herself; and two pieces of
multicolored cloth (*suka*)—one worn around bosom and the other around
the waist—, a drum (*kithembe*), a cane (*ndata*), and a cup, all of which the
other *andu awe* gave her during the ordination ceremony.

Kethi and Mutheu indicated that they treat equal numbers of adults and
children, including free treatment for family members. Most of their clientele
came from their own and nearby sub-locations. As a rule patients come to the
practitioner's residence for treatment, but both TMPs also made house calls
and were available daily at all hours. Fees varied with the treatment required,
including any plants used, and ability to pay. Installment payments were
common, as elsewhere in Africa, and sometimes included noncash goods
such as a goat, chicken, maize, or beans. Both women had also practiced in a
few other areas in the Kilungu Hills.

In the week before we visited her at home, Kethi had treated five
patients (all children): two for gastroenteritis (*nyunyi*) and one each for
pneumonia (*kyambo*), worms (*nzoka*), and "peptic ulcer" (*kavaso*). Mutheu
had treated two adults for madness (*nduuka*) and spirit possession. Although
their combined patient load in this instance was less than one-fifth that of the
herbalists, the time frame is too limited to indicate a pattern of morbidity.

Kethi and Mutheu's sense of their own proficiency and their attitudes
toward various illnesses and therapies are shown in Table 5-13. Both women
treated a fairly wide range of illnesses in addition to those they felt "best
qualified" to manage. With a few exceptions such as *nyunyi*, pneumonia, and

Kamba Culture and Traditional Medicine in Kilungu

Table 5-13. *Andu Awe* Perceptions, Attitudes, and Beliefs About the Treatment of Illness

	Mundu mue	
Questions	Kethi	Mutheu
Three illnesses you are "best qualified to treat"	*Kutavika* (vomiting) *Nyunyi* (infant gastro- nteritis)	*Nduuka* (madness) *Maimu* (spirit possession) *Nyunyi* (infant gastro- enteritis)
Illnesses you cannot treat or cure	N/A	*Mung'athuko* (epilepsy) *Mwimbo* (edema) *Kisonono* (gonorrhea) *Teko* (syphilis)
Illnesses traditional medicine treats most successfully	*Maimu* (spirit possession) Illness due to witchcraft	Mwimbo (edema) *Kwalwa ni ivu* (stomach ache) *Kuua* (continuous men- struation)
Illnesses modern medicine treats most successfully	*Nduuka* (madness) Tuberculosis	*Nduuka* (madness) *Kisonono* (gonorrhea) *Teko* (syphilis)
Illnesses that neither tradi- tional nor modern medicine can cure	All can be cured unless God wants to take the sick person	N/A

Note: From field survey, March 1978.
Abbreviations: N/A, no answer.

stomach ache, these two neighbors also appeared to have an informal division of labor. Whether conscious or not, this complementarity is probably convenient for people who depend on their services. On one occasion I observed Kethi for several hours while she treated a very sick infant for gastroenteritis (*nyunyi*) and examined a toddler brought by its mother for a post-*nyunyi* checkup. Kethi's responses (Table 5-13) indicate self-confidence in her own ability to treat several common pediatric symptoms. On another day we watched as she conducted an elaborate divination session for a middle-aged man who had come from a distance on the initiative and in the company of his elderly mother. The mother was clearly in charge of the case and wanted Kethi to divine why her son had not only lost out to a junior man in his bid for promotion at his place of work, but had been sacked and replaced by this same man—a friend whose initial employment was apparently made possible through a kindness of her son. According to Kethi, the divination revealed the following pattern of events:

The junior man had visited a *mundu mue* in order to interfere with the older man's job performance and gain advantage. This *mundu mue* had seen to it that a foul-smelling substance was magically smeared on the unfortunate man's face and body (unbeknownst to him), and this made him unpopular at work. This is why he was sacked.

Mutheu perceived spirit-caused illnesses of adults (one of which afflicted her before she acceded to the role of *mundu mue*) and digestive diseases of infants as among her greatest therapeutic strengths (Table 5-13). *Nyunyi* refers to a specific pediatric syndrome whose symptoms include a sunken, throbbing fontanelle (*lolotya*), headache, greenish stools, and a white secretion at the corners of the child's eyes. Both Mutheu and Kethi frequently treated infants diagnosed as having *nyunyi*, but their cures are radically different. Mutheu's remedy is powdered soot scraped from the outside of a cooking pot and ash from the fireplace. These ingredients are mixed in water and then spooned into the infant's mouth. Apart from the possibility of aiding recovery by replacement of fluids, the biochemical and therapeutic effects of this treatment remain unknown. Positive results cannot be discounted.[14] In contrast, Kethi mixes juices squeezed from the leaves of *muvou* and the stems of *musoka* with powders made from the crushed stems of *kimala* and the roots of *kithulu* and *muuku*. Sugar is added to this concoction, which is then mixed with honey beer (*ukiwanzuki*) in a tiny calabash and orally administered.

A comparison of the responses of the female *andu awe* (Tables 5-13 and 5-14) and the male herbalists (Tables 5-8 and 5-11) reveals a pattern of discrete primary therapies and a disassociation from spirit-related illnesses by the herbalists.[15] Surprisingly, three of the four TMPs indicated that biomedicine is more successful than traditional therapy in the treatment of mental illness. Biomedicine was also perceived as having more potent cures for tuberculosis and other infectious diseases and symptoms such as pneumonia, fever, and genitourinary infections. None of the conditions against which traditional medicine is seen as the more effective therapy are communicable. Considerable disagreement about cures for epilepsy is apparent, but this mirrors the widespread fear of and ambivalence toward this disease—along with leprosy and tuberculosis—in African societies (Orley, 1970).

Kethi and Mutheu do not treat as broad a range of illnesses as the herbalists. However, the illnesses they reportedly saw most frequently in their work fairly characterized an important part of the overall pattern of morbidity in their local communities. In addition to the three conditions they each felt most competent to cure, Kethi and Mutheu both cited "pneumonia," stomach ache, vomiting, and diarrhea among the "commonest illnesses" their patients present. The other common illnesses Kethi treats are *kavaso/kavoo* ("peptic ulcer"), pain all over the body (*kwalwa ni mwii*), and spirit posses-

Table 5-14. Selected Illnesses and Their Cures According to *Andu Awe*

Illness	*Andu awe*	
	Kethi	Mutheu
"Pneumonia" (*kyambo*)	Prepare a powder from *mu-thika* and *mukolekya* plants and mix with honey beer.	Make cuttings on patient's chest and apply the blackish substance from a battery cell. Juices squeezed from leaves of *muvou* and root of *kyuvi* are given orally.
Stomach ache (*kwalwa ni ivu*)	Wash outside of stomach with soap and water.	Juices from root and leaves of *muvou* are given orally.
Vomiting (*kutavika*)	Extracts of four plants— *muvou, musoka, kimela*, and *kithulu* are mixed with sugar. Given orally to stop the large "worm" in the stomach that causes the upset.	Juice squeezed from *kamusuu* root is given orally.
Diarrhea (*kwituua*)	Same mixture as for vomiting	Same treatment as in vomiting

Note: From field survey, Kilungu, March 1978.

sion. Mutheu added *nyamu* (hard swelling on head or body), boil, fever and chest pains to the list of her patients' recurrent ailments. Neither TMP treated edema, epilepsy, gonorrhea and syphilis, or fractures. Kethi treated infertility when it was not "permanent" and if it was caused by the spirits (*aimu*)—as opposed to witchcraft.

Despite their close physical proximity to each other, their specific cures for these four illnesses illustrate quite remarkably how Kethi and Mutheu functioned in professional isolation from one another (Table 5-14). Indeed, it appears that the concept of standardization of therapy is unknown, and may be consciously avoided by trading on novelty to build and maintain one's reputation. Certainly, not every remedy will be efficacious, and as Table 5-14 suggests, some may be more harmful than neutral. For example, the use of zinc paste or carbon from a "cell" (battery) appears to be one of Mutheu's therapeutic trademarks. In addition to treating pneumonia, she also said she can produce a local anesthesia for pain relief by making cuttings on the skin around the affected area and applying "that blackish substance found in a cell." It is possible that this material is sufficiently caustic to reduce the pain sensation by killing off nerve endings. Moreover, zinc is known to be of value in promoting wound healing if the patient has a prior existing state of zinc deficit.[16] Whyte (1982) describes an application of battery acid in eastern Uganda, where it is an antidote to some of the most powerful types of sorcery.

Snakebite is the only recorded condition for which Kethi and Mutheu employ the same treatment. The method is to "kill a snake, cut the head off, dry the head and then crush it into powder form. This powder is then applied on the wound." This procedure, which suggests homeopathic thinking, differs from those recorded in Lindblom (1969) and Ndeti (1972), the two major sources on Kamba culture. In both latter accounts, the venom is sucked out to prevent its circulation in the blood. Imperato (1977) describes a wide variety of snakebite treatments in his African survey, but none resemble the Kilungu practice.

DISCUSSION

These profiles of *andu awe* and herbalists are representative of TMPs and their therapeutic and social roles in the Kilungu Hills. The local populace depends on them for help with numerous acute symptoms, such as gastroenteritis and fever (naturally caused diseases *are* widely recognized), as well as for the alleviation of chronic physical and mental illness and misfortune. It was not possible to discover, quantitatively, the extent to which TMPs can appropriately and verifiably claim credit for relief of suffering. Nor is it possible to know how much credit their patients' give them for neutralizing the symptoms of illnesses that are typically self-limiting.[17] Furthermore, I can suggest, but not verify, that TMPs (i.e., those *awe* who claim to manipulate supernatural power) who assist some patients at certain times to cope with stress and anxiety in a positive reinforcing manner may also inadvertently promote or find it in their own personal interest to perpetuate the society's often counterproductive beliefs in witchcraft and sorcery. Thomas (1975, p. 278) is probably correct in his argument that the curing powers of Kamba "diviners" (i.e., those among them who consciously manipulate people's jealousies, fears of revenge, and other insecurities) "cannot transcend the very pathologies which they perpetuate—the pathologies associated with the obscurantism of sorcery and ritual impurity." This raises a legitimate question concerning the roles they should be encouraged to play in a developing society.

The Kilungu evidence offers little support for the theory that TMPs are in business because healing for profit is more lucrative than farming or some other rural enterprise. Thomas (1975) asserts that diviners, in particular, have an "exalted economic status" in Ukambani which they occupy together with local government officials and "illegal" ex-government health practitioners (*kitali*). The latter continue to maintain their connections to dispensary and health center drug supplies and operate a private medical business on a fee-for-service basis. While a few of these *kitali* are scattered around the location (they were not part of the census), and may do well financially, few if any local *andu awe* stand out as being relatively more prosperous than their neighbors. However, a few successful herbalists such as Nzeki, who is also a farmer, local government official, and shopkeeper, probably do approximate the rural equivalent of a "petty bourgeoisie" class.

The pattern and rate of change in the traditional medical system of Kilungu is conditioned by several factors, chief among which are the growing acceptance of biomedicine in its various formal and informal guises, and the slow but accelerating changes in educational level and world view. The psychological needs of a rural society under stress are also an important factor in the rate and direction of change in traditional medicine. Rural areas such as Kilungu are not bucolic counterimages of modern city life. While the contexts and symbols are indeed changing, there is no escape from the reality that Third World village life (especially when it is understood as part of a single rural-urban system as evidenced by mobility patterns) is characterized by many stresses, strains, and terrors ranging from vengeful spirits, witches, and sorcerers to jealous kin and angry neighbors (Worsley, 1982). Consequently, people continue to seek out familiar, customary sources of help.

Despite these factors that seem to ensure the continued vitality of traditional medicine in Kilungu, the demographic profile of local TMPs and their apparent inadequate means of social reproduction could lead to the system's atrophy. The most immediate and direct impact of the lack of TMP replacement is to limit peoples' therapeutic options. The average age of all TMPs in the survey ($N = 33$) is about 60. Several in their early to late 70s were infirm and had few remaining years, and a small number were physically disabled by blindness or lameness. By custom there is no formal recruitment or training of apprentices in progress. A number of *andu awe* said that none of their children had as yet manifested the necessary "signs" or omens of supernatural ability (*ue*). More typical of local perception is the response of three old women in Kalongo who were interviewed together. Close friends and neighbors, they are, respectively, a traditional midwife, herbalist, and diviner. I asked them: "Since so many of the TMP's are elderly, and they are not training younger people to take over, what will happen to Kikamba medicine?" After discussing this for a few minutes, one of them offered their consensus. She explained that:

> Many of our sons and daughters are Christians; therefore it is difficult for them to take over. *Ue* is a gift from *ngai* (God) and in order for anyone to practice as a diviner it is necessary that they be "chosen" and granted the power. Herbalists are more compatible with Christianity than diviners.

This observation highlights several aspects of the social reproduction, or sustainability, problem in the traditional sector of the local ethnomedical system. Skills of the midwife and, to a lesser extent, the herbalist are likely to be preserved indefinitely. Both have strong survival value, particularly since official biomedical alternatives are so scarce. Many plant medicines and midwifery techniques constitute folk knowledge and continue to be learned even if they are not formally taught. Even young children can often identify specific medicinal plants growing near the homestead. Unfortunately, the proportion of the local populace that can make effective use of the traditional pharmacopia is apparently diminishing. In general, this sort of knowledge is becoming esoteric.

In the medicoreligious realm, to qualify as a bonafide *mundu mue* a woman or man must still experience the signs—usually a serious illness or prolonged misfortune—of one's God-given supernatural power. These are interpreted and acted upon in a culturally prescribed manner, including an ordination ceremony. Today, many individuals who experience what a cultural conservative would see as the signs of *ue* may prefer to take their chances and avoid interpreting them in the customary fashion. In its authentic form the institution of *mundu mue* is analogous to the priesthood. In the rural setting it can impose constraints on one's life-style, including freedom of association.

Alternatively, there is the option of adopting the TMP role as one's primary source of livelihood. Younger, mobile, adaptive, and innovative *andu awe* may find a lucrative niche in the highly competitive world of "urban" traditional medicine in such places as Nairobi. We met one Kilungu man in his 20s, whom I shall call "Joseph," who was operating as a full-fledged *mundu mue* of the "general practitioner" variety. He maintained his rural home base and family at his farm in Musalala Sub-location but worked as an assistant in the clinic of a private Sudanese doctor in the Eastleigh Estates section of Nairobi. After work he would return to his small, rented room and supplement his income as a *mganga* (Swahili for TMP). He gave me a detailed list of over 50 maladies that he claimed to treat. These ranged from fever, venereal diseases, and epilepsy to impotence, bewitchment, and the adjustment of a menstrual cycle, for example, from twice per month to once per month (*kuvinda nthakame ya mundu muka ethiwa yithiawa mwei keli ikoka savali imwe*). Ingredients of 44 of Joseph's remedies ranged from plant extracts (part of almost all preparations), the black paste from a flashlight battery, and cattle manure (as a hot compress on a dislocated limb) to *chang'aa* (illicit distilled spirits known in Nairobi as "kill me quick"), lion's fat, a glucose–salt–water mixture (to stop vomiting), and penicillin. The treatments revealed a remarkable adaptation and blending of traditional and biomedical procedures and materials.

We met no one else like Joseph who was practicing in Kilungu. The younger *andu awe* such as Kethi and Mutheu were uneducated and female, with children and other domestic responsibilities. There were also very few young *andu awe*. The evidence thus points to a steady decline of new "recruits" for the religious-medical class of specialists within Kilungu, as well as a slower loss of herbalists. Given the skeletal nature and inadequacy of existing biomedical services in the location, and the high fertility levels, resources for individual and community health are probably diminishing in both absolute and relative terms. Some of this growing disparity may be mitigated if unlicensed private practitioners of the biomedical variety are simultaneously expanding their numbers and geographical coverage.

The deteriorating position of the principal health services in Kilungu is directly connected to one of the research aims of this study. This is to explore the feasibility of formal cooperation between TMPs and biomedical practitioners as a means to improve and expand health care. For example, could the existing shortage of (biomedical) primary care providers in rural areas be partially alleviated by a redefinition of the "health team" to include those TMPs who have their communities' confidence? Would TMPs in Kilungu be receptive to such an initiative? To gain initial insight into these issues I asked the four TMPs discussed earlier in this chapter a series of questions. Their reactions are shown in Table 5-15. A much larger group of urban TMPs who practice in Nairobi responded to the same questions (Good & Kimani, 1980). The perceptions and understandings of the Nairobi TMPs are assessed in Chapter 8.

Responses of the Kilungu TMPs indicate a receptiveness to act on opportunities for cooperation with the biomedical sector. Few have had direct contact with biomedical staff. All indicate a respect for the strengths of biomedicine and readiness to learn, if not to teach. To be successful on any scale, collaborative activities would necessarily have to be designed to ensure a two-way process of observation and demonstration, teaching, and learning.

Findings concerning the attitudes and practices of both the rural and urban TMPs indicate that cooperation between TMPs and biomedical workers is possible and that it could be of considerable public benefit. These policy issues are examined more fully in Chapter 10.

NOTES

1. Personal communication, Kikoko Mission Hospital, March 23, 1978. The clinical officer was generally not favorably disposed toward TMPs, but was not hostile either. He reported that some herbal medicines are effective against diseases such as malaria (he referred to a bitter herb "similar to quinine"), and observed that belief in witchcraft (and its resultant psychosomatic disorders) are unlikely to disappear soon.

2. "Client" is used interchangeably with "patient" although one usage is often preferable to the other in a given context. Someone who visits a *mundu mue* for divination in connection with a business venture or love affair, or to obtain an amulet, is appropriately a client. Another who has a fever, abdominal distress, or chest pains is properly a patient in the biomedical sense.

3. *Ngai* is apparently linked to *Engai*, the Maasai word for God, and is a 20th century form used in Christian circles and in the Kikamba Bible (Mbiti, 1971).

4. *Cannabis sativa L. Mbangi* is a Swahili term for marijuana.

Table 5-15. Attitudes and Practices of Kilungu Traditional Medicine Practitioners Related to Biomedicine

| | TMP responses | | | | |
| | Herbalists | | | | |
Issues	Nzeki	Kutenga	Kethi	Mutheu	
Under what conditions would you refer a patient of yours to a clinic or hospital?	N/A	If surgery is needed; if patient does not show improvement after treatment. Some people who respond to biomedicine don't respond well to traditional medicine, and vice versa.	"Tuberculosis" and madness (*nduuka*)	N/A	
Where would you refer them?	Hospital, or in case of epilepsy to (named specialist) in Kavata Nzou, Musalala	Machakos Provincial Hospital	Wherever patient wants to go	N/A	
Do you believe there are areas of medical practice that require collaboration of TMP's and biomedical workers? If "yes," identify the areas you have in mind.	Yes. Bring TMPs and biomedical workers (*kitali*) together in several locations so they can work cooperatively. Doctors would treat fractures and patients could be exchanged. If one in hospital is not improving, he can be sent to TMPs, and vice versa.	Yes	Yes. Because there are illnesses that are treated better by herbal medicine and others by biomedicine. There should be cooperation so that a cure can be achieved easily.	Yes. Because there are illnesses treated most successfully by biomedicine, and others by traditional medicine. Thus there should be collaboration between the two sides.	

Table 5-15. Continued

		TMP responses		
		Herbalists		
Issues	Nzeki	Kutenga	Kethi	Mutheu
How could the government promote cooperation between TMPs and bio-medical workers?	See above	Provide TMPs with licences so that the public knows what they can do. Traditional medical practitioners should be evaluated and tested before issued with licences. They should be brought together, get organized, and practice in a specific place.	N/A	If the TMPs are gathered in one place with their own hospital next to the government hospital it would be very easy to control both groups.
If the government offered a course to assist TMPs to increase their skills, would you enroll?	Yes	Yes	Yes	Yes

If sponsored, what specific knowledge and skills would you want the course to provide?	Learn to treat difficult diseases, e.g., epilepsy. Learn more about causes of diseases so he can "warn" the people. Learn method to protect his medicines so they will not lose potency quickly. He wants to demonstrate to doctors his methods of treatment so that if he is mistaken they can correct him.	Alternative medicines for a given illness when his usual treatment fails. How to preserve his plant medicines so they don't get spoiled by bacteria.	N/A	How to store her medicines so they do not get spoiled quickly. Financial assistance to build a good house in which to treat her patients. Good containers for storing medicines. The training course should be short and near her home because she is a widow and cannot leave her children alone.
Have you ever had professional contact with government or private biomedical workers?	No	Yes. The sister-in-charge at Kikoko Mission Hospital (now deceased) referred patients to him, and vice versa. There is also a private practitioner in Kalongo Market to whom he can occasionally refer patients (former officer-in-charge at a local health subcenter).	No	N/A

Note: From field survey, March–April 1978.
Abbreviations: N/A, not applicable; TMP, traditional medical practitioner.

175

5. International Classification of Diseases. Geneva: World Health Organization, 1978 (9th revision).

6. See Ndeti (1972), p. 117.

7. A total of 65 TMPs (all categories) from four sub-locations were included in the census. Of these, I interviewed 33, 26 of whom indicated that they "regularly use *miti shamba*" in their work.

8. Van Luijk (1982) refers to *wasyungu* as "a disease of the spleen."

9. Kyunuu Kamitu, March 28, 1978.

10. I gathered census data over a period of several weeks by means of personal interviews and home visits, and supplementary registers of TMPs compiled by the sub-location chiefs.

11. This contrasts with Lindblom's (1969, p. 257) much earlier observation that there are "female medicine 'men,' but they are more rare."

12. I visited most of the 37 TMPs at home at least once. The remainder could not be located precisely below the sub-location level because of time constraints, including time needed to send and receive messages concerning appointments with individual TMPs and difficulties of access in the mountainous terrain.

13. The names of individual TMPs have been changed to protect their privacy.

14. Charcoal, the basic substance of soot, absorbs many different substances, including some poisons. Ingested in the manner described, it could tie up various trace minerals such as iron, copper, and zinc and thus deny adequate supplies of growth nutrients to bacteria in the infant's gut. The charcoal may also have a capacity to interact with and quiet an inflamed gut wall and thereby restore recuperative reaction.

15. Only those diseases that the TMPs thought of themselves, without prompting or suggestion, are shown in Tables 5-11, 5-12, and 5-13. Since they did not have an opportunity to revise or add to their answers, the responses may be incomplete.

16. Personal communication, G. E. Bunce, Department of Biochemistry and Nutrition, Virginia Polytechnic Institute and State University.

17. To obtain a more comprehensive picture of patients, TMPs, and biomedical services in Kilungu it will be necessary to systematically gather and evaluate the perceptions and behavior of the help-seekers. It is equally important and even more difficult to design flexible but consistent procedures for assessing the outcome of therapies received. Both objectives require major additional investments of time, expertise, and financial resources that were unavailable to the present study.

Plate 1. Kamba homesteads, Kilungu Hills.

Plate 2. Female filter clinic at Kenyatta National Hospital, Nairobi.

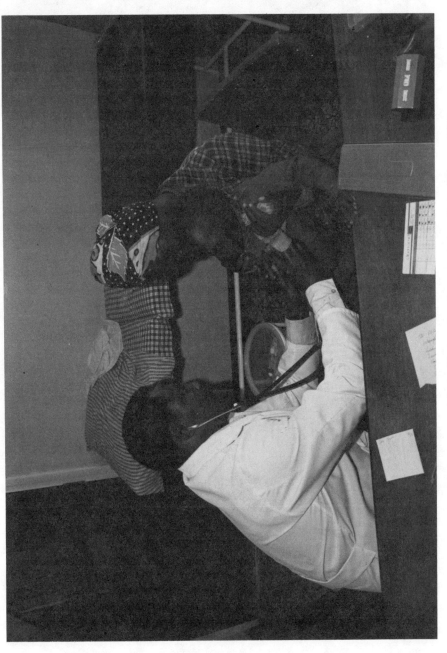

Plate 3. Government clinical officer and patient at mission clinic, Kilungu Hills.

Plate 4. Mathare I, an urban "village" in Nairobi's Mathare Valley.

Plate 5. Sign in Swahili advertises illnesses and personal problems an *mganga* will undertake to cure. Mathare Valley, Nairobi.

181

Plate 6. Digging for *mwaiitha*, a medicinal root, in Kitui District, Ukambani.

Plate 7. Medicines, impala horns, and other ritual paraphernalia in clinic of a Kamba *mganga* at her residence in Ngara, Nairobi.

Plate 8. Herbalist (*mukimi wa miti*) holding *muti* (Aspilla spp.), When crushed, leaves produce a liquid that is applied to the eyelids to treat conjunctivitis, Kilungu Hills.

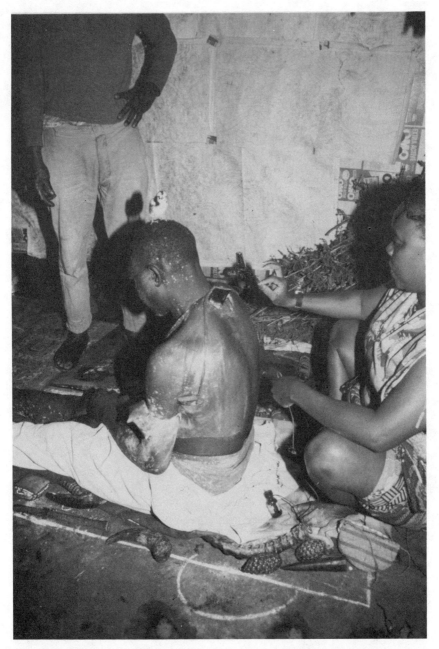

Plate 9. Patient undergoing therapy for chronic chest complaint, enlarged liver, and general malaise, Mathare Valley.

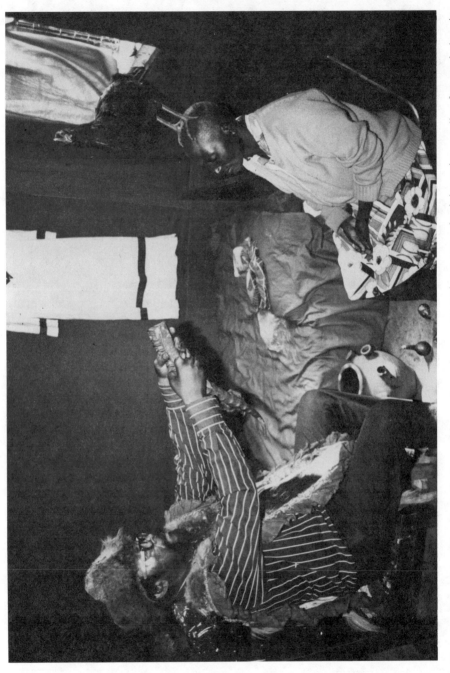

Plate 10. Kikuyu traditional medical practitioners (*mundu mugo*) treating woman for joint pains, swollen feet, and chronic malaise, Murang'a District.

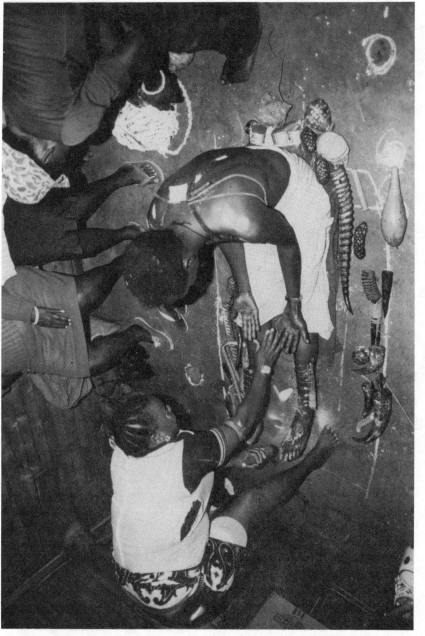

Plate 11. Kamba *mganga* treats young man for impotence, Mathare Valley.

Plate 12. Ritual therapy in Mathare for rural man with malaise attributed to family feud.

Plate 13. Kamba herbalist treating boy for intestinal worms, Kilungu Hills.

Plate 14. Woman undergoing ritual "washing" as part of infertility therapy, Nairobi.

6

Traditional Medicine in Three Urban "Villages": Traditional Medical Practitioners in Nairobi's Mathare Valley

Residents of Africa's burgeoning cities live in close physical proximity to indigenous healers and counselors whose specialized services reflect a continuity with core features of the traditionally based therapy systems. Ibadan's Yoruba *babalawo*, the *marabouts* of Bamako and Timbuctoo, and Kampala's *basawo baganda* are but a few examples of contemporary "traditional" medical practitioners (TMPs) who are key participants in the informal, self-help economy and social services of all African cities in the 1980s. In Dar es Salaam, the Tanzanian capital, traditional medicine "is one of the larger occupational groupings in the city even though it fails to appear on occupational surveys" (L. W. Swantz, 1974, pp. 151–152).

In this chapter, and the two following chapters, I examine the nature, functions, and spatial patterns of traditional medicine and its place in Nairobi's *ethnomedical systems*. The focus is a large cluster of TMPs and their clientele in the "unauthorized settlement" known as Mathare Valley.

Nairobi and many other African cities are expected to continue their rapid population growth due to both rural "push" and urban "pull" factors and natural increase. Because of these demographic realities and the absence of social services that compete with TMPs the urban areas are now and will continue to be the main arena for the growth of traditionally based healing and counseling.

My investigation in Nairobi began with several assumptions based on intuitive arguments and the few existing studies of African urban traditional medicine (Chavunduka, 1978; Maclean, 1971; Swantz, 1974), including Hake's (1977) inquiries in Nairobi. First, the traditionally based healing systems occupy a major, powerful (but not always directly observable) position in social and economic life of modern Nairobi's African population. Second, Nairobi and other urban places in Kenya are the key locations where traditional medicine is growing and adapting in response to population growth and the needs of a society undergoing wrenching cultural conflict and social change. Third, Nairobi's TMPs are disproportionately drawn from a few of the largest Kenyan ethnic groups such as the Kamba and Luo. As later substantiated by the knowledge, attitudes, and practices (KAP) study conducted at Kenyatta National Hospital (Chapter 2), these peoples are particularly known for the continuing vitality and patronage of their traditional medical institutions. Fourth, in urban areas such as Nairobi that are well-served by a high density of biomedical facilities, it was assumed that TMPs would be competing with dispensaries and hospitals for patients—particularly in view of the fact that government health services are free or available at a nominal charge. Finally, studies of African population mobility (Parkin, 1975) suggested that to function adaptively and successfully, urban TMPs would operate spatially within a unified circulation field that includes their own rural village of origin.

As Hake (1977) discovered in the early 1970s, there are hundreds of *waganga* (Swahili) or "self-help" doctors in Nairobi. These TMPs coexist with and form part of the kaleidoscopic array of institutions and individuals who derive at least partial economic support from the health-related services they perform for the population in the city and its circulation region. Within this broad array can be found varied services ranging from the surgical organ transplants and medical opinions of internationally qualified biomedical consultants at Kenyatta National Hospital to the most exploitative and potentially harmful "fringe" practitioners, including hotel-based commercial astrologers and fortune-tellers, back-street "doctors," and bus-stop capsule dispensers (McEvoy, 1976; Omondi, 1977). Biomedicine is publicly and politically the only legitimate and acceptable form of health services for modern urban Kenyans. Hence, they have a tendency to disvalue *bona fide* traditional medicine (which is a logical position only if one assumes a uniformly low intelligence among urban dwellers) and lump it together with the charlatans—and dangerous fringe operators—in whose hands a little knowledge can be lethal. This image of traditional medicine is unfortunate and inappropriate in the majority of cases. Functionally, it also serves to postpone, or eliminate, the added health benefits that could be realized through closer cooperative working relationships between biomedical professionals and those urban *waganga* who render a genuine service. Although this ideal of collaboration seems little closer to realization today than a decade ago, Nairobi's *waganga* nevertheless continue to play a significant role in the city's life. Hake (1977, p. 215) eloquently describes their opaque, equivocal status:

> It is not just that they are an economic factor, handling considerable sums of money between them. Far more important, they provide a crucial element in the 'spiritual', psychological, emotional, psychic inner life of most of the city's population. They're at once feared and ridiculed, sought out and avoided, paid and cheated, being both healers and sources of anxiety, links with the past, powerful influences on the present and seers of the future. Reassuring and yet disturbing, suspect yet trusted, it may be that their essential ambiguity is the key to their role.

Press (1978) suggests that urban TMPs generally offer affordable health care. In Nairobi, however, the services of *waganga* are not cheap. Depending on the problem, fees can range upward to a level of comparability with an expensive private psychiatrist or internist in the United States.

In practice, many urban *waganga* exhibit a strong sense of professionalism. They would like to receive government recognition and legitimization of their work and thereby enhance their prestige and influence. Economic opportunity remains the foremost reason for their presence in the city. Although the vast majority do not become wealthy, all of them are in Nairobi because it is possible to earn a living there as a TMP. At least, the practice of traditional medicine can help to supplement one's income from other sources. Inconspicuous material advancement is a primary goal of many *waganga*. Savings, investment in urban business and property, and improvement of one's rural land and homestead are among the hallmarks of the economically successful *mganga* in Nairobi.

Since the government provides free basic medical services to Nairobi's African population through a system of city council dispensaries and Kenyatta National Hospital, *waganga* obviously are valued for reasons other than cost. Biomedical services are geared to processing "patient units" as rapidly as possible. While the rural person who has recently arrived in the city may find the dispensary or hospital filter clinic a painfully alien environment, seasoned migrants and newcomers alike will often sense them as cold, impersonal places that are not organized to deal with the personal and social adjustment problems that often accompany and precipitate their somatic illnesses. *Waganga*, and TMPs in other African urban cultures, are valued because they can meet such needs, and more, for a large proportion of the city's people. The situation is similar in rural areas of Africa. Even if biomedical services become widespread in the countryside some TMPs will survive because the types of assistance they offer are unavailable at clinics and hospitals.

In his overview of urban folk medicine in the Third World, Press (1978, p. 75) asserts that one of its obvious functions is that of "minimizing the trauma of acculturation" in a situation where "modern medicine can be a threatening solution to an already threatening problem." In general, urban healers have a rural background, or at least effectively display behavior and attitudes with which their patients can identify. When urban traditional medicine incorporates some elements of modern (biomedical) medicine, such as pulse-taking, antibiotic prescriptions, or quasiscientific explanations, it further reduces acculturation pressures by permitting patients to feel more "modern" (Press, 1978, pp. 75–76).

Other African studies (e.g., L. W. Swantz, 1974) amply document the fact that urban TMPs serve as crucial buffers between their clients and the frequently harsh and competitive urban milieu. They go to considerable lengths to adjust their practices to the new, dynamic social and economic circumstances that confront their patients in the city (Katz & Katz, 1981; M. L. Swantz, 1979). For recent arrivals from rural villages, the transition to urban living is typically accompanied by increased individual striving, competitiveness, divisiveness, and anonymity, and often great physical and psychic distance from members of one's primary (rural-based) therapeutic community. In this climate, interpersonal relations often founder and aspirations for affluence and status may fail to materialize. For the individual who responds to such stressful circumstances with dysfunctional emotional and somatic symptoms, urban TMPs can help to ameliorate the consequences by defining the illness in familiar traditional—or "folk"—terms (Press, 1978).

Patterns of rural-urban and urban-rural population mobility—including the movements of TMPs and their patients—substantiate the thesis that East and Central Africa's rural and urban areas are functionally "part of a single field of relations made up of a vast criss-crossing of peoples, ideas, and resources" (Parkin, 1975, p. 3). "Urban" and "rural" represent a unified system rather than functionally discrete elements. Despite the existence of these networks, the "significant others" who normally participate in therapeutic activities in rural areas are often beyond the practical reach of their kinsmen in urban settings (International Development Research Center, (1980). Consequently, depending on the nature of the problem, a patient's social geography may also lead urban TMPs to change from one traditional interpretation of illness to another. For example, a spirit-related cause of illness, which often requires kin-group therapy, may be replaced by a sorcery-related etiology that does not. This may also allow "a displacement of responsibility for failure from self to other sources" (Press, 1978, p. 76). It also redefines and narrows the patient's therapeutic community if the kin group is diffused or not easily accessible. Thus, urban TMPs adapt by diagnosing and treating problems in terms of causes and methods that are possible and credible for them to manage in the urban setting (M. L. Swantz, 1979). Whether in Kinshasha (International Development Research Center, 1980) or Nairobi, most also strive to create a "polyvalent" or general practitioner image, in contrast to the greater specialization found among rural TMPs. Rural TMPs are usually farmers and can rarely afford to practice traditional medicine full time. Moreover, it is not customary for them to engage in self-advertisement or commercial-scale healing activities. Individuals such as Kalii are exceptional (Chapter 4). On the other hand, because traditional healing is their principal source of livelihood urban TMPs try to diversify—working as both diviners and herbalists—in order to attract more patients.

NAIROBI'S RESIDENTIAL SPATIAL STRUCTURE AND "UNAUTHORIZED" SETTLEMENTS

Founded in 1899 as a railhead encampment for workers on the new Uganda Railway, Nairobi rapidly took on significant administrative, communications, and, commercial functions which eventually made it the dominant urban center in East Africa. By World War I it had become a European settler community organized to service a highlands settler economy that continued to expand until the 1950s. Although Africans were "tolerated" for their labor power, and South Asians performed a variety of essential entrepreneurial, service, and technical functions, Nairobi was above all a colonial city designed to serve the exclusive economic, social, and political interests of Kenya's politically influential European population.

Today, Nairobi's residential spatial structure is a striking legacy of the colonial era. To achieve colonial European objectives, including a disease-free urban environment and minimal public responsibility for workers' welfare, the city was "systematically racially zoned in the major plans of 1905, 1927 and 1948" (Amis, 1984, p. 89, citing Van Zwanenberg, 1972). Disproportionately large residential zones situated to the north (e.g., Westlands, Muthaiga) and west (e.g., Upper Hill, Lavington) of the central business district were developed exclusively for Europeans (Figure 6-1). Although affluent African and Asian elite have moved into these attractive, class-conscious "garden" areas since Independence, the residents today are mainly members of the foreign business and diplomatic communities—and Europeans. Every evening these neighborhoods are "invaded" by legions of uniformed and other private security guards. Virtually all residents have hired at least one of these *askaris* to protect their house and possessions from burglars and thieves, who often operate as gangs.

A second ethnically and culturally distinctive residential zone—a landscape of flat-topped "Asian" architecture and large extended South Asian families—emerged in the cluster of communities known as Eastleigh, Ngara, Pangani, and Parklands, areas situated north and northeast of central Nairobi. At least into the 1970s, about four of every five of Nairobi's considerable Asian population were housed in these areas (Ferraro, 1978–1979).

Traditional working-class housing for Africans is located in Eastlands—which is the area east of Nairobi's downtown and north of the industrial area. Well-known settlements here include Kariokor and Pumwani (the oldest housing areas convenient to the railway and industrial district), Shauri Moyo, and Kaloleni. Farther east are relatively newer working-class "estates" such as Bahati, Ofafa, and Makadara, which were primarily built by the Nairobi City Council or large employers (Halliman & Morgan, 1967). However, the inertia of long-entrenched, ethnically segregated zoning in residential areas and rapid expansion of the urban population means that land for housing the poorer African majority in Nairobi remains in critically short supply.

Between 1962 and 1979 the population of Nairobi—now a "million city" with an area of 684 square kilometers—grew from 343,500 to 827,800. This growth has been both rapid (about 5% annually) and poorly planned. Consequently, the city's housing market has recently been characterized as one that is "in a permanent condition of an excess of demand over supply" (Amis,

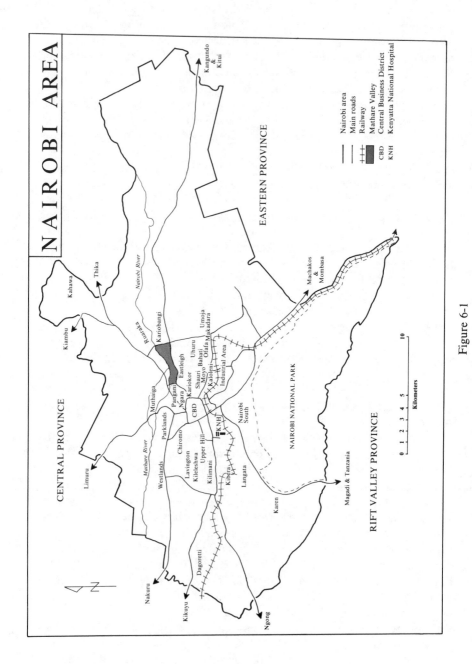

Figure 6-1

1984, p. 89). People have responded to the dearth of legal housing by constructing thousands of units of so-called "unauthorized housing"—structures and areas that do not meet government building standards and are technically not supposed to exist. In 1962 there were an estimated 500 unauthorized housing units, almost all of which represented noncommercial "subsistence shelter." By 1979 the inventory had expanded to some 110,000 unauthorized units, which then housed about 40% of Nairobi's population (Amis, 1984). "Subsistence shelter," which Amis sees as a transitional urban development, had virtually disappeared by 1979. In its place, a vigorous capitalist rental housing market was, and is, operating within the unauthorized settlement areas such as Kibera, Kariobangi, and Mathare Valley. Amis (1984) asserts that the extraordinary scale and swiftness of this development was spurred by its "exceptional profitability," which was itself conditioned by Nairobi's inappropriate (unrealistic) housing standards, the high demand for shelter, and the limited availability of land. All Kenyan occupational groups, ranging from doctors, managers and high-ranking civil servants to the urban poor, are involved in the unauthorized commercial housing phenomenon as landlords, tenants, or both.

MATHARE VALLEY: A SETTING FOR URBAN TRADITIONAL MEDICINE

Mathare is Nairobi's largest and, probably, most densely populated area of unauthorized housing. It was selected as a research site because it is a physically well-defined community, is representative of Nairobi's poorest majority, and was reputed to have a large number of resident *waganga*. Mathare symbolizes the consequences and nature of inequality in modern Africa, and it reflects the failure of political leaders in almost every country to invest a fair share of national resources in the rural areas. If Mathare is attractive to tens of thousands of Kenyans, what is the level of despair in the countryside?

Mathare is home, at least temporarily, to an estimated 100,000 of Nairobi's inhabitants—people who are willing to trade the disadvantages of life in the city's most notorious settlement for a chance to carve a niche in its transitional, self-help economy. It is a sprawling, congested maze of high-density shanty and former squatter settlements—organized as a chain of nine distinct, named "self-help villages" (Hake, 1977)—which have mushroomed along a 4-kilometer stretch of the Mathare Valley. The villages occupy some 160 hectares, including stretches of a steep-sided escarpment and hazardous abandoned quarry pits. Each exhibits a degree of social cohesion (Ross, 1973) and has an elected elders committee of the national party [Kenya African National Union (KANU)] that provides liaison with city and national government officials, mediates minor disputes, provides night security guards, and occasionally organizes local self-help projects (Nelson, 1978–1979). So-cial services are greatly undersupplied and the few physical facilities, such as community halls and nursery schools, are crude.

Mathare is about 4 kilomters northeast of Nairobi's central business district. Juja Road, its southern boundary, is a major city bus route. The Eastleigh communities and Kenya Air Force base are adjacent to Mathare on the south, while the northern periphery is bounded by the Mathari Mental Hospital (government), Mathare Primary School, and the Nairobi "police lines." Population growth in the Valley from 1971 to 1976 was estimated at nearly 7% annually (Waweru and associates, 1976). Two-thirds (52,500) of the estimated 78,600 persons in the nine villages in 1978 lived in the three oldest settlements known as Mathare I, Mathare II, and Mathare III (Figure 6-2). These oldest villages are the focus of this case study.

Development of unauthorized housing in Mathare began about 1939. By 1952 there were five villages in the valley and more than 4,000 settlers, primarily Kikuyu. Most of the African residents rented the plots for their mud and wattle houses from Asian landowners. No service facilities existed. None of the original villages could be seen from the main road since they had been intentionally situated below the edge of the escarpment. Each village had organized an internal political leadership to resist attempts by the Nai-robi City Council to evict the residents or demolish their houses. Neverthe-less, in late 1952 at the start of the Mau Mau "Emergency" in Kenya, all of these villages were bulldozed by the colonial government and the mainly Kikuyu residents were sent to rural areas or placed in detention camps (Ross, 1973; Waweru and associates, 1976).

Self-help housing quickly reappeared in Mathare in 1961, following the end of the Mau Mau Emergency and just before national independence. Four villages were reestablished on their original sites, and five were built in new locations east of Mathare III. Most of the houses were again built out of mud and wattle, and many were sited on hazardous slopes or at the bottom of the valley. A large proportion of these units was still in use in 1980, providing a source of income to residents who sublet space to new arrivals in Nairobi. Residents of these "squatter" settlements tend to be older and form a more permanent, socially cohesive and organized community than those in other types of housing.

The remainder of Mathare's housing stock is mostly high-density timber barracks with galvanized, corrugated iron roofs and concrete floors. This type of housing has been constructed since 1965 by numerous companies, including some *bona fide* cooperatives (Waweru and associates, 1976). Such "company" housing is normally laid out on a rectangular grid, in contrast to the irregular, unplanned pattern of the squatter-type housing. Company residents are also more transient, and few are owner-occupiers.

Overall housing density reaches 70 structures per net hectare in Mathare I, II, and III—the older and most densely populated villages at the western

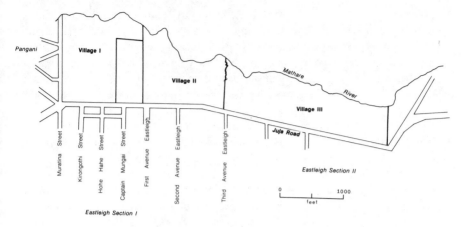

Figure 6-2. Mathare I, II, and III (Nairobi).

end of the settlement. Monthly rents in 1978 ranged from approximately $8.00 for a room in a mud and wattle structure to $23.00 for a room in company housing. Few rooms in either type of housing exceeded 9 square meters in area. Sheltered living space was often less than 2 square meters per person, and as many as six to ten people might share a small room.

Mathare's residents are generally under 35 years old and have, at most, a few years of primary education. Although every Kenyan African ethnic group is represented, about two-thirds of the population is of Kikuyu or Kikuyu-related (Embu, Meru) background. Nelson's (1978–1979) survey revealed that over half, and possibly 60%, of the total population is female—which is disproportionately much greater than Nairobi's overall female/male ratio of 1:4. The majority of these Mathare women are unmarried and have several children. Their emerging family systems tend to be female-centered, which represents a radical departure from the traditional, patrilineal forms found in rural areas. Economically and status-wise, Mathare women are much more independent from men than are women who remain in the rural areas (Nelson, 1978–1979). In a structural and emotional sense, Mathare womens' ties with their sisters, mothers, and daughters (all of whom may be coresidents) are more important than their ties with males. Relationships with men often have a commercial orientation, and women speak of their pride in and need to control their own family, sexual, and economic lives (Nelson, 1978–1979).

Most of Mathare's residents are unemployed, or self-employed in the "informal sector." Only about 20% of the men, and probably less than 10% of the women, hold wage jobs. The few women who earn wages are typically house servants (outside of Mathare), barmaids, or nursery school teachers. Some women earn money by offering cooked food, sewing, selling fresh fruits and vegetables in a local market, by prostitution, or by exchanging sex and domestic services in consensual unions of varying duration. More than half of the *waganga* in our TMP census were female. Overall, however, Nelson (1978–1979) found that most women in Mathare—certainly in the older villages where this traditional medicine study was conducted—earn at least part of their income from illegally brewing and selling *buzaa*, a local maize beer. *Buzaa* is a profitable, popular drink sold by the glass or tin at various open-air bars, or "joints," such as "Kijiji's Mathare Valley."

Men, on the other hand, engage in a wider variety of informal sector enterprise. They operate small kiosks and restaurants; sell clothes and durables; fashion scrap metal into utensils; sell charcoal; and work as tailors, carpenters, water carriers, butchers, shoemakers, and TMPs. Men (and women) also distill *chang'aa*, an illicit gin jokingly referred to as "kill me quick," in large oil drums along the banks of the Mathare River. Some of this *chang'aa* is "exported" to other shantytowns in Nairobi. Others undoubtedly gain some support from petty theft. Young boys are hired to serve as sentries for *buzaa* brewers and their customers. They are known for their distinctive whistling, done to warn brewers and customers of approaching plainclothes police.

Although the city has constructed a few permanent public latrine houses, and people can obtain potable water from scattered standpipes, Mathare's environmental quality is exceedingly poor—especially in Villages, I, II, and III. In truth, these villages, which have a human density in excess of 1,200 persons per hectare all at ground level, are by any standard teeming, filthy, and malodorous. Open earth drains, scattered human and animal excrement, and huge mounds of uncollected garbage are the norm. Young children, particularly, and adults are continuously a risk of exposure to pathogens and other health hazards, including fire.[1]

There were also no government dispensaries or first-aid posts within Mathare. A small, Baptist-operated clinic operated part-time, but treatment was dependent on the availability of an American doctor. With this exception, residents of Mathare who sought government biomedical services had to go outside the settlement to a city council health center, such as the one less than a kilometer away in the neighboring Eastleigh community; to the maternity hospital at Pumwani; or to the Kenyatta National Hospital some 7 kilometers to the west in Upper Nairobi. The Nairobi City Council also operates seven maternity units and 40 child welfare clinics. Numerous private hospitals and clinics are scattered through the city, including a "surgery" located in a modern-style, four-story hotel-shops complex situated near the eastern end of Mathare III adjacent to the Kenya Air Force base at Eastleigh. A private maternity home on Juja Road is immediately adjacent to Mathare II. Nairobi thus has a large and varied assortment of public and private dispensaries, clinics, and hospitals which are easily accessible—for the cost of a *matatu* ride or bus fare—to Mathare's residents. Mathare's *waganga*, as well as their own patients, also utilize Nairobi's biomedical facilities.

MATHARE'S TRADITIONAL MEDICAL PRACTITIONERS: ETHNICITY AND LOCATIONAL PATTERN

A census of resident *waganga* who practice in Villages I, II, and III revealed 62 TMPs whose location is shown relative to their ethnicity and respective village populations in Table 6-1. Their internal spatial distribution is shown in Figures 6-3 to 6-5. The overall ratio of *waganga* to population in Mathare (1:833) contrasts with L. W. Swantz's (1974) estimate of 1:350–450 for Dar es Salaam, Tanzania, and also with the 1:146–1:345 values for rural Kilungu. In comparison, the estimated population per university-trained biomedical doctor in Kenya is 1:987 for urban areas and 1:70,000 for rural areas (Family Health Institute, 1978).

Although Kikuyu formed the majority (nearly 70%) of residents in the three villages, (Nelson, 1978–1979), Kamba comprised approximately two-thirds of the *waganga* in each village (Table 6-1). Kamba practitioners had thus created an important niche for themselves in the informal sector as the dominant providers of traditional medical services in the Mathare study area. Strikingly, the rural (home) villages of 90% of these Kamba were not in Machakos but in the western locations of Kitui District (Figure 6-6). Agriculturally, much of Kitui is marginal, drought- and famine-prone. The area is renowned in Kenya for its many so-called "witch doctors" and powerful medicinal plants. Nearly half of the Kamba TMPs from Kitui had migrated from Mutonguni and Mwingi Locations. However, evidence of clustering disappeared almost entirely at lower levels (sublocation and village) in the settlement hierarchy.

Kikuyu formed nearly a third of the other *waganga* in Village I—the oldest and most crowded settlement—but few of them lived in villages II and III. Traditional medical practitioners from western Kenya—predominantly Luo and a few Luhya—comprised about one-third of the *waganga* in village III. All of the TMPs attracted a heterogeneous clientele. In this urban setting many patients "cross over" and choose a TMP whose ethnolinguistic and regional background (e.g., Bantu, Nilotic) is very different from their own.

Following the census a large sample (68%) of *waganga*, stratified by village, was used to gather extensive information concerning their personal characteristics and their professional KAP. This survey included 35 Kamba practitioners (70% of all Kamba and 53% of the total TMP population) who served as the urban counterparts of the *andu awe* and *akimi wa miti* in rural Kilungu.[2] Eighty percent of the urban Kamba referred to themselves as

Table 6-1. Mathare Valley Study Area: Villages, Population, and Traditional Medical Practitioners

Village (area in hectares)[a]	Est. Population (1978)	Population Density per hectare	No. TMPs recorded		Ratio TMPs/Population	Ethnicity (%)		
			O	E		Kamba	Kikuyu	Other
Mathare I (12.1)	17,716	1,464	20	21	1:886	65	30	5
Mathare II (10.6)	14,698	1,386	14	17	1:980	67	13	20
Mathare III (20.7)	20,085	970	28	24	1:717	64	4	32
	52,499	1,210	62	62	1:833	65	14	21

Note: From field survey, 1977–1978.
Abbreviations: TMP, traditional medical practitioner.
[a]Estimates based on Waweru and associates, 1976.

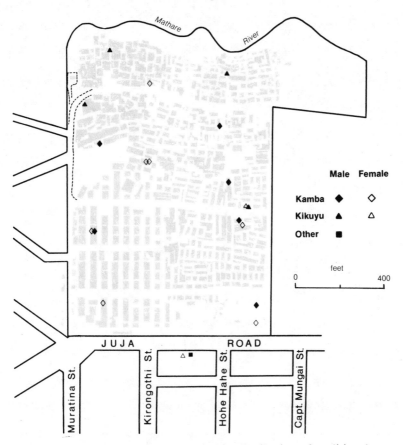

Figure 6-3. Mathare village I (Nairobi): distribution of traditional healers.

diviners and herbalists, that is, general practitioners. The remainder added midwifery to their repertoire. Similar proportions of specializations emerged across the entire sample of TMPs.

A SOCIAL AND SPATIAL PROFILE OF MATHARE'S TRADITIONAL MEDICAL PRACTITIONERS

Overall, male *waganga* (53%) slightly outnumbered females in Mathare I, II, and III. Among the Kamba TMPs, however, women (54%) outnumbered men. In contrast, women formed less than 30% of the Kikuyu, Luo, Luhya and other non-Kamba ($N = 10$) TMPs. The youngest TMP was 23 years old, while the oldest exceeded 80 years. Half of all the practitioners exceeded 47 years of age, although the average woman (45 years) was seven years younger than her male counterpart.

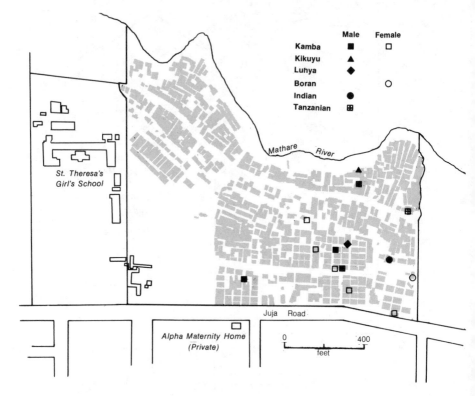

Figure 6-4. Mathare village II (Nairobi): distribution of traditional healers.

Taken as a whole, more than nine of ten TMPs were partnered—either married or in other forms of temporary or more permanent unions with a member of the opposite sex. Overall, the *waganga* had an average of 1.7 partners, with a range of up to five among the Kamba—95% of whom were "partnered." Approximately 20% of the TMPs in Mathare III did not have a partner. This anomaly is related to the relatively higher proportion of male practitioners (63% overall and 58% of the Kamba) in that village. Overall, and for the Kamba, the mean number of persons defined as "adult family members" who actually lived with a *mganga* in Mathare was 0.9, although 40% of the Kamba had no one they considered "adult family" within their household. Some 42 of the 45 TMPs (93%) claimed approximately 247 children, or an average of 5.9 per practitioner. However, *waganga* had only 1.3 children physically present and living with them in Mathare. This difference suggests the importance of a supporting family infrastructure to care for the children left behind in the TMP's rural home village. It is generally perceived to be both expensive and difficult to raise young children and adolescents properly in Mathare. Nelson (1978–1979) examined the organization of the mainly female-centered, Kikuyu families in Mathare. She found that over 40% of the women in her sample (most were beer brewers) preferred to have one or more of their children fostered by a rural relative. The mother's mother was the fostering adult in 80% of these transactions. Observation of Kamba TMPs who are also mothers suggested that they too choose fostering out as a solution to the conflicts between child-rearing and a city-based livelihood.

Few of the Mathare *waganga* had a formal education. Overall, 69% had never attended school—and the proportion rose to 83% among the Kamba. Average time at school for the two groups was 1.2 years and 0.5 years, respectively. Such evidence leaves no doubt that Mathare's TMPs had almost no chance for alternative employment in Nairobi's formal sector.

Most *waganga* had an average of two decades of professional experience (Table 6-2). More than half of their experience as TMPs occurred outside Nairobi, and the typical practitioner had been in Mathare for about 5 years. Kamba TMPs had come to Nairobi and Mathare more recently than the group as a whole. Women, being younger and still less autonomous, had a shorter urban experience than men, although a few very old and experienced male TMPs tended to increase the difference between the experience of men and women. The most common places of professional experience outside of Nairobi among the Kamba included Kitui District, Mombasa, Kikuyuland, Machakos District, and Kisumu.

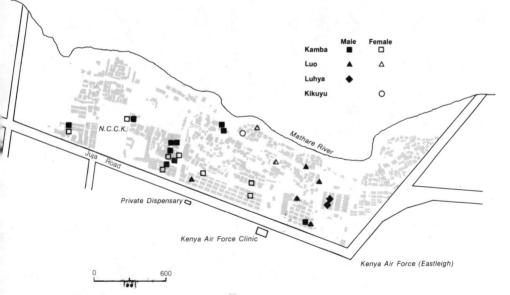

Figure 6-5. Mathare village III (Nairobi): distribution of traditional healers.

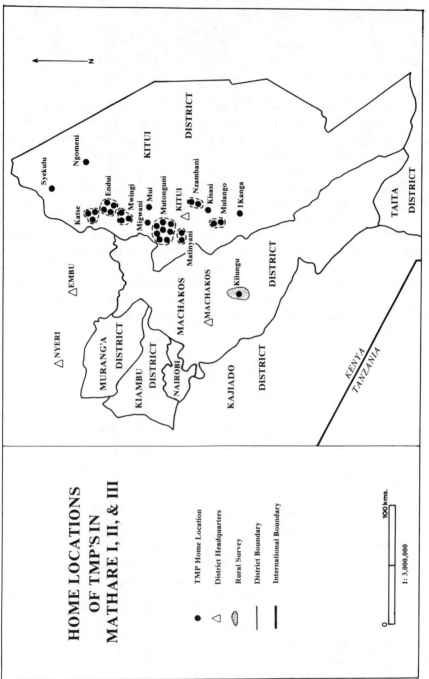

Figure 6-6

Table 6-2. Mathare *Waganga*: Length of Professional Practice

	Overall practice (years)	Practice in Nairobi (years)	Practice in present Mathare location (years)
All waganga ($N = 45$)	$\bar{X} = 20.0$ Median $= 18.0$ Range $= 3–45$	$\bar{X} = 8.4$ Median $= 5.5$ Range $= <1–26$	$\bar{X} = 5.1$ Median $= 5.0$ Range $= <1–20$
Akamba waganga	$\bar{X} = 20.4$ Median $= 17.0$ Range $= 4–45$	$\bar{X} = 7.4$ Median $= 5.0$ Range $= <1–22$	$\bar{X} = 4.3$ Median $= 4.0$ Range $= <1–11$
All male waganga	$\bar{X} = 23.0$ Range $= 7–45$	$\bar{X} = 8.9$	$\bar{X} = 6.1$ (half 5 years or less)
All female waganga	$\bar{X} = 16.4$ Range $= 3–36$	$\bar{X} = 7.8$	$\bar{X} = 4.0$ (half 3 years or less)

Note: From field survey, 1977–1978.

Mathare *waganga* conducted their practices from rented rooms that also served as their residences. A few TMPs rented two contiguous rooms in order to accommodate both their household members and their patients. Only one *mganga*, a 35-year-old Kamba woman, maintained a clinic in Mathare and a residence physically outside Mathare proper. Charismatic, enormously popular, and affluent by local standards, she easily afforded a more prestigious residence in a contemporary "Asian" style duplex located within a 5-min walk from her clinic. She employed a house servant and supported a young male trainee—also a Kamba—who participated in many of her therapeutic activities.

Over 90% of the *waganga* in each village identified traditional healing as a full-time occupation while they are in Nairobi. Less than one-quarter reported an income source other than their TMP fees. A few of those who did report other income were night watchmen. A husband and wife, both TMPs, operated a "shop" within the same small (3 × 3 meters) rented room they used as a residence and clinic. Other jobs represented included birth attendant in a private maternity home, fish trader, cobbler, traditional dancer, and petty landlord. Men were more likely to have other sources of income (29%) than women (19%). Only 12% of Kamba women said they had other income opportunities, while Kamba men (31%) had more sources of supplementary income. This male/female disproportion apparently reflects real differences in the sense that women had fewer alternatives and strong competition from men in many "informal sector" activities.

Since the great majority of TMPs retained strong connections with families and land in their rural villages, the figures above certainly underestimate those who had other sources of support. For example, 80% of the Kamba practitioners left Mathare to visit their home villages *at least* once per month. Over half reported that they made at least two visits each month. Overall, less than 7% of the *waganga* said they "only rarely go" to their home village. Thus the high frequency of contact between the urban-based practitioners and their home areas enabled TMPs to supervise their family farms; attend to family business, which is supported by income earned in Nairobi; collect fresh medicinal plants (*miti shamba*) for their urban practices; obtain foodstuffs for personal use when they return to Nairobi; and, occasionally, to treat rural patients. Such mobility and transactions provide strong evidence of the "single system" character of Nairobi and the rural regions that contribute a large share of its population. Indeed, as Ross (1975) also discovered in his study of Africans in the Shauri Moyo and Kariokor neighborhoods, Nairobi is not "home" for most of its residents. None of the Mathare TMPs ($N = 45$) were born in the city. Moreover, most of them saw their life in the city as instrumental—a means to an end—to achieving an increasingly higher socioeconomic status. My observation of TMPs in Mathare suggests that they shared the same values and model behavior found among the general population. Rural life with urban amenities was and is the ideal. Ross (1975) found that intensity of contact with rural areas increased as individuals moved into a higher socioeconomic status. Urban success thus promotes a strengthening of rural ties, and urban resources assist people to maintain a status in their home areas. Those who break their connections with home, or who maintain them at a low intensity, seem to "do so because of economic necessity rather than . . . social choice" (Ross, 1975, p. 49).

Aside from a minority who posted signboards outside the door to their room that advertise their services, outwardly Mathare *waganga* generally remained indistinguishable from the other village residents. They lived close together and saw their patients in small, rented rooms. The mean spacing between the *waganga* clinics was less than 70 meters, and ranged from 68 meters in Mathare I to a mere 24 meters in Mathare III. In general, the distributional pattern revealed marked clustering where several pairs of *waganga* worked from the same or adjacent rooms (Figures 6-3, 6-4, and 6-5).

Given the continuing scarcity of shelter in Mathare and elsewhere in Nairobi (Ross, 1975), the most important determinant of the spatial pattern of practitioners' clinics was quite simply the availability of living quarters. Few had the luxury of choosing from a variety of neighborhoods or housing options. They rarely considered a particular village or room in terms of a "rational" framework of decision-making, that is, one determined according to economic criteria such as the situation of competitors' clinics or market potential.

Whether moving to Nairobi for the first time, or relocating in Mathare, *waganga* relied extensively on direct help and information provided by relatives. Nonkin who may be more casually connected to a *mganga* also provided assistance (Table 6-3). Introductions and personal vouchers by relatives and other established contacts in a neighborhood enhanced one's prospects for locating housing. There was also considerable changing of residence among people already living in Mathare or elsewhere in Nairobi. Over two-thirds of the TMPs who explained a reason for their last move were not newcomers (Table 6-3).[3] Such moves occurred in response to a wide variety of factors, including rent increases, troublesome neighbors, fear of thugs, a desire for more space or privacy, a preference to live in closer proximity to other members of a *waganga* association, or a desire to move to a less objectionable physical environment (e.g., away from the communal toilets). Two TMPs had previously lived in rooms in Mathare that had been destroyed by fires, while another moved because his previous room had been demolished by authority of the Nairobi City Council.

Table 6-3. Mathare Traditional Medical Practitioners' Sources of Assistance in the Acquisition of Local Housing

Response	N	Moving from another place in Mathare or Nairobi?		
		Yes	No	?
Relative(s) assisted	7	4	2[a]	.
Contacted landlord personally	7	6	1	–
Friend assisted	6	4	–	2
Assisted by relatives of former patient	2	1[a]	–	1
Assisted by former client	2	1	1	–
Joined husband/wife/sister already renting a room	4	–	3	1
Assisted by *waganga* association	2	1	–	1
Located by Kikuyu Society	1	–	–	1
Located by Kiambu-Murang'a Society	1	1[b]	–	–
Provided by committee of Mathare elders	1	–		1[d]
Informed of room by neighbors in rural area	2	–	2[a]	–
Informed by occupant of nearby room	1	1	–	–
Room self-constructed	1	–	1	–
Details of acquisition unclear	7	4	1[c]	2
No account provided	1	–	–	1
	45	23	11	11

Note: From field survey, 1977–1978.
[a] Rent paid in advance by benefactor.
[b] Previous room demolished by city council.
[c] Move forced by accusations of sorcery or witchcraft.
[d] Indigent woman (80 years old): Cares for her orphaned grandchildren. Has no living children. Husband deceased.

In view of the high average population density in the villages (>1,200 persons per hectare or 120,000 persons per square kilometer), and the close proximity of tens of thousands of potential patients in the adjacent Eastleigh estates, virtually any location in Mathare is central. On the other hand, the relatively high density of TMPs underscores how important it is that a practitioner have personal charisma and an in-place, supportive social network as he or she strives to build up and maintain a city clientele.

Large numbers of patients are naturally attracted to the most famous and reputable *waganga*. It is to the patient's advantage that at any one time the population of TMPs available for consultation is large and accessible. Of course, this information is useful only if therapy managers and patients can obtain it. For their part, patients of Mathare TMPs were not limited to the fluid Mathare population. Many came from other parts of Nairobi and from up-country.

Although the bulk based their practices in Mathare, most *waganga* occasionally treated patients when they visited their own home villages, and some made "outcalls" to patients in other parts of Nairobi and adjacent districts. Only one-quarter of all practitioners said they "often" traveled to their home areas to treat patients, while 10% to 20% reported they never went home for this reason. The majority (70%) said it is only "occasionally" that people from their home areas come to them for treatment in Nairobi. The remaining TMPs included those who reported they saw rural patients in rural areas often (20%) and "never" (10%).

For various reasons patients may, of course, seek out a trusted *mganga* from their village who has moved to Mathare. Regardless, as Kilungu illustrated, TMPs are numerically well represented in the rural areas. For example, Mathare *waganga* identified an average of five and a mode of eight active practitioners in their home villages. However, these rural TMPs were seldom consulted or sent referral cases, and vice versa.

PROFESSIONAL ASSOCIATIONS

Traditional healers lost status and power throughout Africa during the colonial era, and little of this has been recouped since Independence. They were challenged (and often but inconsistently suppressed) by colonial administrations and Christian missions (Janzen, 1978), and indirectly by the biomedical community. They gave up authority not only in individual and public health matters (narrowly defined), but also over broader community concerns such as the agricultural calendar, land use, warfare, adolescent training, and jural matters (Feierman, 1984; Lindblom, 1969).

Over the last 25 years, or since Independence, the mushroomlike growth of Africa's urban populations through migration has included a surge of individuals who seek to support themselves by practicing forms of traditional healing. While an urban need for the services of *bona fide* TMPs is documented in this and many other studies, their activities are in the "informal sector" and there is little firm evidence of how many individuals are actually gainfully occupied as urban traditional healers. If the ratio of TMPs to population in the Mathare villages (1:833) is representative for all of Nairobi, the expected number of TMPs currently in the city is approximately 960. L. W. Swantz (1974) estimated the number of TMPs in Dar es Salaam at more than 700. A Zairian authority on traditional medicine estimated that in the late 1970s there were at least 10,000 TMPs in Kinshasha, the national capital.[4] One important consequence of the greater freedom, opportunity, and numbers of TMPs in urban areas is the desire of many of them to associate and cooperate for mutual advantage and external recognition. In Ibadan, Nigeria, for example, Maclean (1971, p. 77) found in a survey of 100 herbalists that "all but six were members of some kind of association or professional society; Ewesanmi, the Medical Herbalists' Association, being repeatedly cited by both *babalawo* and herbalists."

In Zaire, professional associations of TMPs represent a new form of cooperation in rural areas that is spreading to urban centers (International Development Research Center, 1980). Typical objectives of these healers' associations include enumeration, training, cooperation among members, promotion and regulation of customary practices, and cooperation with official health services. These aims form but part of the Zairian and other African TMPs' more fundamental purpose of association, which is a "desire to define who they are with respect to the rest of society" (International Development Research Center, 1980, p. 23).

In their organization the Zairian associations generally have a regional and local base that keeps them closely connected to their original rural locus. Each association tends to be structured around a well-known TMP. While most of the associations represent a limited area of the country, the names of those in Kinshasha frequently reflect members' desire for national recognition (International Development Research Center, 1980).

In Zaire and other states there is always the potential for some leaders to misuse association resources for personal gain. Thus the enthusiasm that leads to the creation of official associations is usually tempered by fears of losing one's investment of personal funds and identity. In his essay on traditional medicine and psychiatry in Africa for the World Health Organization, Koumare (1983, p. 29) notes that "mistrust among the healers is still prevalent and one can hardly speak of national associations comprising a substantial body of practitioners. There is likewise considerable suspicion between the health professions and the healers. The public therefore needs to be given statutory protection against malpractice."

Currently traditional healers' associations have official recognition in at least ten African countries, including Senegal, Mali, Niger, Togo, Benin, Cameroun, Congo, Zaire (Bannerman *et al.*, 1983), Swaziland (Green &

Makhubu, 1984) and Zimbabwe (Kronholz, 1982). Numerous TMP associations also operate in Kenya. None have official approval and most appear at
present to have marginal viability even if their membership has high expectations.

In Mathare, 69% of the *waganga* in the sample (*N* = 45) belonged to one
of four professional associations. The most prominent association was the
Waganga wa Kenya wa Miti Shamba Society, a Kamba-based organization
which was first registered as a corporation in 1971. Three-quarters of the
Kamba TMPs reported membership in at least one association. Over 80% of
this group belonged to Miti Shamba. The other associations represented
included the Jamii-Tiba Society (1973), which had 19% of the Mathare
membership; Kitile and Syokisinga Itamba (Kenya) Co. (6.5%); and African
Repairs-Tibibwa Kienyeji Co. All three maintained Nairobi postal boxes.

Mathare I had the highest (82%) and Mathare III (56%) the lowest level
of membership in these associations. This difference is explained in part by
the presence of Luo and Luhya TMPs in Mathare III, who tended not to
belong to an association. Also, one-third of the Kamba TMP's in Mathare
III—the only village where male Kamba practitioners form a majority—
had no such professional affiliation. A distinct gender difference was also
apparent in that over 70% of women, but only half of the men, in the three
villages belonged to a professional association. Women may have placed
greater emphasis on membership because they are more prone to perceive it
and rely on it as a form of insurance policy—as a means of risk-reduction in
an urban setting characterized by competitiveness, unpredictability, and the
insensitivity of strangers.

Acceptance into a professional association was contingent upon one's
ability to pay the required fees (the Kenya Miti Shamba Society levied a one-
time Shs. 100/- (approximately U.S. $13) initial membership fee and a
monthly fee of Shs. 5/-). Presentation of a letter from the applicant's local
chief at home that vouches for his or her character and reputation as a
mganga was also a standard practice. These documents, and the certificate
issued by a society to each member, are typically framed and displayed inside
the *mganga's* clinic. In addition, virtually all prospective members are
quizzed, in at least nominal fashion, about their particular skills and practices. No evidence was available concerning the rejection rate for those who
can afford the fees.

Mathare *waganga* identified several benefits of membership in professional associations. The two most prominent advantages included corporate
assistance in the case of police harassment and or arrest; and financial
assistance, particularly loans to help with payment of childrens' school fees, a
court case, or some other genuine need. When necessary in case of a
member's death, the Kenya Miti Shamba Society had committed itself to take
care of the TMPs body and arrange for its removal to the TMP's rural home
for burial.[5] Other perceived benefits of membership included assistance in

case of an accident or sickness. Several *waganga* believed that the certificate itself—with whatever margin of legitimacy it provided—was the principal benefit of membership in a professional association.

Like many other associations of TMPs in Africa, the Kamba-based *Waganga wa Kenya wa Miti Shamba* Society aspired to be a national organization.[6] It boasted a certificate of corporation (July 23, 1971) issued under Kenya's "Rules of Societies." Its headquarters were located in a room in the "Swahili village" section of Machakos (Masaku) town, the up-country administrative and market center of Machakos District located about 60 kilometers southeast of Nairobi. A post office box was also maintained in Machakos. The Society had three "branch" offices located in Nairobi, Nakuru (a large town in the Rift Valley), and Mombasa. Apparently the Mombasa branch had become more or less autonomous, and issued its own certificates. The chairman traveled regularly to Nairobi and Nakuru and returned to Machakos with the names of prospective members working in those areas. He then prepared the formal certificates and took them back to Nairobi and Nakuru to be exchanged for membership fees. This system is similar to the pattern in Zaire (International Development Research Center, 1980) in that the society sought to maintain its locus of authority and control in the local, ethnic area of origin.

The chairman and female treasurer of the Miti Shamba Society, both Kamba, lived in Machakos. Both owned cars and represented part of the African lower "middle class." According to the treasurer the Society's "Trustee" was an Indian who practiced both traditional healing and biomedicine. Functionally, the organization's leaders suggested that it was designed to provide mutual aid and opportunities for self-improvement. A bank held the members' fees. Attempts by the society's leadership to purchase houses to rent out had not received government approval. Another idea discussed but not acted upon by the Society was the possibility of sending some of their younger, brighter members abroad for training in biomedicine. Upon returning to Kenya they would be expected to share their new knowledge with the membership at large. While this last aspiration may be unrealistic, it signified a progressive attitude and, as such, a resource that could be developed locally for the betterment of community health care.

THE CALLING OF URBAN TRADITIONAL MEDICAL PRACTITIONERS

In rural areas such as the Kilungu Hills, TMPs are known by their particular skills and functions. A richly varied range of terms that describe these specialties remains in use today (Chapter 5). In contrast, when TMPs leave their home areas to practice in the city they tend to adopt the polyvalent image of a general practitioner—an apparent concession to their urban clientele and need for economic survival. Practically all of the urban TMPs said that they are proficient at both divination (which always requires supernatural powers) and curing illness with herbal medicines. This last observation raises the issue of how urban TMPs are recruited to their role. Did the individuals who worked as TMPs in Mathare bypass the customary signs and rituals that continued to legitimize and sanctify the role of the religious-medical specialist (*mundu mue*) in rural areas such as Kilungu and Kitui? How and where did Mathare TMPs acquire their calling and expertise? Was there evidence of significant differences between men and women in terms of their entry and preparation for the role of *mganga*?

Most of the Mathare practitioners, whether Kamba, Kikuyu, or Luo, came from families whose membership included other TMPs. Although there is little evidence of inherited roles, there was a distinct predisposition for certain kin groups to reproduce TMPs. In the great majority of cases chronic personal suffering and a visionary experience with the spirit world through dreams or possession served as key elements in the transition to the individual's private acceptance and public recognition of his or her role as a *mundu mue* (Kamba), *mundu mugo* (Kikuyu), or *ajuoga* (Luo) (cf. Janzen, 1978). Various forms of mental illness, physical and psychosomatic illness (particularly abdominal symptoms, edema, vision impairment, and pain all over the body), and death of children are common examples of the miseries Mathare TMPs reported they experienced before (and until) they accepted the call. These behavior changes and illnesses usually did not manifest themselves until the late teenage years (and for many women not until after marriage, pregnancy, and/or giving birth). Cultural rules dictate that, sooner or later, such signs must be interpreted by a traditional religious-medical specialist within the sufferer's own ethnic group. In the Kamba cases, some individuals were taken to a *mundu mue* for divination within a few days; others waited for months or years, all the while with persistent misery. Misfortune continued until the candidate acquiesced to the call and was actually ordained as a *mundu mue*. In this sense acceptance of the TMP role is mandatory; to ignore or reject the call is to insure a life accompanied by chronic sickness or other misfortune. In 1979 the Kamba continued to observe these customary signs and rules for the recruitment of new TMPs into their ranks in Mathare.

The following cases illustrate the variety and underlying pattern of experiences that led several Mathare TMPs into their professional callings.

1. N. S. is a female Kamba *mundu mue*, age 45, who was born in Kitui and has lived in Mathare for seven years. She did not serve an apprenticeship, and was taught how to treat by the spirits. "My parents told me I was born with small pieces of herbs (*miti*) in my hands. Later, as a teenager, I saw spirits in my dreams at night. The spirits told me about things that would happen, and also about which herbs to use for different illnesses.

After I married and during my first pregnancy I dreamt I would give birth to a boy. This dream came true. I had the same dream during my second pregnancy, and the child, a boy, was born with 'herbs' in his hands. Another time a medicine bag appeared to me in a dream. In the morning I found a medicine bag without knowing where it came from. I told my parents about this and they told me to keep the bag in a safe place because it is equipment used by a *mundu mue*. One night my grandmother came in a dream and told me that my father had died. I told my husband about this the next day but he refused to believe I could just wake up and say my father is dead without receiving information from some one. About 11:00 A.M. that day some people came to the village and told them her father had died.

After some time I became very sick with madness and pain all over my body. My husband and I went for divination to know what the problem was. The *mundu mue* said I had a call to be a *mundu mue*, and that my husband was refusing to allow me to be one.

My sickness lasted 5 years. After I recovered I attended a circumcision ceremony and became possessed. The spirits told me to gather the equipment a *mundu mue* needs to practice or they would take me away. My mother and a *mundu mue* she sent for both asked the spirits not to take me away. An ordination ceremony was arranged. My mother provided a goat and a medicine bag for the ordination. Afterward, the first person I treated was a woman who had abdominal pains and severe pain after menstruation which had caused her to be barren. After I treated her she delivered a baby boy. The second person I treated had continuous menstruation, and she also recovered. From this time I began to treat other illnesses also."

2. M. T. is a female Kamba *mundu mue*, age 38, who was born in Kitui and has practiced in Mathare for nine years. She did not serve an apprenticeship, but was taught how to treat by the spirits. "I was born with *ue*. There were small 'herbs' on my palms. Between the ages of 10 and 12 the spirits would wake me up around 1:00 A.M. and tell me to sit on the hearth. Another night, after the spirits had ordered me to sit at the fire, I saw a goat coming down to me from the roof. It stood on the hearth and the spirits told me to hold the goat, which I did until morning. The spirits told me not to tell anyone how the goat came to me, and that it should be killed, and I should bury the head but keep the skin to use when treating people.

Two days later, after I had gone to the garden, I saw the image of a woman who had one eye, one leg, and one house. This was a well-known spirit called Mwiitu wa Lala [a prophetess who looks and dresses like an Indian and lives in Kitui District]. This spirit told me she is the one who brought the goat. She also taught me how to treat different illnesses, the herbs to use, and where to find them. I started treating people from that time.

After I got married my husband persuaded me to become a Christian. When I agreed we had to take my tools and burn them before one of the Catholic fathers. After becoming a Christian I changed from the Catholic church to the Africa Inland Church. I finished catechism class and passed the test. It was then high time for me to be baptized, so I was taken to a dam at a river. The pastor stepped into the middle of the river, and when he dipped me I got lost in the water. People tried but failed to find me in the water. I stayed there in the water for a week. During the second week they found me sitting on a rock on the river bank. After I was taken home I was mad and disappeared into the forest for 10 days. When my husband, parents, and other relatives found me I was very thin because of not eating and I was talking and acting like a mad person. The spirit said my tools should be given back to me so that I could begin treating people again. My parents and husband returned my things so that I could resume work as a *mundu mue*. From that time the madness was healed, and I started treating madness and wounds."

3. K. W. is a female Kamba *mundu mue*, age 59. Born in Kitui, she has practiced traditional medicine for about 32 years, including the last 6 years in Mathare. She did not serve an apprenticeship, having learned her art from the spirits. "My father and grandfather were *andu awe*. I started to practice several years after I got married. I gave birth to four children and they all died. I then went back home to my parents and the elders discussed my case. They told me I must practice *ue* like my father and grandfather are doing. A goat was slaughtered and a piece of its skin with some *ng'ondu* inside was put around my wrist to show that I was a *mundu mue*.

My fifth child was born alive. However, I was still young and ashamed to work as a *mundu mue* because the others would laugh at me. I had seven more children who died, and I became mad. I was examined at the hospital but they could find no cause for my illness. Later, I went back to my parents and was told I must be a *mundu mue*. I finally agreed and a goat was slaughtered. Later I had a live boy, my last born."

The histories of these Kamba female practitioners reflect the two basic paths along which the Mathare women TMPs entered the profession of *ue*, or *uganga* (Swahili). Over half were born with herbs or "seeds" in their hands, which is a sign that a child is destined to become a *mundu mue*. Another sign of *ue* at birth is ritual objects or markings (*masyawa*) in a child's afterbirth, especially on the umbilicus. In a few cases, women born with *ue* did not experience illness or other misfortunes during or after adolescence. They attributed this to their parents' early detection and verification of the signs, and their own active acceptance of the call as young women.

The second path to becoming a *mundu mue* (*mundu mugo* or *ajuoga* in Kikuyu and Luo cases) is through an encounter with spirits and misfortune after puberty, and particularly after marriage. Women's personal experience of high infant mortality, or barrenness, can produce enormous emotional strain and suffering. A culturally approved strategy that brings healing is for a woman's family to take her to a *mundu mue* to have the signs and symptoms of her malaise interpreted—and for her to accept and act on the call.

In at least 40% of the cases there were other TMPs, both male and female, in the immediate or extended family of Mathare's female practitioners. However, among the Kamba—ideologically at least—these other *andu awe* have little role in training neophyte female practitioners. Less than one in five of the women practicing in Mathare said they had apprenticed under a *mundu mue*. Instead, the great majority of women received instruction from the spirits.

As the following cases reveal, the career experience of Mathare's male TMPs offers a striking contrast with their female colleagues.

1. N. M. is a male Kamba *mundu mue*, age 55, who was born in Kitui. He had practiced *ue* for over 40 years, and had been based in Mathare for 1 year at the time of his interview. He was a highly respected and popular TMP in Mathare. His father was a *mundu mue*, and N. M. apprenticed with him for 15 years at home in Kitui.

"I was born with *ue* in me, holding marbles (*mbu*, or divining counters) in my hand. One morning when I was 10 years old I awoke and saw that a small piece of wood had dropped on my bed. My father said it would be mine to use in practicing *ue*. To this day I do not know the name of this thing. On another day when I was preparing sticky gum (*lembwa*) for catching birds, I heard something drop onto my bed. I picked it up and took it to my father. It is a spongy-looking thing [N. M. showed it to us] that my father called *ng'ondu* [a magical substance that may consist of parts of one or more of several plants used in ritual therapy, but that must include the plant *ndatakivumbu*]. Father told me that it might be God (*Engai*) who brought this to me, and that it should be kept in a safe place. My father held it for me and handed it over to me when I grew up.

The thing I got from my bed as a boy I use today to treat some cases. I just touch the painful part with this *ng'ondu* and the pain goes away. I sometimes grind this thing and mix the powder with water to make medicine. Whenever it is cut it always returns to its original size and shape.

When I was young I used to see things like people carrying fire. They would disappear if I tried to go near them. I also used to accompany my father when he went to collect herbs and treat people, and I would treat people in his presence. When I grew up people started asking me to come and treat their families to prevent their riches from being spoiled by a clan with the evil eye.

2. K. N. is a male Kamba *mundu mue*, age 48, who was born in Kitui. He had practiced *ue* for about 20 years, and had been in Mathare for 10 years. His paternal grandmother in Kitui was a *mundu mue*, and he apprenticed with her for 3 years.

"I was born with *ue*. When I was still a small boy I became sick and my parents took me for divination. They were told that I had a call to be a *mundu mue*. When I was about 18 I became very sick with severe abdominal pain. After nine months I was told I should be practicing *ue*, and I decided to respond to the call. I went to a *mundu mue* with my parents to be ordained. The ceremony began very well. My parents and some *andu awe* gave me some equipment, including a magic bow (*ota*) and divining counters (*mbu*). However, when it was time to slaughter a goat the goat's neck was too hard to cut. People tried to kill this goat for three days without success. Finally, the ordination was stopped. The *mundu mue* who had been called to lead the ceremony said there was nothing he could do because he could not discover the problem. After the goat was taken home it disappeared and was never seen again. Later, I thought of going to church and becoming a Christian. So one day I went to church. Nothing bad happened while I was there, but when I came home my original illness worsened. I continued to attend church for a year, and also got married during this time. My wife gave birth twice, and both children died. Another ordination ceremony was planned, and this time it was successful. The first illness I treated was *nyuunyi* (gastroenteritis)."

3. N. O. is a male Luo *ajuoga*, age 34 who was born in Homa Bay District in western Kenya. He has practiced traditional medicine for 7 years, and had been working at Mathare for 5 years. He came from a family of TMPs and was apprenticed to a Tanzanian Luo practitioner for 2 years.

"My grandfather and grandmother were both TMPs. When I was in Form One of secondary school I became mad. I started to run around the classroom and misbehave. The teacher started beating me with a ruler. I had a small knife in my pocket. I took it out and stabbed the teacher on the hand. I was taken to jail. Later, my brother took me to Tanzania, where I stayed for 2 years and was treated by a Luo *mganga* who had emigrated from Uganda to Tanzania. This man trained me and I used to treat patients in his presence. After I started to dream at night and see the herbs used to treat patients it was decided I was qualified to practice on my own. When I returned home from Tanzania I was treated well."

The career experience of male TMPs provides a strong contrast with their female counterparts. Only about 20% of the men were "born" with the signs of *ue*. The stories of the two Kamba men presented above are thus exceptions to the general pattern for males. Also, 87% of the male TMPs reported a period of apprenticeship. In comparison, 83% of the women did not apprentice.

Men also reported a much higher prevalence (80%) of other *waganga* in their extended families than did female TMPs (35%). Male Luo *waganga* in the study pointed out their membership in clans known to have a large number of TMPs. These findings are also of interest in relation to the results of the traditional medicine KAP survey conducted among 331 outpatients and inpatients at the Kenyatta National Hospital in Nairobi (Chapter 2). Overall, 13% of these patients, who represented 22 ethnic groups, reported at least one TMP in their family. Kamba patients reported the highest prevalence (28%) followed by the Luo (21%).

Among Mathare TMPs, the great majority of both sexes were already recognized as TMPs in Kitui prior to initiating a practice in Mathare and establishing a divided rural-urban household. Thus most of these Kamba *andu awe* had legitimated claims to their profession and few could be accused of entering the profession from a position of cynicism or greed. As M. L. Swantz (1970, p. 209) also found among the Zaramo people in Tanzania, the fact that becoming a *mganga* (religious-medical specialist) is not customarily a personal choice, and necessarily involves the approval and assistance of other *waganga*, "prevents the office from becoming only a tool of personal ambition."

What conditions explain the distinct "career differences" between male and female TMPs in Mathare? One unusual aspect is that Kitui females form a slight majority of all Kamba TMPs in Mathare. As a mobile occupational group these women are part of a selective process that typically favors males, and are thus different from the majority of women who tend to be disproportionately represented in the rural areas and underrepresented in urban areas.

One of the salient differences between the sexes concerns the onset or manifestation of *ue*. The proportion of women "born" with signs was 2.5 times greater than was the case for men. Almost half of the women manifested physical and mental symptoms of *ue* after they married, whereas men "at risk" generally had to confront the signifying illness and behavioral changes before marriage—during adolescence, in the late teens, or in their early 20s. A large majority of the Kamba men underwent an apprenticeship prior to practicing on their own, while most of the women denied they received any formal training in Kitui. The women emphasized that the transmittal of medical knowledge and powers to themselves came from the spirits alone, regardless of whether *ue* was diagnosed in infancy or after they had borne children.

Similar spiritual paths to the ranks of healers, often accompanied by personal suffering, are reported from other culture areas. Janzen (1978) draws a parallel between such African healers whose careers follow a spiritual or visionary experience—who begin to practice healing arts without apprenticeship—and Arctic shamans. Romanucci-Ross (1978, p. 133), writing about curers in Melanesia, describes how a long illness becomes a "journey of self-discovery for the sick person," who eventually finds his own cure and is thereafter sought out as a healer. However, the literature offers few clues to gender differences in recruitment to the healer role.

In the case of Kamba from Kitui, I do not know what proportion of the total female population is born with the signs of *ue*, or later develops a chronic mental or physical difficulty that leads to a specific therapy through which their affliction is validated and they are recruited to the role of *mundu mue*. Nor is it possible to know, in the case of those said to be born with *ue*, how many never experience a problem that is serious enough to require a course of therapy that ultimately leads to ordination as a *mundu mue*. My sense is that Kamba women have fewer outlets than men for the expression of chronic illness and misfortune, yet their well-being, reproductive success (the principal criterion of status), and labor are crucial to the health and maintenance of the family system. Other studies are needed in areas of Africa such as Ufipa, Tanzania, where TMPs (*asing'aanga*) "seem to be exclusively male" (Willis, 1978), to determine the extent to which comparable "safety-valve" institutions exist for women.

It is well known that African women of child-bearing age are everywhere a high-risk group, particularly in regard to nutritional needs and the effects of nearly continuous pregnancy. For those who are also barren, or suffer repeated deaths of their children, the psychological and social effects can be traumatic and lead to chronic mental and physical dysfunctions. In principle, validation of the "cause" of these problems by a *mundu mue* initiates a process of healing, or at least social acceptance, of the sufferer's condition. For women especially, recruitment to the role of *mundu mue* is not by choice but provides a culturally approved symbol of and explanation for their affliction. It places them in a position to assist others and, as a result, offers an opportunity to regain status.

Moreover, there are apparently fewer constraints on the mobility of a woman who becomes a *mundu mue* than on other females. Today, with the increasing commercialization of traditional medicine, female *andu awe* also have a chance to achieve a degree of economic and domestic independence when they move from Kitui to Nairobi. As in the case of male TMPs, this strategy enables more creative women to earn and invest incomes that permit them to move themselves and their families up to what Collier and Lal (1984) term "the middle peasantry." Opportunity costs are large if the younger female TMPs choose to stay home in remote and drought-prone Kitui, where high rates of male out-migration and poverty place even greater day-to-day domestic responsibility on women. Although other studies focused on women TMPs are needed to establish the accuracy of this scenario, I believe the pattern is widespread in Africa today.

DISEASE CLASSIFICATION ACCORDING TO ETIOLOGY

As in rural Kilungu, Mathare TMPs readily identified a system of disease classification that is specific to their own ethnic group. Although on comparison these group-oriented typologies reveal variation, the most salient feature is the similarity of response and predominance of three major categories of causation (Table 6-4). The ranking of responses is of interest for several reasons.

First, although the importance of "God-given"/natural diseases appears unambiguous, the perceived ultimate cause of an illness can change. Thus in certain circumstances some afflictions such as mental illness or sterility may initially be attributed to human agency, such as witchcraft. It is only later, when the illness becomes chronic and is perceived as incurable, that it is said to be in God's hands. It is crucial to recognize that in actual practice the assignment of cause to illness is subject to broad interpretation. For example, three TMPs might observe a case of generalized edema and independently attribute it to three different causes such as witchcraft, "natural" factors (e.g., circulation of bad blood, or eating spoiled meat), or heredity. Such variation

Table 6-4. Classification of Diseases: Mathare *Waganga*

Origin/cause	Rank all Waganga	%	Rank by ethnic group		
			Kamba	Kikuyu	Luo
"God-given"/natural diseases	1	98	1	1	1
Witchcraft	2	86	2	1	1
Curse	3	44	3	2	2
Seasonality	4	12	4	—	—
Hereditary	4	12	5	—	3
Spirit possession	5	9	5	—	—
Poisoning	6	7	6	3	—
Broken social taboos	6	7	7	3	3
Infection	7	5	6	—	—
Evil eye	8	2	7	—	—
Sorcery	8	2	7	—	—
Other	—	28	—	—	—

Note: From field survey, Mathare, 1977–1978.

occurs across all medical systems because of differences in symptoms, length of illness, and the training experience and opinions of the therapist (see Chapter 5). Table 6-5 indicates the range of etiologies Mathare TMPs attributed to ten of the illnesses they perceived themselves "best qualified to treat." They usually gave evidence of multiple etiologies. Nevertheless, the Mathare findings support those who insist that in Africa generally traditional therapies are employed to treat a wide range of *naturally* caused illnesses which have little if any connection with witchcraft or other "magico-religious conceptions" (Feierman, 1984, p. 10). The evidence also suggests that natural causation is more widely recognized in the city than in rural Kilungu (and probably Kitui), which is due in no small measure to TMPs' observation of and borrowing from biomedicine made possible by more intensive contact with dispensaries and hospitals. Thus there are indications of a shift toward "modern" interpretations of disease involving such concepts as germ theory and contagion (see entries "God-given/natural diseases, seasonality, and infection, Table 6-4).

A second feature of the disease classification by the Mathare TMPs is the minor rank given to spirit-caused illnesses relative to the rural study area (Table 6-4). Possible explanations for this include a diminishing concern with ancestral and clan spirits among the more sophisticated urban population. In addition, it is possible that the urban location has become an impractical venue in cases of spirit possession (and other situations) where the patient's kin or age group—located at a distance in the rural area—would customarily be expected to participate in the therapy (M. L. Swantz, 1979).

Table 6-5. Varieties of Causation of Selected Illnesses According to Kamba Traditional Medical Practitioners in Mathare

---

Madness (*Nduuka*)
    No blood/anemia
    Blood "boils"
    Bewitchment/witchcraft
    Curse from parents
    Excessive thinking
    Complication of malaria
    Drug use (e.g., *bhangi*)
    Jealousy
    God-given
    Studying too much
    The worm in head becomes strong
    The worm swims in the cerebral spinal fluid
    Heart failure

Epilepsy (*Kizala* or *Munguthuko*)
    Blood not moving and remains weak
    Brought by God—mostly natural
    Witchcraft of enemies and relatives
    Hereditary, passed from generation to generation
    Worm in the head turns and patient falls down
    Overly rapid heart beat

Infertility (*Ngungu*)
    Natural—born to be barren
    Breaking a family rule, e.g., failure to pay dowry
    Witchcraft—person "tied" by jealous friend who got hold of her menstrual blood
    Disease, e.g., venereal disease, would destroy ovaries
    Diseases from prostitution
    Bad blood (black clots) during menstruation
    Curse

General Edema (*Mwimbo* or *Kithakame*)
    Witchcraft
    Hereditary
    Originates from liver and kidney defects
    Bad blood circulation, some thick and some thin
    Eating infected meat
    Caused by meat diet
    Hot, strong sunshine
    Lack of blood or too much blood

Infant Gastroenteritis (*Nyunyi*)
    God-given
    God-given only
    Hereditary from mother
    Eating soil (infants who crawl)
    Coldness
    Teething
    Unclean food or impure water
    Natural causes—infections passed among children in a family

Impotency (*Ndewa*)
    Witchcraft—caused by a woman who wants a man who already has another woman
    Lack of blood

Gonorrhea (*Kisonono*)
    Infection via sexual contact

Table 6-5. Continued

Worms (*Nzoka*)
    Uncooked food, bad meat
    Contaminated water and food
    Seasonal, when it becomes very hot

Abdominal pains (*Kwalwa ni ivu*, etc.)
    Witchcraft
    God-given
    Coldness
    Contaminated/poisoned food
    Change of diet
    Ulcer
    Worms
    Menstrual period

Pneumonia (*Kyambo*)
    Coldness, e.g., during rains
    God-given
    Witchcraft

If spirit-related illness etiologies were on the wane in the urban environment, witchcraft and sorcery remained alive and well as causal agents of unusual or abnormal illness. "Witchcraft" (merged conceptually with sorcery) had become a "catch-all" explanation for myriad disorders in interpersonal relations. It was also a widely acceptable cause of personal failings ranging from lack of success in business, love, and school to a sudden illness or strange disease, and especially barrenness and impotency.

PERCEPTION AND TREATMENT OF DISEASE AND ILLNESS: TOWARD AN ETHNOMEDICAL EPIDEMIOLOGY

Mental illness and "madness" head the list of illnesses that Mathare *waganga* perceived themselves as "best qualified to treat" (Table 6-6). Half of all the TMPs identified these disorders as one of their three main areas of expertise. Since depression, anxiety, and unusual behavior are also frequently associated with many other conditions such as infertility, alcoholism, drug abuse, curses, and bewitchment, the magnitude of mental health problems surely exceeded all others in the view and experience of Mathare's TMPs. Mental illness was also proportionately greater in Mathare than in rural Kilungu. Although not a list of illnesses actually treated, Table 6-6 is an important illustration of how urban TMPs oriented their practices. Females showed a greater propensity than males to view themselves as best prepared to treat abdominal problems, infant gastroenteritis, and epilepsy. Men, on the other

Table 6-6. Three Illnesses Mathare *Waganga* Are Best Qualified to Treat

Illness	Percentage of all *waganga* ($N = 45$)		Percentage of Akamba *waganga* ($N = 35$)	
Mental illness/madness	49		51	
Abdominal problems	38	♀[a]	40	♀
General edema	27		31	
Infertility	22	♂[b]	23	♂
Infant gastroenteritis	16	♀	14	♀
Epilepsy	16	♀	20	♀
Pneumonia	13	♀	17	
Impotency	11	♂	11	♂
"Witchcraft"/bewitchment	11	♂	9	
Gonorrhea	11		14	
Bloody diarrhea	7		9	
Worms	4	♂	6	♂
General body pains	4	♀	3	

Note: From field survey, 1977–1978.
[a]♀ Female percentage > twice male response.
[b]♂ Male percentage > twice female response.

hand, were twice as likely as women to claim competency in the treatment of infertility, impotency, and worms, and in methods of neutralizing witchcraft.

Problems associated with the abdominal tract and sexual organs—which uneducated rural folk may perceive as "the most important part of their bodies, sustaining the functions of nutrition and reproduction" (Ndetei & Muhangi, 1980: 249)—also loom large in the urban environment. In spite of Kenya's very high fertility (Sindiga, 1985), I saw daily evidence in the clinics of Mathare *waganga* that large numbers of urban women are deeply troubled by their inability to conceive. In Nelson's (1975) Mathare project, for example, barrenness was reportedly characteristic among about 14% of the resident women.

Two other measures provided direct evidence of the nature and frequency of illnesses treated by Mathare TMPs. Each practitioner in the village surveys ($N = 45$) specified the problems his or her patients presented during the week immediately preceding our interview with them. Analysis of the Kamba TMPs (Table 6-7) reveals important areas of agreement in comparison with the rural study area (Table 5-10). The major differences include the greater frequency of edema and mental illness, and the high visibility of divination sessions related to problems of personal and social adjustment in the city; and the prominence of spirit possession cases—as well as maternal and child health concerns—in the rural area. The latter observation is particularly important because it highlights the actual population structure and thus the prominence of maternal and child health needs in an area characterized by high male out-migration and scarce biomedical services.

Nine Kamba *waganga* received instruction from my research assistants in the keeping of written patient records. They also received supplies of printed forms in their own languages. They completed 468 usable forms (cases) on a discontinuous basis during a 3-month period, often with help from my research assistants. This information provides a third perspective to the types and pattern of illnesses handled by Mathare's TMPs (Table 6-8), about which more will be said later. The five categories of illness shown account for 70% of all conditions recorded. The prominence of personal and social adjustment problems in Tables 6-7 and 6-8 highlights what Hake (1977, p. 322) correctly terms "the essential insight," which is to say that many afflictions are traceable to broken relationships. "While a Western-trained doctor will ask a patient if she has opened her bowels, a traditional African doctor will investigate the hereditary factor by asking about ancestors and if she has quarreled with a neighbor, or is at odds with her society and its culture, finding out if she is expressing guilt" (Hake, 1977, p. 222).

As I outlined in Chapter 1, the *ethnomedical approach* emphasizes the personal and social meaning contexts of illness. Its main goal is to provide a more realistic and comprehensive understanding of illness, and through that process enhance the effectiveness of therapies in medically pluralistic settings. As such, the information about the identity, causes, and relative frequency of the illnesses shown in Tables 6-4 through 6-8 (Tables 5-7 and 5-10 through in Chapter 5) represents a necessary and important first step in develop-

Table 6-7. Illnesses Presented to Kamba Traditional Medical Practitioners During Week Before Their Participation in Survey

Illness/problem	Rank
Abdominal problems	1
General and specific edemas	2
Personal and social adjustment problems requiring divination, e.g., home problems, unpopularity, medicine for love, desire to improve business success, chronic unemployment	3
Mental illness/madness	4
Headache	5
Infertility/impotency	6
Sexually transmitted diseases	6
Palpitations/"heart attack"	7

Note: From field survey, 1977–78.

Table 6-8. Illnesses and Problems Recorded by Nine
Mathare Traditional Medical Practitioners on Client
Record Forms ($N = 468$)

Illness/problem	Rank
Personal and social adjustment problems	1
Abdominal problems and body pains	2
Infertility and female reproductive/GU tract	3
Sexually transmitted diseases	4
General and specific edemas	4

Note: From field surveys, 1977–1978.
Abbreviations: GU, genitourinary.

ing an ethnomedically informed epidemiological approach to illness and misfortune in urban and rural communities in Kenya. It will be obvious that practically all of these illnesses are seen to occur, and are therefore approached by TMPs, in relationship to patients' personal histories, physical environments, and social networks. For biomedical practitioners, this is a new and discrete kind of epidemiological knowledge of illnesses. When coupled with information about the prevalence, perceived seriousness, and the stressors and circumstantial factors of illness, such knowledge helps to create a clearer, less mystical and less prejudicial image of patients, their needs, and the TMPs whom they consult.

NOTES

1. For example, on Christmas Eve in 1982 a Kikuyu woman's kerosene cooking stove burst into flames causing a fire that spread wildly to neighboring shanties. By the time a city fire brigade arrived, 10,000 families had reportedly lost their homes and belongings to fire and looters. J. Muchunu, "A Kenyan Remembers," *World Vision*, December 1984–January 1985, 28, 10–11. In the case of similar fire in 1977, the city fire brigade arrived some 30 min after the inferno began [*Standard* (Nairobi), October 29, 1977]. Flooding and landslides caused by heavy rains are also seasonal hazards in Mathare and other "unauthorized" settlements in Nairobi [*The Weekly Review* (Nairobi), April 24, 1981, 10–11].

2. The original random sample was enlarged to provide greater representativeness and, as a concession to the human feelings reflected in *waganga* politics, to maintain a climate of participation and equanimity. In effect, the sample sizes are as follows: Mathare I (85%), Mathare II (57%), and Mathare III (57%). Comparable village figures for the Kamba subset were 92%, 78%, and 67%, respectively. Four of the 35 Kamba resided physically outside Mathare—3 in adjacent Eastleigh and 1 in nearby Ngara. Kinship and functional ties of TMPs in Mathare I lead to their inclusion.

3. The following questions about housing were part of the standardized survey (p. 7) administered to the 45 TMPs in the Mathare sample. "How did you acquire this room? (What factors influenced your decision to live and practice your medicine in this particular village and in this specific house?) Please explain."

4. Personal communication, Mauvumi Sa Kiantandu, January 23, 1977.

5. One female Kamba TMP expressed her dissatisfaction with the society because members were asked to raise additional funds, particularly for costs connected with a TMP's death.

6. Information in this discussion was obtained during an interview at Swahili Village, Machakos, with the treasurer of the Kenya Miti Shamba Society, June 9, 1978.

7

Patterns of Illness and Patient Behavior in Mathare

METHODOLOGY FOR THE STUDY OF PATIENTS, ILLNESS BEHAVIOR, AND THERAPY

There are two basic approaches to the study of patients' interactions with health care resources. Population-based studies focus on the features and behavior of all members of a sample population—the healthy and the sick—in a specific area. Such studies are costly but valuable because they can differentiate individual and community patterns of illness, well-being, and choice of therapeutic alternatives. Chavunduka's (1978) study of illness behavior in a random sample of 200 residents of Highfield, an African township in Salisbury (Harare, Zimbabwe), is a useful example of a population-based study. It focuses on how and when patients choose between biomedicine and traditional medical practitioners (TMPs). One advantage of this approach is that it provides an opportunity to observe illness behavior throughout a course of therapy—from the time an illness is first perceived, through its various episodes, and within a framework of therapeutic alternatives. It also distinguishes the sick from the nonsick; thus it has the potential to describe and quantify health care needs in a given population during a specific period.

Utilization surveys are the second method of investigating illness behavior. They focus on the population that actually visits a particular therapist and are thus provider- or facility-based in terms of information gathering. A liability of this method is that it does not adequately identify the true extent of the unmet need for health care in the community at large. Another disadvantage of utilization surveys is that a patient's therapeutic choices and outcomes that occur prior to contact with an interviewer at a TMP's clinic or a hospital outpatient ward, for example, must be determined retrospectively and may therefore be more likely to suffer from bias and incomplete reporting.

Advantages of the utilization survey include: (1) it permits a broader locational analysis of patient origins and characteristics (the population-based approach is limited to individuals in the sample who become sick, and to the TMPs or facilities they choose to visit); and (2) patients contacted in a TMPs clinic, for example, have "voted with their feet" to put faith in a kind of therapy that may be stigmatized by others. In a population-based study, in

228

contrast, the patient may deny that he or she would make such a choice or, perhaps more often, would obscure his or her actions by making surreptitious visits to TMPs—as reported from northern Machakos District (van Luijk, 1984). Due to the influence of Christian missions, Western education, and colonialism in general, many people in Kenya[1] are ashamed to admit that they consult with *waganga*. No such stigma attaches to associations with biomedical practitioners, however, even though many patients in TMPs' clinics exclaim that they have been sorely frustrated by the poor results from earlier hospital and clinic visits made in connection with their current problem.

In the present study, information about patient characteristics and illness behavior was developed mainly through utilization-type surveys at TMPs' clinics and by surveying the TMPs' general knowledge of their clientele. However, formal questionnaires on their own are inadequate (and sometimes disruptive) tools in surveys of health behavior, whether in a divination session or a hospital ward. Thus patients were also interviewed informally, and observed in a wide variety of therapy sessions together with the "significant others" providing support. My research associate, research assistants, and I were participant observers in scores of therapy sessions involving patients and TMPs in Mathare. Some of these sessions were spread out over 2 or 3 consecutive days followed by a return visit after an interlude of days or weeks. In several instances one or more of us was obliged to play a peripheral or, in one case described later, key role as a co-opted member of a patient's therapy management group.

IDENTITY OF PATIENTS WHO CONSULT MATHARE'S TRADITIONAL MEDICAL PRACTITIONERS

Studies in Lusaka, Zambia (Leeson, 1968) and Kinshasha, Zaire (International Development Research Center, 1980) indicated that proportionately greater numbers of women than men attend the clinics of urban TMPs. Leeson also found that children are most often taken to a biomedical facility. In contrast, the Kinshasha study revealed that while TMPs treat small numbers of patients aged 6–15 years, "a large number of children under age 5 in Kinshasha were consulting herbalists and spiritualists. This finding was quite unexpected because of the numerous medical services available for both mother and child" (International Development Research Center, 1980, p. 9). In his population-based survey in Salisbury, Rhodesia, Chavunduka (1978) compared men and women with regard to their first choice of treatment, that is, whether it was "scientific" or "traditional," and found no statistically significant difference based on gender.

Evidence from Mathare patients corroborated most of the Lusaka and Kinshasha findings with respect to the gender and age structure of patients. A majority of patients placed in the 20- to 40-age group. Women accounted for 55 to 60% of the adults recorded by the nine *waganga* who agreed to keep patient record forms for us. This compared well with Leeson's (1968) finding that females made up 63% of the adults attending the *ng'anga* in Lusaka. In Mathare, children accounted for <5% of TMPs' patients. In sharp contrast with Kinshasha, children under 5 presented for treatment infrequently.

Since traditional healing is usually their main source of livelihood in Nairobi, it is not surprising that many *waganga* said they treated "all age groups" (Table 7-1). While many treated children on occasion, or would agree to do so, adults formed the core of their practices. Table 7-2 records the great variety of patients and conditions treated in the course of one day by a popular, male Kamba TMP (pseudonym "Ngali") based in Mathare III. Only adults, mainly women (59%), presented for treatment on that day. Remarkably, Jaluo—a people whose language and culture history are fundamentally different from the Kamba—made up over 40% of Ngali's patients. Although it provided only a "snapshot" view, the information in Table 7-2 coupled with later observations showed that Ngali's charisma and compassion outweighed his own ethnicity as a decision factor when his patients needed a TMP.

In the urban environment most patients seeking treatment from a TMP are referred by a relative, friend, or other associate who has some prior knowledge of the practitioner's reputation. In Nairobi, as in Kinshasha (International Development Research Center, 1980), ethnicity was not unimportant but neither is it the dominant motive for selecting one TMP over another. Client record returns from Mathare indicated that while some Kamba TMPs saw other Kamba more often than Kikuyu or Jaluo patients, as Ngali's case illustrates, there is also wide variation in the ethnic proportions of clientele. This suggests a weakening of strict ethnic reference and probably a growing awareness of the interethnic character of African tradi-

Table 7-1. Age Groups and Sex Structure of Patients in Categories Reported by Mathare Traditional Medical Practitioners

Age/Sex	All TMPs ($N = 45$) (%)	Akamba TMPs ($N = 35$) (%)
All age groups	44.0	43.0
Adults: males and females	20.0	23.0
Adults: males	7.0	5.5
Adults: females	20.0	20.0
Children	4.0	5.5
Children and pregnant women	5.0	3.0
	100.0	100.0

Note: From field survey, 1977–78.
Abbreviations: TMP, traditional medical practitioner.

Table 7-2. Patient Load of "Ngali," Mkamba Traditional Medical Practitioner, August 5, 1977

No. patients	Sex	Ethnic group	Home area	Problem presented
1	M	Mkamba	Machakos	Abdominal pains
2	F	—	West Kenya	In 12th month of pregnancy(!), and occasional vaginal bleeding
3	M	Luo	Kisumu	Divination requested due to problems at home
4	M	Kikuyu	Murang'a	Failure to find employment
5	F	Kikuyu	—	"After-pains"
6	F	Mkamba	—	Abdominal pains
7	F	Mkamba	Kitui	Itchy body
8	F	Mkamba	—	Wants husband to return home
9	M[a]	Luo	Kisumu	Rectal pains
10	F[a]	Luo	Kisumu	Vaginal pains
11	F	Luo	Kisumu	Divination requested
12	M	Mkamba	Machakos	Divination requested
13	F	Luo	Kisumu	Requested medicine to make her successful in trade, and for love affairs
14	M	Mkamba	Machakos	—
15	M[a]	Luo	Kisumu	Complained of "extravagance"
16	F[a]	Luo	Kisumu	Infertility for 20 years of marriage
17	F	Luo	West Kenya	Abdominal pains

Note: From field observation, Mathare III.
[a]Patients 9 and 10 and 15 and 16 are married couples.

tional healing engendered in the heterogeneous urban milieu (International Development Research Center, 1980).[2] The Mathare sample of TMPs provided confirmation of this cross-over phenomenon (Table 7-3). Two demographic factors also influenced the apparent patterns of interaction between patients and TMPs. The Kikuyu, who form about 60% of Nairobi's African population, and the Luo, are the largest ethnic groups in Kenya. On the other hand, Kamba formed a disproportionate share of the TMPs in Nairobi (to the extent that the Mathare findings can be extrapolated). Consequently, there is reason to expect that "supply and demand" factors also encouraged the practice of interethnic traditional medicine.

A strong pragmatic quality characterized the actions of both patients and healers in Mathare. The emphasis was increasingly on doing what seems to work, and testing out strategies that might work, which is why the therapeutic process involved many kinds of actors. The latter sometimes included TMPs of a different ethnic group than the patient; *mabwana mkubwa* (Swahili for "men of great influence") such as Kalii in Machakos (Chapter 4) or Kabwere at the coast (he operates the "Anti-Witchcraft

Table 7-3. Ethnic Groups Treated Most Often By Mathare Traditional Medical
Practitioners

Ethnic groups (rank order)	All TMPs ($N = 45$) (%)	Akamba TMPs ($N = 35$) (%)	Kikuyu TMPs ($N = 35$) (%)
Akamba, Kikuyu, Luo	36	46	—
Akamba, Kikuyu, Luo, Luhya	27	—	—
Kikuyu, Akamba, Luo, Luhya	9	34	—
Kikuyu, Akamba, Luhya, Luo	—	9	—
Kikuyu, Luo, Luhya	—	—	40
Kikuyu, Akamba, Luo	—	—	20
Luhya, Kikuyu, Akamba	—	—	20
Other	28	11	20
	100	100	100

Note: From field survey, 1977–1978.
Abbreviations: As in Table 7-1.

Clinic" in Malindi); and biomedical practitioners. All formed part of the same *ethnomedical* system. For better or worse, openness, flexibility, and innovation emerged as key features that characterized the attitudes and behavior of both patients and TMPs. Contemporary ethnomedical systems are thus multiethnic and polymedical! In the Kinshasha study, for example, many patients had psychic disorders for which a ritual treatment (e.g., Zebola) provided by a specialist is required. Despite the deep-rooted cultural foundations of such therapy, only 45% of the urban patients consulting a TMP chose one from their own ethnic group (International Development Research Center, 1980, p. 11).

In general, the sex of a TMP is not an important consideration in the choice of a practitioner in an urban or a rural area. Katz, Katz, and Kimani (1982) also observed this in their study of trends in urban traditional medicine in Nairobi. The evidence suggested that male TMPs are no less likely than females to treat gynecological disorders such as infertility, menstrual difficulties, or miscarriage (Table 6-6). However, definitive evidence concerning patients' sex preference in TMPs for certain kinds of conditions may be confounded by illness labels. "Abdominal pains," for example, which female TMPs tended to identify as a specialty at least twice as often as the men (Table 6-6), may in fact frequently refer to irregularities of the menstrual cycle. Often, only direct questioning of the patient can satisfactorily determine the nature of the problem in a biomedical sense (Katz *et al.*, 1982).

Evidence concerning the educational level, marital status, and occupation of the patients of *waganga* in Mathare was gathered informally and through numerous case histories. Most patients had at least a few years of primary school education and an estimated one-fifth had received some secondary training. Individuals with additional vocational or professional education also consulted *waganga*. Although we met several such persons in Mathare clinics including, for example, a Kenya Air Force officer, a nurse from Kenyatta National Hospital, and a former medical assistant, it was difficult to know their overall proportion among patients. Not infrequently their embarrassment at discovery pressured them into making a quick exit from the TMP's clinic. It was and is widely believed that persons of higher socioeconomic status who patronize TMPs usually attempt to make themselves inconspicuous, preferring to come after dark or to arrange a consultation at a private venue.

Collectively, patients of TMPs reflected the full spectrum of domestic and family-living arrangements characteristic of the contemporary African city. Some had chosen a church wedding and lived in a monogamous relationship with a spouse. With their children, they maintained a "Western"-type, nuclear household. In a majority of cases, however, it was difficult to certify the "marital status" of patients without invading their privacy. While a stable marriage and family life remain the ideal standard, consensual unions, some long-term, many ephemeral, are commonplace. Some are based on mutual affection and are relatively secure, others on convenience or assured profit. Members of many intact families, both polygamous and monogamous, are separated from their spouses by distance and time because they work or are looking for employment in Nairobi. Many other young people and adults, women and men, are products of broken homes or are otherwise on their own and attempting to find ways to survive and satisfy basic needs in the city. The upshot of these varied personal and social circumstances is that use of the terms "married," "single," and "divorced" is problematic and can lead to inaccurate assumptions. We did not attempt to formally determine the status of patients' current living arrangements, although in the case of TMPs we confronted reality by simply identifying those "partnered" (91%), their average number of partners (1.7), and those "not partnered."[3]

Occupations and employment status of patients who consulted Mathare *waganga* showed wide variety. Many maintained self-employed status in the "informal sector," and some had wage or salaried positions in the formal economy. Many others were seeking jobs and remained more or less dependent on relatives, friends, or their own devices for support. The range of "occupations" among patients included the following:

—Unemployed	—Kenya Air Force Officer
—Student	—Kenya Army noncommisioned officer
—Peasant farmer	—Medical assistant
—"Businessman"	—Dressmaker
—Beer brewer	—Hairdresser
—Prostitute	—Enrolled nurse–midwife

—Sales clerk   —Housewife
—Butcher   —Cleaner
—Driver   —House maid
—Mechanic   —Waiter

In Zimbabwe (Rhodesia), Chavunduka (1978) divided the patients in his study into two income groups: those making under (46.5%) and over (53.5%) Rh $42 per month. He found no significant difference between the two groups with respect to whether their first choice of therapy was "scientific" or "traditional" when they became ill. Education level was similarly not an important determinant in therapy selection. Three-quarters of the people Chavunduka (1978) described in 20 case studies consulted TMPs at some stage in their illness. Their occupations included police sergeant, builder, office manager, laborer, shop assistant, housewife, shoe repairer, supervisor of a railway work gang, teacher, vegetable seller, and dressmaker.

SOURCE AREAS OF PATIENTS AND ACCESS TO TRADITIONAL MEDICAL PRACTITIONERS

Interviews of TMPs, patient record forms, and sustained observation indicated that approximately 60% of the patients of Mathare TMPs originated from within Nairobi, including Mathare. Asked to specify where most of their patients live, 64% of the *waganga* ($N = 45$) reported "Mathare and Nairobi." Only 22% specified a particular village in Mathare. In addition, although I had assumed that a small proportion of patients would originate from outside Nairobi, an unexpected 37% of *waganga* indicated they received substantial numbers of patients from rural areas and other towns. While these TMP reports do not represent the actual proportions of patients who came from the urban and rural areas, they do provide a useful surrogate measure. Supportive evidence comes from an origins analysis of 271 patient record forms filled out by five TMPs themselves or with the assistance of research assistants. Some 58% of the patients came from within Nairobi, including 19% from Mathare. Analysis of the origins of 125 patients recorded by one female Kamba TMP without our assistance revealed the following locational pattern: 5% from within Mathare, 38% from other places in Nairobi, 51% from outside Nairobi, and 6% undetermined. The most frequent sources of patients from outside Nairobi included Kiambu, Machakos, Murang'a, Kitui, Embu, and Meru districts. The distances of these areas from Nairobi ranged from 5 miles to over 100 miles. At a later time, my research assistants interviewed 27 patients who consulted this same woman. None of the patients originated in Mathare; 48% came from other areas of Nairobi such as Eastleigh, Kariobangi, Pangani, and Muthaiga; and 52% came from other Districts, especially Machakos, Murang'a, and adjacent

Kiambu. A similar pattern of "distance-decay" has been identified in many other studies of health services utilization undertaken in a wide variety of social and cultural environments (Gesler, 1979a,b; Girt, 1973; Joseph & Phillips, 1984; Stock, 1980).

Perhaps the most remarkable finding concerning the patient origins in Mathare is the strong inward flow of patients from the rural areas. It is likely that such movement was (and is) occurring, at least in part, in response to the declining numbers of TMPs in rural areas—such as Kilungu—that I postulate in Chapter 5. This pattern is also supported by the evidence cited earlier in this chapter that 20% of Mathare's TMPs "often" receive patients from their own home village.

Comparative data concerning patient origins are not available or do not exist. Chavunduka's (1978) study focused on patient behavior and was not designed as a utilization survey of urban TMPs' patients. Nevertheless, what he has presented is of interest because it appears counterintuitive to the findings for Mathare and Nairobi discussed above. Of the 132 people in the Harare (Salisbury) study who consulted TMPs, 70% chose to consult *rural* TMPs, 5% chose TMPs resident in other towns, and only 20% selected a practitioner in Harare. The remainder provided insufficient information (Chavunduka, 1978, pp. 42–43).

In contrast, the Nairobi findings highlight the popularity of urban TMPs among rural as well as urban residents. However, since the present study is provider-based I did not determine the proportions of Mathare and Nairobi residents who traveled outside the city when they wished to consult a TMP. It is not inconceivable that a population-based study would reveal patient behavior in Nairobi and Harare (Salisbury) in a comparable light. More research is a prerequisite to understanding this important spatial dimension of illness behavior in Africa. In the Zimbabwe (Rhodesia) study, rural TMPs had a higher reputation than their urban counterparts, and people were willing to travel long distances to contact them. Patients believed that urban TMPs are not as well qualified as rural TMPs. They also took exception because urban practitioners expected to be paid *before* a patient was cured, and also tended to publicize their availability—thereby breaking an "unwritten code in Shona medical practice that no healer may advertise or promote his own competence" (Chavunduka, 1978, p. 23). Many Shona also preferred to be treated near their rural home (especially if the illness is life-threatening) because of the desire to be amongst a supportive group of kinsmen, the ceremonial requirements associated with death, and the desirability of burial at the family home together with ancestors. Since traditional medicine carries a stigma, some city-dwellers also consulted rural TMPs because they wanted to avoid the risk and shame of being seen by their urban friends, neighbors, and employers, and the possible effects of such discovery on "their job, family, religion or reputation in their urban neighborhood" (Chavunduka, 1978, p. 24).

Some of the above factors certainly must also be considered by people in Nairobi when illness arises. Yet I sense that the situation is different in Nairobi. The city's population (including its TMPs) and its functional region are apparently characterized by much greater ethnic diversity than Harare (Salisbury) and its surroundings. These factors provide a greater sense of anonymity for patients who wish not to be seen in a TMP's clinic. Also, although there is extensive and continuous circulation of people *from* Nairobi to the rural areas and back again for social and livelihood purposes, a large proportion of folk who consulted TMPs in Mathare (and elsewhere in the city) remained essentially rural-based. This suggested that the Nairobi pattern is not the same as Harare's.

DETERMINANTS OF THE CHOICE OF THERAPY

Residents of Africa's cities generally have a much broader range of choice of therapeutic options than their rural kinfolk. Potentially all of the alternative strategies identified in Figure 2-1 are available and physically accessible to city folk, whether they happen to live in Nairobi, Harare, Kinshasha, or Ibadan. In contrast, although rural residents generally have a more restricted choice of therapy and a "hierarchy of resort" (Schwartz, 1969) is often apparent, physical distance may nullify the practical value of a particular option—especially when the sickness is acute, the individual's mobility is impaired, and transportation is unavailable.

Given the pluralistic character of ethnomedical systems in both rural and urban settings, what factors determine the choice of therapy? Who has authority to decide and thus exercise control over the patient and his or her course of therapy? Satisfactory explanations of these important theoretical issues are crucial to comparative understanding of the search for health and wholeness, how diagnoses are made, and thus the healing process itself. The debate has been joined by scholars from various disciplines, particularly medical anthropology and medical sociology. Much of the literature on choosing a particular therapy explains such behavior in terms of some characteristic(s) of the patient (e.g., education, income, ethnicity) or some feature of the illness itself (Colson, 1971; Yoder, 1982b). Others argue that since patients frequently consult traditional, biomedical, or other specialists for the same malady, decisions are based on perceptions of what each therapist can contribute to restoring well-being (Gonzalez, 1966). However, such explanations are usually individualistic and have generally overlooked the linkage that exists between therapy choices and structural aspects of the health system. Key areas of such linkage include the social management of illness (Chavunduka, 1978; Feierman, 1984; Janzen, 1978) and the relative location (urban or rural) of the patient and therapist. Relative location underscores the importance of accessibility of all kinds of therapy options

(Lasker, 1981). Thus choosing therapy also has a significant geographical dimension, involving interactions of specific groups of people in an array of locations whose attributes may reflect a "hierarchy of resort." Every course of therapy is, therefore, shaped by place and time relationships. These factors also act as determinants of therapy, or "independent variables," and help to create social geographies of the therapeutic process (Chapter 1). These geographic dimensions to healing have received little attention in the literature on medical pluralism in Africa or elsewhere.

In his recent seminal review entitled *The social origins of health and healing in Africa*, Feierman (1984) observed that the literature on African therapy systems offers three different categories of answers concerning how therapies are selected. These focus on (1) cultural interpretations, (2) control by healers, and (3) lay control, especially the authority of "significant others."

Cultural interpretations of therapy selection emphasize shared values rather than who is in authority or in control of the therapeutic process. Each society or ethnic group maintains its own specialized knowledge or set of responses to disease and illness which are vested in a more or less integrated medical system. There is, however, much debate over whether it is in fact even possible to speak of a "core" medical tradition, particularly in the face of the multitude of therapeutic forms in Africa today.

I believe it is more realistic and productive to think of African therapeutic traditions as constituting open systems. Concepts and practices specific to the traditions of a person's own ethnic group typically form the "core" reference points when illness arises. Today, however, these traditional approaches are usually insufficient as guides to or explanations of healing strategies. I have argued throughout the present study in favor of conceptualizing a society or community as having an "ethnomedical system" (EMS), defined as the entire set of strategies and traditions that are relied upon to maintain and restore well-being. An EMS incorporates *all* practices acceptable by the community, including biomedicine. It also accommodates the readily observable "cross-over" behavior of different ethnic groups and social classes, and allows (not without tension and misperception) the coexistence of different causal explanations of illness and misfortune.

Chavunduka (1978, p. 40) found among the Shona people of Zimbabwe that the definition attached to an illness by the sick person and his or her social group *"at any given time"* (my emphasis) is the key determinant of the choice between TMPs and biomedical practitioners. Furthermore, Shona distinguish between "normal" illnesses, which are self-treated with herbs or referred to biomedical practitioners, and "abnormal" illnesses, which are usually referred to TMPs. Abnormal illnesses tend to have chronic or deviant characteristics, such as persistent headache, abdominal pains, or inability to conceive. Most Shona "assume that an illness is a normal one until subsequent events prove otherwise. Thus when in need of professional help, most patients nowadays make scientific medicine their first choice" (Chavunduka,

1978, p. 40). This assumes, of course, that appropriate biomedical facilities are nearby. Overall, "scientific" medicine was the agency selected by 78% of 132 Shona patients before they consulted TMPs (Chavunduka, 1978, p. 45). There is evidence of a similar hierarchy of resort in Zaire (International Development Research Center, 1980).

The unfolding of an illness can thus serve as a guide to therapy. In this regard Davis-Roberts (1981, p. 315), who studied the traditional medical system of the Tabwa people in Zaire, seems very close to the mark for Africa in general in asserting that "the relationship of experience to time is at the core of the cultural construction of illness."

The second focus in the literature concerning the choice of therapy in Africa examines the kind of competition that occurs among healers who represent the different therapeutic traditions available today (Feierman, 1984). Professionalization and the variety of criteria by which the efficacy of a therapy is judged are key areas of concern. Because centralized authority over all but public sector biomedicine is generally lacking, and professional or bureaucratic referral enforcement is impossible (Janzen, 1978, p. 130) the "practical effect . . . is to leave the crucial decisions on therapy in the hands of patients and their relatives" (Feierman, 1984, p. 30).

Lay control is the third approach to the study of how therapeutic decisions are made. The focus here is on the sick person's kinfolk, neighbors, and other associates in the individual's social network who choose among the therapies available. Chavunduka's *Traditional healers and the Shona patient* (1978) and Janzen's *The quest for therapy in lower Zaire* (1978) are pioneering studies of lay control of the therapeutic process in Africa.

Janzen's central organizing concept is the "therapy managing group," (TMG) a previously neglected phenomenon in African healing which he skillfully depicts in a case study of lower Zaire. Janzen argues that the process and pattern of illness management and the choice and interrelatedness of therapies selected are difficult to recognize if the focus of attention is limited to the patients alone or the specialists who treat them. Instead, close kinsfolk and occasionally friends and other associates—the ever-present "family" of significant others who surround the sufferer once an illness or major problem materializes—are the key to understanding why, how, and when alternative therapies are chosen. In essence, the TMG rallies together

> for the purpose of sifting information, lending moral support, making decisions, and arranging details of therapeutic consultation. [It] . . . thus exercises a brokerage function between the sufferer and the specialist, whether this be for a hernia operation by a Western doctor or a plant cure known to a traditional practitioner for the treatment of sterility (Janzen, 1978, p. 4).

Membership in the TMG may change in composition and, hence, "personality" during the course of an illness in response to a particular diagnosis or new information.

As in Zaire, among the Shona of Zimbabwe it remains customary for a wide range of kin, both educated and uneducated, to become involved in a person's illness, such that "he does not often act on his own" (Chavunduka, 1978, p. 45). Therapy decisions are made jointly. Money is usually not a major limitation when a person becomes ill, and kinfolk are ultimately accountable for the medical fees incurred. On occasion kinfolk may also be held responsible for a person's illness. In such cases they may be viewed as "extended patients." Kin participation and unity is also promoted by individual fears of being accused as a witch. Since all relatives are expected to show concern and visit the sick person, failure to meet this obligation can raise suspicion (Chavunduka, 1978). In Mathare, for example, it was quite common for a mother-in-law to accompany a woman seeking treatment for barrenness.

Lay control of therapy is thus a core feature of ethnomedical systems in many areas of contemporary Africa. The degree to which the Kenyan evidence further validates the "pan-African" character of this therapeutic institution is examined below. The preceding discussion affirms the argument that the process of choosing therapy lies at the heart of African therapeutics today. However, I believe Feierman (1984, p. 12) is essentially correct in saying that if we want to understand the central historical dynamic of change in ethnomedical systems—that is, "the direction in which the whole system is moving"—we must look beyond cultural interpretations, practitioner control, lay control, and therapeutic pluralism *per se*. To the extent that the therapy managing group, for example, plays a role in health care behavior, African ethnomedicine is integrated into local household, kin, and community organization. This set of institutions involved in making health care decisions also happens to be the framework of economic activity. Consequently, changes in therapy-related practices (e.g., decreasing size of or diminished kin responsibility within the TMG due to cash cropping and/or male migration and dispersal) will necessarily reflect ongoing changes in the peasant economy in both its rural and urban dimensions, and vice versa. Changes in mode of production and, hence, productive relations will cause changes in levels of health (e.g., through the emergence of new conditions of risk) and in approaches to healing. If we wish to determine the locus and direction of change in African ethnomedicine we must, therefore, examine the broader political economy (Feierman, 1984).

CHOICE OF THERAPY IN NAIROBI: EVIDENCE FROM MATHARE

Observations and case studies of patients at TMPs' clinics in Mathare indicate they presented three basic but not always mutually exclusive categories of health problems. Traditional medical practitioners saw only a minority of *medically acute* cases. Some women brought infants or toddlers with gastroenteritis or worms to a TMP who was perceived to have an effective remedy available. Occasionally adults with diarrhea also consulted a *mganga* for a strong herbal remedy. Gonorrhea and other sexually transmitted diseases with compelling symptoms ranked among the leading acute disorders seen by TMPs, although many cases probably resulted from repeated exposure and had become chronic for all practical purposes. Urban TMPs only rarely treated conditions such as fevers, tonsilitis, abscesses, wounds, and trauma.

In practice, most TMPs today are not eager to see patients who present conditions which can be, at least in theory, handled more effectively at a dispensary or hospital. Urban TMPs often attempted to determine if a patient has visited a biomedical facility, and 90% indicated they refer patients to the hospital for a wide variety of maladies ranging from tuberculosis and accidents to fractures and asthma. Katz *et al.* (1981, p. 4) argue that this practice of referrals "places much of the responsibility of triage on the medical profession" in that "most acute emergencies, acute infections amenable to antibiotics, and surgical cases are filtered out." To the extent that it is true this process probably serves the interests of TMPs by minimizing unsuccessful treatments, and, indirectly, limiting legal complications.

Patients in search of *counseling* and *divination* ranked among the three largest categories of clientele (Tables 6-7 and 6-8). Comparable services did not exist at biomedical institutions. Many people desperately sought for guidance and resources that could improve or stabilize the course of their lives. Frequently their problems reflected the disaffection of a close relative, a broken marriage, or loss of a job or lack of success in obtaining one. Some patients were habitual alcohol abusers and appeared disoriented. One woman asked a TMP to provide her with herbal medicine that would keep thieves away from her shop. A female secretary who had apparently been fired from her job and not paid the salary she was due came to a TMP to be ritually "washed" so that her luck would improve. Other individuals with less savory life-styles also presented themselves. In one case a prostitute reported that she had come to the same Mathare TMP more than 20 times in 2 years as part of a strategy to increase her "fortune." When we observed her she was given medicine to place on her tongue and palate to enhance her chances of finding men with money.

Women and men with *chronic, refractory conditions* whose illness was complicated by significant social and personal etiologies such as witchcraft, depression, anxiety, and conflict with kin formed the third and largest class of patients observed in Mathare. These cases, ranging from abdominal complaints and infertility to edema and mental illness, best illustrate peoples' use of multiple therapeutic strategies and the existence of a preferred sequence or hierarchy of resort for different therapies.

Evidence gathered from participant-observation, discussions with patients, and client record forms indicates that at least two-thirds of the patients seen by Mathare TMPs had first visited a biomedical facility for the same illness. Most of these patients presented illnesses of the chronic, refractory type whose etiologies were in the process of being redefined in terms of personal and social circumstances. Thus in Nairobi as in other African cities, towns, and rural areas that are well served with the services of dispensaries and hospitals (Chavunduka, 1978; Janzen, 1978; Lasker, 1981), biomedicine was and is respected and more or less systematically utilized as a first line of defense because of its power to treat the physical symptoms of disease, to perform surgery, and to repair wounds and fractured limbs. Such conditions and needs are natural, unexceptional, and even "normal" occurrences (Chavunduka, 1978). If, however, the illness progresses or otherwise fails to respond, despite professional biomedical treatment, frustration sets in and suspicions about extenuating circumstances such as witchcraft begin to emerge. Mathare TMPs "confirmed" via their diagnoses that one-half to two-thirds of their patients had been bewitched. Relatives, including affines, acted as the most frequent perpetrators of witchcraft, but neighbors, co-workers, prostitutes, and other *waganga* were also frequently implicated.

In Harare, Chavunduka (1978, p. 49) found that over 53% of the people in his Highfield Township sample who received biomedical treatment later visited a TMP for the same illness. The following ten capsule descriptions represent a small sample of cases recorded in Mathare that illustrate the kind of involvement patients had with biomedicine before they came to a resident TMP.

- *Male* Kamba, late 30s. Second visit to same TMP. Diagnosis: *nduuka* (mental illness). Ill for 2 years. Prior treatment at Aga Khan Hospital, Kenyatta National Hospital (KNH), and inpatient at Mathari Mental Hospital (government) adjacent to Mathare Valley.
- *Female* Kamba in 30s. A traditional birth attendant, diviner, and specialist in infant health and "protection." Married to Kamba TMP. Both live and work in Mathare. Gave birth to her seventh child at the Pumwani Maternity Hospital operated by Nairobi City Council.
- *Female* Kikuyu, age 30. Hairdresser. Separated from husband. Three children. Diagnosed by TMP as having general malaise: "no strength in her body; blood is moving too slowly; result of witchcraft; hospitals can do nothing about it." Patient reported that before visiting the TMP she had been to Rhodes Avenue Clinic, "all the private doctors in Ngara," and M. P. Shah Hospital.
- *Female* Kamba in 30s. Divorced with two male children, ages 8 and 7. Self-employed dressmaker. Diagnosed by TMP as bewitched by maternal aunt who obtained some of patient's menstrual blood and has

used it to prevent her from getting married again. One year before visiting this TMP the patient was treated at Thika General Hospital, where they took blood and urine samples and performed a pelvic examination.

• *Male* Kamba, late 50s. Former medical assistant. Cancer patient. Admitted to Machakos Provincial Hospital and within 10 days was referred to KNH in Nairobi. Discharged from KNH on seventh day due to terminal diagnosis. Died 8 weeks later in care of TMP and his son.

• *Female* Kikuyu, 28 years old. Treated at Rhodes Avenue Clinic "over ten times without improvement" before visiting TMP in Mathare.

• *Female* Kamba, 37 years old. Continuous headache attributed by TMP to "family quarreling." Had initially attended KNH (twice), Bahati City Clinic, and Bahati Immaculata Clinic without improvement.

• *Female*, age 30. Unable to conceive. Visited several private hospitals and KNH without improvement. The woman and her relatives "got tired of hospitals" and decided to try a TMP.

• *Female*, age 52. Kikuyu. Abdominal pains and apparent old case of elephantiasis. Treated initially as an out-patient and then inpatient at Kiambu Hospital outside Nairobi. Her condition did not improve. Decided to visit a TMP because she was "fed up with modern medical treatments."

• *Female*, age 40. Kikuyu. General malaise. Visited Kiambu District Hospital and Nazareth Mission Hospital but was not cured. Said she "had lost patience with hospital treatments."

In Africa as elsewhere, most illnesses begin with symptoms that the sufferers and those around them perceive as natural, or unexceptional. If this was not so, why would the herbalists (*akimi wa miti*) of rural Ukambani— who do not perform divination and who specifically disavow *religious*-medical powers—have such an important role in the traditional medical system? In urban areas where biomedical services are relatively abundant people have little need to rely on herbalists for primary care. Hospital medicine is generally more efficacious (at least potentially) in coping with physical complaints. Instead, the demand for traditional healers in urban areas centers on TMPs endowed with divinatory and healing powers received from God—women and men who double as priests, social workers, and psychiatrists.

As noted earlier, to function successfully as a TMP in Nairobi one needs to be a generalist, that is, to be a diviner and also have special knowledge of plant medicines. Some 80% of the Kamba TMPs in Mathare fit this category. Most of their patients presented physical and mental problems that for one reason or another had not been adequately resolved through biomedical therapy. Because their illness had not responded to treatment, or their expectations of therapy had not been satisfied, these patients and "significant others" who shared in their concerns had essentially redefined their illnesses by the time they had arrived at the TMP's clinic (Chavunduka, 1978). Frustrated by persistent symptoms and by what they perceived as the failure of hospital medicine to cure them (many Mathare patients said they had become "fed up" with the ineffectiveness of hospitals and doctors) they began to view their condition as abnormal. At this point concern with ultimate causation, about which biomedicine was seen to have little to offer, moved to the forefront and the patients typically consulted TMPs for guidance and therapy.

Chavunduka (1978) seems fundamentally correct in suggesting that the most important determinant of illness behavior in the African milieu, particularly therapy selection, is the definition of the malady by the sick person and those who participate in therapy management. Considering the Kenyan evidence, including patient interviews and testimony from TMPs, I believe Chavunduka is also accurate in his assertion that it is the reassessment and *redefinition* of illness from a "normal" to an "abnormal" condition that is most influential in motivating urban dwellers to shift from the physically oriented biomedical system to the more holistic, psychosocial domain of traditional healing. In the latter domain the ultimate causes of illness can be addressed. Such concerns are perceived as the special competence of traditional therapy, and they serve to emphasize a fundamental dichotomy that characterizes African thinking about illness causation. Janzen (1978, p. 8) characterizes the distinction as "illness of God," which is natural, unexceptional, and "unrelated to human intentions"; and "illnesses of man," which arise from human malice, social strains, or ritual error.

In practice, the exercise of therapy selection is neither more nor less rational and mechanical than most other kinds of social decisions that must be made on the basis of imperfect knowledge. We now know that health behavior in general, with therapy selection a critical element of such behavior, is less a function of "belief" and more a variable that depends on the perception of the benefit derived from changing behavior (Foster, 1982). As in the economic sphere, international health research demonstrates that people are pragmatic—that new forms of behavior will be adopted if such action has perceived advantages. On the other hand, as Foster (1982, p. 194) observed, "the acquisition of new knowledge will not automatically lead to changes in behavior; individual priorities and personal preferences override strict rationality."

The evidence from Kenya confirms that fear, anxiety, and frustration often lead to impulsive or idiosyncratic therapy choices, including concurrent consultation with specialists in different systems. This was particularly evident in the urban setting where "significant others" may be a smaller, weaker, and less authoritative force than in the rural areas.[4] In Mathare, we recorded many cases of patients engaged in a peripatetic, costly quest for a "cure" (Figures 7-1 and 7-2).

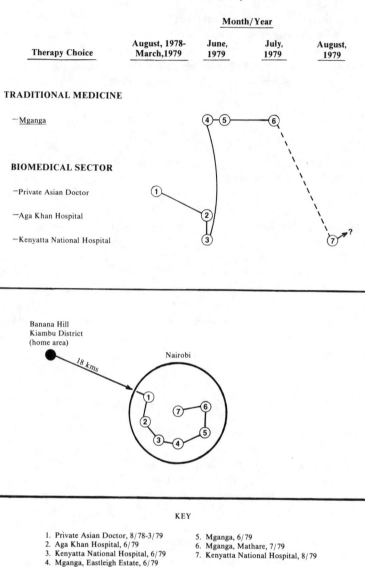

Figure 7-1. Kamau's course of therapy: time and place relations.

In Nairobi, my evidence suggests that the popularity of urban TMPs was bolstered by the conceptual and structural weaknesses of biomedical services. For various reasons, including the narrowness of biomedical training, authoritarian communication patterns that often appear to result in poor understanding and noncompliance by mostly unsophisticated patients, and proce-

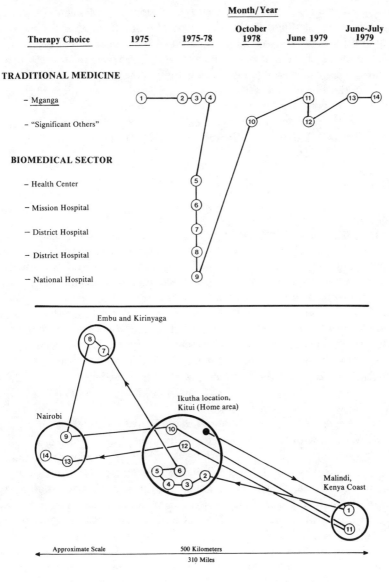

Therapy Choice	1975	1975-78	October 1978	June 1979	June-July 1979

Month/Year

TRADITIONAL MEDICINE

– Mganga

– "Significant Others"

BIOMEDICAL SECTOR

– Health Center

– Mission Hospital

– District Hospital

– District Hospital

– National Hospital

Embu and Kirinyaga

Nairobi

Ikutha location, Kitui (Home area)

Malindi, Kenya Coast

Approximate Scale   500 Kilometers
310 Miles

KEY

1. Mganga, Kabwere's Anti-Witchcraft Clinic, Malindi (Kenya Coast), 1975
2. Mganga, Ikutha Location, Kitui District (home area)
3. Mganga, Ikutha Location, Kitui District (home area)
4. Mganga, Ikutha Location, Kitui District (home area)
5. Health Centre (government), Ikutha Location
6. Mission Hospital, Ikutha Location
7. District Hospital, Embu

8. District Hospital, Keruguya, Kirinyaga District
9. Kenyatta National Hospital, Nairobi
10. Negotiations in family and with location chief, 10/30/78
11. Kabwere's Anti-Witchcraft Clinic, Malindi, 6/2/79
12. Location Chief: Ill person reported to him on 6/5/79 following Kabwere's treatment
13. Mganga, Eastleigh Estate, 6/79
14. Mganga, Mathare Valley

Figure 7-2. Mwithi's course of therapy: time and place relations.

dures that encourage minimizing the time considered "necessary" for routine diagnosis and consultation, a broad variety of illnesses presented to biomedical practitioners appear to go unrecognized (no diagnosis), are misdiagnosed, or are inadequately diagnosed and followed (Good, 1986). These kinds of patient experiences served to increase the flow of the ill toward TMPs. Added to this is the fact that most conditions brought to TMPs are chronic in nature, often have a strong psychosomatic component, or are maladies that have not responded as anticipated to earlier therapy—including cases where patient–doctor relations have broken down (Katz et al., 1981).

THERAPY MANAGEMENT: MATHARE EVIDENCE

Although men often arrived alone, most people who visited a TMP in Mathare came with at least one other person. Close kin such as a parent(s), sister, brother, husband, wife, son, or daughter most often fulfilled this support function. Also, friends (especially a woman's close female friend), and to a lesser extent co-workers, often accompanied patients to a TMP's clinic. Knowing about the persons who accompany the patient is usually crucial to identifying the social context and nature of the patient's illness or dilemma. In one case, a married girl, 19 years old, arrived at a clinic in the company of her own mother and father. The girl had not been able to conceive and also complained of pains in her joints. Although the girl's husband was paying for the treatment, he reportedly was unable to inform *his* parents about his wife's visit to a TMP because they are "converts to Christianity" and would raise strong objection.

In the case of the more popular TMPs, five or six other patients might also be waiting, with their supporters, for treatment or consultation. Everyone squeezed into the same room where the TMP was working (in Mathare room size is usually about 2.7 X 2.7 meters). Some patients had to wait for hours before their turn came. Consequently, more often than not therapy was an activity conducted in a public setting. If the TMP found it necessary for the patient to disrobe, he or she drew curtains to divide the room for visual privacy.

Undoubtedly in the present study those persons who actually accompanied the patient to consult the TMP represented only part of the patient's TMG. Persons responsible for the patient or otherwise implicated in the illness and/or its cure might show up at one time and be absent another, particularly when employment or distance intervened, or in response to their relative personal importance in successive episodes of an evolving illness.

One of Janzen's (1978) most significant theoretical contributions to the study of the therapeutic process is the product of a research design centered on the patients and their therapy managers from the beginning of an illness. Although this approach requires extraordinary investments of time and other resources, he was able to trace the evolving, changing cast of characters involved in therapy decisions as an illness progressed (but Janzen, too, was unable to follow all of his cases through to their final outcome). Thus, he could simultaneously chart the pattern of selection among therapy options, ranging from "Western therapy" (village nurse, dispensary, and hospital), African therapy specialists (various *nganga* and prophets), kin group meetings, and nonkin social groups. Janzen based his observations and analysis on a rural area, and his attention focused on the dynamics of interpersonal and group relations in therapy. His recognition of place–time patterns is implicit, but he did not develop the social geography of therapy, which is so crucial for seeing how and where the varied elements of health-seeking behavior and therapy fit together to form a whole.

S. Feierman (1979) argued in an insightful analysis that the growth of wage labor and the accompanying geographical dispersion of kin groups in Africa have considerably reduced the degree to which the extended family can affect, or even comprehend, the daily working lives or living conditions of relatives who are sick or in distress. Consequently, the active membership of groups traditionally responsible for therapy management has diminished along with their capacity for unified action on matters relating to the health status of kinsfolk. The problems the bureaucrat experiences in the office, Feierman observes, cannot be truly appreciated by his or her peasant kinsmen. Neither can the schoolteacher influence the working conditions of his or her kinsman the factory worker. On the other hand,

> the very fact that a therapy managing network can be spread across town and country, spread among farming, bureaucratic, and factory occupations, leads to a balancing of resources. The peasant farmers provide a place for their city relatives to retire to when sick or old, and the bureaucrat provides cash to help with the peasant's illness. This does not impede the emergence of the ultimate problem of reduction in control over the daily working and living environment. But the comfort of group assistance, and the vitality of the private kinship-based welfare system, must be given full weight (S. Feierman, 1979, p. 283).

There is now ample evidence from African cities that treatments that customarily required ritual therapy are being adapted to the new realities of urban forces. A report on *Traditional medicine in Zaire* (International Development Research Center, 1980) concluded that although traditional principles of therapy are still followed, their length and complexity have been reduced due to such factors as shortages of time and space, unavailability of family members, cost of living, and the materialistic nature of urban life. As noted in the case of Dar es Salaam, rituals are also increasingly focused on the patient. Family members may still assist in therapy management, but are less likely to participate directly in actual treatment.

Throughout Africa the increasing division of kin groups between rural and urban areas which are often widely separated necessitates special arrangements for bringing people together at a time of illness. In this sense illness presents opportunities for social travel and participation in one's society. It fosters social contacts that cut across all other categories and levels of social groupings and activities and it links countryside and city (M. L. Swantz, 1970). In Kinshasha, Zaire, for example, the healer, patient, and family members living in the city may travel to the village and convene a family council to seek the root cause of the illness. In other cases, as I also observed in Kenya, family members in a rural village may be called to the city to help identify the underlying causes of an illness (International Development Research Center, 1980). As the following discussion illustrates, the search for healing can become a lengthy and costly process that involves contact with several forms of therapy that are widely separated geographically.

COURSES OF THERAPY: MOBILITY, PLACE, AND TIME

Numerous patients of Mathare TMPs consented to intensive interviews which permitted me to reconstruct the prior episodes of therapy received for their current illness. The two cases discussed here exemplify chronic illnesses, show the patttern of resort within and between the biomedical and traditional medical sectors, and highlight the role of geographic mobility in health care behavior (Figure 7-1). In both instances the sufferers made repeated efforts to obtain relief from therapists in one sector before switching to the other. While no two illnesses or individuals behave alike, the cases of Kamau and Mwithi illustrate the theory and practice of therapy-seeking in a medically pluralistic society.

Kamau was a married Kikuyu businessman who was about 55 years of age at the time he entered our study in July, 1979. He was illiterate and lived in a rural area known as Banana Hill in Kiambu District, about 18 kilometers from Nairobi, and thus had gone to some inconvenience to find a reputable TMP in the city. By his own report, Kamau became ill in August 1978 with chest problems and swollen feet. (When we first encountered him in the clinic of K., a female Kamba TMP, he had suspected clinical asthma, a palpable liver, and pains extending to the lower end of the sternum/xiphoid process. He also complained of a general weakness, coughing, and insomnia.) Kamau's initial choice of therapy was an Asian doctor in private practice in Nairobi, whom he visited several times over the next 8 months (Figure 7-1). Disappointed with the results of those visits, he went to the Aga Khan Hospital in Nairobi in June, 1979. There he was examined, received a chest x-ray examination (for which he was reportedly charged K. Shs. 315/= or U.S. $42.00), and admitted as an inpatient. He stayed in this hospital for 5 days, which he said was all he could afford. Shortly after he left the Aga Khan Hospital in June, Kamau visited the outpatient clinic at KNH, and upon examination he was admitted in Ward 6 and kept under observation for 5 days. He was then discharged and told, without understanding why, to report back to KNH on August 1, 1979. (He complained to us that while a patient in KNH he was "not given even a single tablet.")

By the time of his discharge from KNH, Kamau had begun to lose hope that he could be cured by doctors and hospitals. Thus his next strategy was to consult a "Tanzanian Swahili" *mganga* in Eastleigh, a settlement adjacent to Mathare Valley. Because he had no medicines available, this so-called *mganga* prescribed and supplied Kamau with three bottles of "pure water," charging him K. Shs. 200/= (U.S. $26.00). Kamau said he drank the water over a 2-day period but again experienced no change in his condition. Annoyed and frustrated, he turned to a second TMP who also was unable to help him. Thus during June 1979, Kamau sought therapy from four specific sources in two medical sectors, but afterward felt he was no closer to a cure than before.

I first encountered a tired and rather desperate Kamau in July 1979, when he arrived unaccompanied to consult with K., a popular TMP in Mathare with whom our research team was already engaged in intensive participant-observation. K. agreed to treat Kamau. After igniting a sweet-smelling incense that would "invite other clients," she initiated a rigorous, detailed process of diagnosis and treatments that extended over a period of 3 days. K.'s initial action was to determine the cause of Kamau's illness through divination. This lasted about an hour, during which time K. revealed correctly and to Kamau's amazement that he was involved in a land dispute that included his sister and two unrelated parties. Following the divination, which K. accomplished by playing the "magic bow" (*ota*) and making numerous counts of small colored pebbles shaken from her special divining gourds (*ketiti*), K. instructed Kamau to write down the names of all people who might bewitch him. Since he could not write, my research associate took his dictation. Kamau suggested the Kikuyu names Njoki, a female; and Njeri, a male, and said that these two people wanted to purchase his sister's land. He informed K. that he was against the sale. He objected to the fact that his sister had already received K. Shs. 1000/= (U.S. $133.00) for the land from Njoki and Njeri, which permitted them to use the land on a rental basis. They were now pressing to buy the land outright. Kamau told K. that he was ready to hire an advocate to deal with the case. K. suggested that Njoki knew that Kamau was wealthy and could afford to pay a lawyer for his sister; consequently, Njoki was determined to kill him, using witchcraft, in order to obtain the land.

K.'s diagnosis also determined that Kamau was suffering from a nearly fatal illness caused by the witchcraft. To symbolize the seriousness of his illness, Kamau had to be ritually "resurrected" from his graveside. When he returned for therapy the next day he was therefore instructed by K.'s young male assistant to disrobe to his undershorts. Meanwhile, an elderly male assistant to K. mashed three special plants (*miti*)—*mumbaa, ndatakivumbu*, and *maguta na thenge*—on the concrete floor with a stone and then mixed them together with water. The assistant used a plant dipped in this mixture to "wash" (purify) Kamau from head to toe. Kamau was then instructed to lie down on the floor with his head resting on a pile of impala horns. Medicine bags were placed at his sides. K.'s elderly male assistant then proceeded to wrap Kamau with an underlayer of red cloths, on top of which he was covered and closely bound with mummylike white sheeting from his chin to his toes. The other assistant then covered Kamau with numerous cuttings of a green aromatic herb. Numerous dabs of an oil were then placed on the floor around the patient's body and set aflame. Periodically during this process Kamau was given sips of a liquid medicine (_muyulu_). After about a half hour, Kamau was raised up and unwrapped. K. then performed various ritual movements around him. Soon, Kamau was allowed to dress and was instructed to return the next day. He did not leave immediately, but stayed to observe K. work with her next patient.

Kamau's second day of treatment involved elaborate, extensive ritual therapy. K. revealed that Kamau had not had sexual relations for the past 5 months, and he agreed. K. said this was a further sign of Njoki's nefarious deeds. Ceremonies included the use of a baby chick "less than 10 days old" whose blood, drawn from a small cut made on its crest, was rubbed into small cuts K. had made on Kamau's back with a razor blade. The chick was also made to ingest some of Kamau's blood, thus transferring the evil power of the witchcraft from Kamau to itself. The chick was then released outside K.'s clinic and left to go its own way. Before Kamau left on this second day of treatment he was given plant medicines (_miti_) to take with him: _muyulu_ to drink, and _mwaiitha_ to rub on his swollen legs. He returned to see K. 3 days later and paid her an installment of K. Shs. 200/= (U.S. $26.60), and indicated to us that he was pleased with the results of his treatment. She instructed him to return a week later for more medicine.

Because fieldwork ended at this point, I do not know if Kamau kept his appointment to see a doctor at KNH on August 1, 1979, 10 days after his last scheduled visit to K.'s clinic. But after 12 months of searching for help, his condition appeared to improve under K.'s management. Was K.'s therapy efficacious? Did she succeed in reducing Kamau's anxiety and ameliorate his physical symptoms, such that he could function with less stress and perhaps resolve the land dispute? These issues underscore the necessity of appropriate follow-up of patients in the field, including medical monitoring, as a first step toward defining and evaluating the efficacy of therapy.

Mwithi's case provides an illustration of therapy-seeking behavior by an individual whose apparent primary problem was a chronic but generally nonincapacitating form of mental illness (Figure 7-2). This case was an extreme example of the habitual therapy "shopper," and differed from Kamau's case in several respects. First, it extended over a period of 4 years instead of one. Second, it involved visits to therapy sources in widely separated locations in Kenya and is thus a particularly interesting case of geographic mobility. Third, Mwithi apparently began his peripatetic quest for a cure with a visit to a TMP rather than to a biomedical practitioner, and he related details of 14 episodes in his course of therapy.

Mwithi was a Kamba male from Kitui District who had 6 years of formal education. He was 24 years old when he entered our study at K.'s clinic in July, 1979. Once again, the patient's social background is crucial to interpreting illness behavior. Mwithi's father, who had three wives, died in 1974, leaving his eldest sons to manage the home. The first wife had two sons who, according to Mwithi, seemed to lead relatively successful lives. Mwithi's own mother died in 1963, while his late father's third wife—Mwithi's second stepmother—was a *mganga* whose sons were reportedly "not doing anything worthwhile."

Mwithi had sought therapy for mental illness and concurrent joint pains in his ankles, knees, and hips. He blamed his younger stepmother for his suffering. Despite the fact that the rest of the family reportedly was very kind to this woman, attempting to comfort her and assist her in everyday living, she did not respond positively. During a period of friction at home in 1975, Mwithi concluded that his stepmother was capable of bewitching him at any time. Consequently, he decided to take countermeasures and traveled to the renowned Kabwere's Anti-Witchcraft Clinic in Malindi, at least 300 kilometers away on the Kenya coast, to purchase a protective charm.

Despite the anti-witchcraft treatment, Mwithi started to experience joint pains after he arrived back at home in Kitui. Moreover, while in his bedroom one day he stepped on a thorn that inflicted a deep prick in his right foot, causing him chronic pain. To Mwithi, this was "nothing less" than his stepmother's witchcraft. Over the next 3 years he visited three local *andu awe*, a health center, and a mission hospital—all in his home area—but did not get relief.

In 1978, Mwithi moved over 200 kilometers to Kerugoya in Kirinyaga District near Mr. Kenya, where he was employed as a junior electrician at a local tea factory (he had trained with a self-study course at home). While there his joint pains worsened. With help from friends, he hired a taxi to take him to Embu Hospital. That hospital was "too congested" so he was taken immediately to Kerugoya District Hospital. It too was overcrowded, but because of his poor condition he was given a corner space in a corridor of the hospital. Later, he was admitted to a ward where he was "plastered and treated for 3 months." When he failed to recover as expected, he was referred to KNH in Nairobi and admitted there for observation. He was discharged after 2 days because his doctor thought he had improved, although he said he still did not feel well.

After he returned to his home in Kitui, Mwithi said he "started to have mental problems on October 28, 1978." He then ran away from home and into "the forest" for 2 days. When he returned his brothers, acting now as therapy managers, held a family meeting and decided that Mwithi's younger stepmother was in fact causing his health problems. The case was thereupon reported to the sub-location chief, who instructed Mwithi's two older brothers to go to Kabwere's Anti-Witchcraft Clinic in Malindi because Kabwere could prove whether the young stepmother was the one causing their brother's illness. Reluctantly, the two brothers set out for Malindi on June 1, 1979, carrying an official letter which they had to return to the Chief with Kabwere's signature as proof of treatment.

Upon the brothers' arrival back home in Kitui on June 5 the Chief declared the case settled "once and for all." Mwithi, however, still did not feel "perfectly well." Later in the month he decided to visit a *mganga* in the Eastleigh section of Nairobi, but this TMP did not seem to offer much help. Having heard about K., he decided to try her as well. Our first contact with Mwithi came on the day of his first session with K., which we were able to observe. Afterward he said he had already realized "tremendous improvement." Perhaps K. was able to see into and touch Mwithi's troubled psyche with her recognizably formidable intuitive skills and personal charisma. Indeed, K. amazed all of us in attendance that day. She had never seen Mwithi before, and had no foreknowledge of his condition or the circumstances leading up to his visit, yet she described the brothers' long journey to Kabwere's in Malindi and noted that they carried a letter about the case from their sub-location chief.

Details are inadequate to determine with certainty if the young stepmother also directly and habitually upset other members of Mwithi's family, or whether they became involved because of the intensity of his conflict with her. Mwithi's suffering appears to reflect pathology in the corporate sense. To the extent that this is true he was a "symptom of his family's sickness."[5] He adopted a sick role that gradually involved a widening circle of people, including nonkin such as his friends, co-workers, and the sub-location Chief, as therapy managers. The chief became directly involved on behalf of the family for a second time in an attempt to achieve closure in the case. Nevertheless, 4 years after the problems began—and after Mwithi and others in the case had traveled hundreds of kilometers and made substantial financial outlays—a "final" resolution was still problematic.

The preceding discussion illustrates the geographic and social correlates of illness behavior. It is notable that at the local scale distance was apparently not a determinant of therapy selection for any episode of the search for a cure by the two men.

Mwithi and Kamau are neither isolated nor exceptional cases. They certainly defined their own illnesses as abnormal. And they mirror the large burden of chronic, nonincapacitating illness and social pathology present in Kenya, and Africa generally. Although patients tapped biomedical resources, these alternatives proved ineffective. Ultimately, this permitted the TMPs (as nonauthoritarian specialists) to have a central role in the therapeutic process. In the next chapter I discuss patient–TMP relationships in regard to consultation and treatment of other conditions such as infertility, uvula syndrome, and terminal cancer.

NOTES

1. As in Zimbabwe (Chavunduka, 1978).

2. In the initial census of Mathare TMPs, approximately 15% emphasized that they had on occasion personally treated "Asians" or "Indians."

3. Chavunduka (1978) surveyed his sample Shona population in Highfield and classified them, without further explanation, according to the conventional categories of married, single, divorced, or widowed. Over 75% of the males, and 76% of the females, were "married."

4. Concurrent resort to different therapies, for example, visits to biomedical and traditional practitioners during the same episode of an illness, may occur more often in urban than rural areas. However, Janzen (1978, p. 37) argues against the likelihood of such behavior since, at least in a rural context, "the beliefs and practices constituting these systems rest upon different premises. For most people translation from one to the other is difficult, and individuals usually manipulate them separately rather than synthesize them."

5. "The sufferer is but a symptom of his family's sickness" [Dr. Denis Bazinga, quoted by Janzen (1978, p. 114)].

8

Environment, Diagnosis, and Treatment of Disease and Illness in Mathare

CLINICS OF TRADITIONAL MEDICAL PRACTITIONERS

Housing stock in Mathare is essentially of two types: old, "squatter"-type housing made of mud and wattle, supplemented with an assortment of scrap metal and wood, and cardboard; and "company" housing, generally of clapboard construction on a concrete floor and with a corrugated iron roof. As noted in Chapter 6, in both types the standard room size is roughly 3 meters to a side. Since these rooms double as a residence and a place where TMPs see their patients in the great majority of cases, space is intensively used and may be extremely congested during consultations. It was often difficult to separate the internal proceedings of the clinics from the external press of local residents, strangers, smells, and noise in Mathare. Since Mathare is also situated adjacent to the Kenya Air Force Base at Eastleigh, there were occasionally startling reminders of the paradoxes that punctuate life in Third World cities as F-4 jet fighters screamed overhead in strafing-style runs barely 100 meters above the rooftops. Children screamed and scampered, and covered their ears with their hands. Within this environment, three examples of individual clinics suggest their character and ambience—and the personalities of the resident TMPs.

M. is a male Kamba TMP who lived in Mathare III. His room was in a multiunit "company"-built structure consisting of eight adjacent rooms laid out four to a side, and it has a rough dirt floor. There was one window with shutters that opened out. The room contained a wooden single bed, a combination charcoal cooker/heater, several aluminum cooking pots (sufurias), two small, crude tables, a chair and four low stools for guests, and a metal drum for water storage. M. kept his divining instruments, including a magic bow (*ota*), hand mirror, calabashes, and pebbles underneath his bed. Ten tins of various powdered medicines (Swahili = *dawa*), numerous bottles of liquid medicines prepared from plants, and small hand-wrapped packages of *dawa* stood in one corner of the room. They included remedies for madness and dizziness, infertility ("to open blocked fallopian tubes"), general edema, palpitations, impotency, constipation, and joint pain. A large commercially printed poster on one wall featured a map of Kenya showing photos of former President Mzee Jomo Kenyatta and government ministers. According to the framed certificates on the wall, M. was a paid-up member of the New Kamba Union and the Kamba professional association of TMPs known as the Kenya Waganga wa Miti Shamba Society. A wall calendar for Kenya Breweries showed a woman serving bottles of lager beer in a "Salute to Sports." Upon arrival for our first visit M. was talking with a former patient, a Kikuyu woman, who had "come from Murang'a to greet" him. He was dressed in a light print shirt with a red and blue-striped tie, gray-striped trousers, and black leather dress boots. Inside the room a professionally painted sign advertised his practice as that of a "Mganga wa Kienyenji." It included his name and home address in Kitui District, other places he had practiced (Kitui and Machakos), and four illnesses for which he claimed special competency: *kithakame* (edema), *kiathi* (vomiting with blood), *mutumbuko* (joint pains), and *nduuka* (madness).

K.'s clinic was unusual because she did not live there. She preferred instead to rent an Asian-style house located about a 5-min walk from Mathare. The floor at the back of K.'s clinic was crowded with numerous bottles of assorted shapes, sizes, and colors that once held whiskey, beer, Chianti, and fruit drink mixes. She had filled the bottles with water or some medicinal preparation. K. had also spread a cloth on the floor on which she arranged an assortment of small medicine gourds, medicine baskets (*ngusu*), talc powder, cups, impala horns, amulet cases, and incense. Three paper bags contained fresh medicinal herbs brought back from rural Kitui. Large sheet maps from the Kenya Meteorological Survey covered two of the walls. A framed photograph of K. and her spouse and a certificate of membership in the Waganga wa Miti Society also adorned a wall. Wooden benches and chairs lined three sides of the room, and the clinic was typically full of patients waiting their turn. If privacy was required (when it was necessary for a patient to disrobe) a colorful curtain was pulled across the center of the room on a guy wire. For divination sessions K. sat on the floor and spread out a red cloth on which she rapidly counted and classified pebbles into lots after she had shaken them from the divining calabash.

D.'s clinic/residence was also "company"-type housing. A handpainted blue and white signboard attached to the outside wall of his room identified him as a "Dr." from Kitui District and a "Native Medicine Consultant" for "incurable stomach pains" and "all sorts of body pains." The sign also stated that D. "can make a barren woman conceive" and "can cure a bewitched person." On the wall inside his clinic a framed document certified him as a member of the Kitile & Syokisinga Itambya (K.) Co. of TMPs. It identified D. as a "witch-doctor" who "travels to all Provinces of Kenya, Uganda, Tanzania, Sudan, Somali, and Ethiopia."

D.'s room was exceptionally clean and well-organized. It contained three single beds, a table and folding chairs, and space for his medicines and paraphernalia—including a wooden carving of a bare-breasted woman. His bed was covered with a bright red and black blanket, and a new transistor radio sat beside it. D. kept his medicine for bloody diarrhea in a White Horse liquor bottle, and decoctions for general edema and fever in Haig whiskey bottles. Smaller bottles and gourds held powdered medicines for mental problems, madness and numbness, love problems, male impotence and unemployment, teething diarrhea, tapeworm, and a medicine that D. said regularizes menstrual flow and promotes conception.

"M.," "K.," and "D." are rural folk who had managed to establish themselves as financially viable TMPs in the informal economy of Nairobi's poorest and least attractive neighborhood. Since their original legitimization as *andu awe* in rural Kitui District they had shrewdly adapted the rural values and procedures of traditional medicine to the needs and expectations of many city residents and in support of their own aspirations for a higher standard of living.

URBAN TRADITIONAL MEDICINE AS A LIVELIHOOD

In time, some amenities of urban life must come to possess a more or less permanent allure for people who have migrated to the city from rural areas. Yet we also know that for a large share of its African citizens Nairobi will never be "home" (Ross, 1975). If it is true that "rural life with urban amenities is the ideal" for most of the African population, residence in the city thus becomes instrumental to achieving that goal (Chapter 6).

Available evidence certainly provides confirmation of such motivations among the Mathare TMPs, 90% of whom were Kamba people from Kitui District (Figure 6-6). The more densely inhabited, western sections of this large area are within a day's journey from Nairobi. Although portions of western Kitui have a high agricultural potential, subsistence farming predominates and cash-earning opportunities are scarce. Consequently large numbers of men and, at least in the case of TMPs, women migrate to Nairobi as a strategy for improving living standards for themselves and their families.

In contrast to many other poorly educated and unskilled migrants, rural TMPs from the same socioeconomic background who left Kitui and came to Nairobi were unique because they *already* possessed a highly specialized set of skills which could be readily adapted in Nairobi's "self-help city" (Hake, 1977). With certain other conditions satisfied, they soon used their traditionally based medical skills to produce a viable income in Nairobi's extensive, informal economy. A TMP's income potential was determined in large measure by his or her personality force and social skills, coupled with secondary enabling factors. These included an opportunity to find living space in a locale such as Mathare in which one's own ethnic group was well established and where some kin and friends already formed part of the community fabric.

After arrival in Nairobi the Kamba TMPs and those from other ethnic groups emerged—as Press labels the urban *curanderos* of Latin America—unambiguously as "professionals for hire" (Press, 1971, cited in Worsley, 1982). Payment in kind for services rendered, which was still common in rural areas, was the rare exception to cash transactions. Although such cash-earning opportunities seemed likely to attract numerous pretenders, apparently almost all of the Kamba practitioners in Mathare had been legitimate *andu awe* (religious-medical specialists) in rural Kitui prior to moving to Nairobi. In cultural terms, their original "calling" to the profession was prompted by a religious experience and would not reflect simply personal choice or ambition (Chapter 6).

Evidence of the economic incentive to practice traditional medicine in Nairobi was obtained from a question put to the 45 TMPs who formed the Mathare study population. They were asked: "Are there differences between traditional medicine or midwifery as practiced in urban and rural areas today?" Their responses to this question (Table 8-1) highlighted the importance of the monetary reward available in the city. "More money" and "more patients" together accounted for 43% of the responses from all TMPs, and

Table 8-1. Differences Between the Practice of Traditional Medicine in Urban and Rural Areas of Kenya According to Mathare Traditional Medical Practitioners

Responses	Percentage of responses: all TMPs			Percentage of responses: Kamba TMPs		
	Male	Female	Total	Male	Female	Total
More money available in city	17.6	29.4	23.5	20.8	32.3	27.3
More patients in city	17.6	20.6	19.1	25.0	22.6	23.6
City provides valuable experience	11.8	11.8	11.8	12.5	12.9	12.7
No difference	14.7	8.8	11.8	12.5	9.7	10.9
No livestock payments in city	2.9	8.8	5.9	4.2	9.7	7.3
Different tribes in city	5.9	2.9	4.4	8.3	—	3.6
More diviners in city	2.9	—	1.5	—	—	—
Fewer herbal medicines available in city	2.9	—	1.5	—	—	—
More medical facilities in city	2.9	—	1.5	—	—	—
Cheating by TMPs in city	—	2.9	1.5	—	3.2	1.8
Other	20.6	14.7	17.6	16.7	9.7	12.7

Note: From field survey, 1977–1978.
Abbreviations: TMP, traditional medical practitioner.

51% of the Kamba responses. Female Kamba practitioners mentioned these two factors more than any other group.

The goal of most Mathare TMPs was to improve the living standard of themselves and their families (Good & Kimani, 1980). It is important to keep one's material advancement as inconspicuous, locally, as possible. Conspicuous consumption is subject to traditional leveling mechanisms, and socially disruptive jealousies are easily kindled among TMPs who live in the same locale. For many, the practice of traditional healing was accompanied by (quiet) professional pride, compassion for and respect from patients, and a striving for recognition and legitimization by the government. Savings, and investment in an urban or rural business (e.g., shops, private transport services) or property (e.g., on which to construct flats), served as the hallmarks of the most successful urban TMPs. For the minority successful enough to earn Shs. 500–1,000/ (U.S. $67–$133) per working day,[1] a substantial rise in standard of living was possible within a few years. This represents an extraordinary accomplishment in a country where the annual per capita GNP is $390 (Population Reference Bureau, 1984). Agunga, a Luo *mganga* who was also based in Nairobi (but who is not a Mathare TMP) reportedly earned "as much as $3,000 a month" (Darnton & Corbett, 1980).[2]

AVAILABILITY AND COST OF TRADITIONAL MEDICAL PRACTITIONERS' SERVICES

Nine of ten Kamba TMPs in Mathare said they treated patients 7 days a week. Over two thirds claimed that they would consult with a patient any hour of the day or night, although another quarter (mainly women, who often mentioned nighttime security as a concern) preferred to work only mornings and afternoons. Most patients willingly endured long waits for a consultation, although many appeared to spend less time at the TMPs practice than at a city dispensary or hospital outpatient ward.

While some TMPs maintained a list of fees for common treatments, standard fees—and upper and lower limits on fees—were rare. The most common determinants of fee level included (1) the severity of the illness; (2) type of disease and the time required to treat the patient (e.g., treatment of infant gastroenteritis or a simple divination to determine if someone is likely to find employment versus a case of madness or infertility); and (3) the amount the TMP believed the patient could afford. Some TMPs said they determined ability to pay by the patient's appearance; others insisted that the spirits "told" them the appropriate fee during divination. The fee was also affected in some cases by the amount and rarity of the plant medicines required for treatment.

Many TMPs emphasized their desire to show leniency with the poor, elderly, and children. Some complained that patients took advantage of them by failing to pay after treatment. Whereas goods in kind such as goats, cows, chickens, and foodstuffs such as maize and beans remained an acceptable form of remuneration in the rural areas, cash payment was the standard in the city. In cases that involved more serious conditions patients at first usually paid no more than half the fee agreed on, and arranged to bring the balance at a later date. Partly for this reason TMPs found that keeping an accurate account of actual payments was problematic.

A composite list of fees charged by eight Kamba *waganga* for selected conditions is provided in Table 8-2. The mean fees shown may be on the low side due to the relatively small number of cases for which data are available. [In a rural environment on the fringes of the town of Murang'a, a district headquarters some 80 kilometers north of Nairobi, we witnessed "installment" payments to a local *mganga* that ranged from Shs. 100/ (U.S. $13) to a high of Shs. 800/ (U.S. $103)]. The fees for home delivery in Mathare may be compared with the then current Nairobi City Council fees of Shs. 120/ (e.g., Pumwani Maternity Hospital and Ngara Maternity), Kenyatta National Hospital (Shs. 60/), and a private facility such as Alpha Maternity, immediately adjacent to Mathare (Shs. 60).

Table 8-2. Fees Charged by Kamba *Waganga* in Mathare: Selected Illnesses, 1977–1978

| Illnesses/condition | Fee charged (Kenya shillings)[a] | | |
	Range	Mean	Comments
Epilepsy	—	200	1 case
Gonorrhea	10–70	30	7 cases
Impotence	10–80	47	3 cases
Infertility	18–100	52	20 cases
Abdominal problems	2–150	33	35 cases
Diarrhea and vomiting	5–10	6	5 cases (mainly infants)
Excessive menstruation	10–55	33	5 cases
"Home"/family problems	1.75–230	25	60 cases
Madness	15–217	82	7 cases[b]
Medicine for love	5–31	17	13 cases
Headache	10–120	42	6 cases
Uvulectomy	—	25	Fixed rate
Body ache/malaise	5–60	15	31 cases
Unemployment/business problems	2–120	22	12 cases
Delivery	20–60	37	Reported by five resident TBAs

Note: From patient record forms.
Abbreviations: TBA, traditional birth attendant.
[a]U.S. $1 = 7.5 Kenya shillings (1978).
[b]Usually also requires a hen for sacrifice.

DIAGNOSTIC METHODS

Diagnosis by TMPs was based on one or, usually, a combination of procedures including *divination, observation, case history, clinical examination,* and *evaluation of body substances.* Of these, divination, "the endeavor to obtain information about things future or otherwise removed from ordinary perception by consulting informants other than human" (Rose, 1911) is probably the principal method of diagnosis in all African traditional medicine. It is also frequently used as a guide to prognosis (Koumare, 1983).

Among Kamba TMPs in the city, the "props" for divination typically involved vigorous (solo) singing or chanting accompanied by a tune played on the "magic bow" (*ota*)—a wooden bow with a wire string stretched taut across a half-calabash which forms a sounding board. Playing the *ota* set the scene for divination, enabled the practitioner to request help from his or her spirits, and could also lead into a possession or trance state. The TMP often poured out libations, and offered tobacco (cigarettes) or snuff to "feed" the spirits. Patients were expected to place an offering of money in the TMP's *ngusu* (small medicine basket) before the divination began.

During divination proceedings some TMPs also offered prayers for their work. Ngali (pseudonym), a charismatic male Kamba practitioner, often used the following prayer:

> Oh God (*Ngai*) on high, give me power if you really meant me to be a *mundu mue*. For this patient I am treating I pray for healing from you. The work I am doing is not mine. I pray that you give me healing power. The herbs that you showed me I will use them to bring your healing power on him.

After the *ota* is played the TMP continued the divination by casting and counting lots of pebbles (*mboo*) shaken from a divining calabash (*ketiti*). She might also look into a small mirror to "see" images which provide relevant diagnostic information (one TMP compared this to switching on a television set). Another diagnostic tool used by some TMPs was to "read" a passage from a book. This technique was apparently borrowed from Koranic medicine, but any book with printed pages sufficed. One female TMP consulted up to four books, including a Methuen biochemical monograph (title page missing) on the structure and functions of lipids (1957); Jonathan Edwards' *God glorified in man's dependence*; Book One of *The Beacon infant readers*, by James Fassett (1933); and *Akili ni mali* ("Intelligence is Riches"), Tanzania Publishing House, 1975. Some (illiterate) TMPs also followed a practice of "writing" in a blank notebook or receipt book as part of their effort to gain insight into the patient's problem. A few others found meaning in the reflections of water in a cup.

Whatever combination of techniques they used, after conversations with their spirits and "throwing" pebbles TMPs usually followed-up with leading questions and statements to their patients. These dialogues concerned their social history and personal behavior, and usually required the patient to offer either a true/false or substantive response. The TMPs also carefully observed patients' attitudes, gestures, and other nonverbal cues. Each TMP strived to impress, even astound, the patient with his or her knowledge of the case in order to build up the rapport and hope necessary for successful therapy. Depending on the nature and seriousness of the problem, divination sessions took as little as 15–20 min or extended for an hour or more.

Mathare TMPs also resorted to other diagnostic methods, usually as a complement to divination. Depending on the particular case, these might involve a clinical examination—including sensing the pulse rate and temperature; palpation (e.g., depressed fontanelle); an inspection of the mouth, tongue, and eyes, an examination of urine of adults (gonorrhea and infertility) and stools of infants ("teething diarrhea"); or a vaginal smear to test for "gonorrhea." Some TMP's said they could "test" blood vessels and blood flow, and locate an offending spirit inside the body, by "reading" a mirror.

Traditional medical practitioners' descriptive symptomologies for various illnesses reflected areas of consensus as well as idiosyncratic interpretations. Most of the practitioners had similar, rural socioeconomic backgrounds. However, the absence of a common "syllabus" for traditional medicine, individual differences in exposure to biomedicine and other systems, and the custom of practicing in relative isolation had produced diversity in understanding, perception, and labeling of symptoms as well as in ideas about causation and methods of therapy. Representative, actual responses of Kamba TMPs who were asked to describe the "visible physical or emotional signs" of a particular illness are as follows:

Madness (*Nduuka*)

1. "Abnormalities in speech, odd behavior (jumping, tearing clothes). Loss of memory. Has red eyes that look like fire. May say 'an animal is eating me.' "
2. "Singing and running around; some beat other people."
3. "Has red eyes; speaks to self; sways head."
4. "Twitching of eyes; abnormal behavior while talking and moving. Loss of memory and fierceness."
5. "Confused; talks nonsense; collects rubbish and may wear it; looks dull; patient is not straightforward in what he or she does."
6. "Laughs to self, shouts; can fight and even remove clothing. Usually tied up when they come."
7. "Speech changes; talks to self; wild and restless behavior."
8. "Eyes look yellowish; shifting eyes and misbehavior; stronger than normal."

Epilepsy (*Kizala or Munguthuko*)

    1. "You look at the eyes and see shadows under them. The eyes look very red. Patient then gets an attack."

    2. "Reddish eyes, twitching of eyes; body is flabby and a worm goes around in patient's head."

    3. "Foreign blood in body."

    4. "Shivering unconsciousness; foam from the mouth. Very stiff. Observes patient's reactions."

    5. Pulse rate and heart beat perceived as abnormal (blind TMP).

    6. Stares continuously at one place; saliva constantly on the lips."

Infertility (*Ngunga*)

    1. "Lack of mentrual flow."

    2. "Patient looks strong. Abdominal muscles tight."

    3. "No symptoms."

General Edema (*Mwimbo*)

    1. "Body and eyes are swollen; constipation."

    2. "Legs are swollen; there is pitting edema (*kiathi* or *kithakame*) when flesh is pressed."

    3. "Overweight and pitting edema on touching."

    4. "Pitting edema; armpits swollen; uneven swelling; shiny skin and whitish eyes."

    5. "The whole body is swollen."

    6. "Body is extra large/overweight. Patient falls down when astonished. Likely to have nosebleeds and heart failure."

    7. "Skin looks shiny; body swollen; body looks transparent because of lack of blood; other fluid tends to be more than the blood."

    8. "Swelling; hard stool."

Infant Gastroenteritis (*Nyunyi*)

    1. "Sunken fontanelle; looks wasted on chest; body swollen and appears reddish or whitish; vomiting and diarrhea; red eyes."

    2. "Child cries much; body is hot; greenish-yellowish diarrhea; vomiting."

    3. "Vomiting and greenish diarrhea; sunken fontanelle; palpitations; taut skin."

    4. "Sunken fontanelle; body feels heavy (like stone); vomiting and greenish diarrhea."

5. "Sunken fontanelle; green stools; breathes very hard; vomiting and diarrhea; blood is like water; sometimes whole body is swollen."
6. "Greenish diarrhea; sunken fontanelle; stools look like greenish fermented milk."
7. "You look at the eyes. Vomiting and diarrhea; the body is swollen and gets white."
8. "Sunken fontanelle; greenish diarrhea; stools look like spoiled milk."
9. "Diarrhea; no interest in food; severe headache; coldness."

Impotency (*Ndewa*)

1. "He cannot sit down properly. Head of the penis is swollen."
2. "The patient says he is unable to 'go with women.' "
3. "Test ability of men by having sex with them" (female TMP).
4. "No symptoms" (two TMPs).

Gonorrhea (*Kisonono*)

1. "Patient feels pain when passing urine. Some blood and pus also pass out."
2. "Patient complains of lower abdominal pains and backache. Urine is yellowish and contains 'mucous'. Smear shows blood and pus."
3. "Abdominal pains, painful urination, pus discharge."
4. "Painful urination; pus comes out of sexual organ."

Worms (*Nzoka*)

1. "Itching and scratching over whole body; swelling of body which later disappears; generalized weakness and rashes."
2. "General weakness; patient very dull; abdominal pains, especially during the night."

There was much variation in the treatments individual TMPs in Mathare provided for various diseases and illnesses. They used medicines prepared from plants nearly universally, except in consultations restricted to divination. Traditional medical practitioners administered plant medicines (*miti shamba*) primarily orally in liquid form, although in certain instances patients inhaled the steam from a boiling pot of herbs. Sometimes they rubbed powdered forms of medicine into small cuts made in the patient's skin (subcutaneous injection and counterirritation). Massage was also used in cases involving muscular and skeletal discomfort. Forms of psychotherapy, including the use of rituals, development of the case history, and prayers served as integral elements of all therapy. None of the Kamba TMPs practiced surgery. Two Kikuyu TMPs in Mathare I provided uvulectomies on a "walk in, walk out," fee-for-service basis.

TRADITIONAL MEDICAL PRACTITIONERS AND THERAPY

A Note on African Pharmacopeia

Scientific research to discover the pharmacodynamic properties and therapeutic potential of many plant medicines is a common activity today in African universities and certain government-aided agencies. Such work is often undertaken in cooperation with foreign research centers such as the National Cancer Institute in Washington, D.C. (Aikman, 1977; Duke, 1985). These activities may result in discoveries of great medical and social value and should be encouraged and supported. However, because this kind of research is separated from the behavioral and religious context of African therapeutics it may do little to advance understanding of how African cultures interpret the functions of or use plant medicines.

As a rule, Western observers show intense interest in learning about the nature, applications, and efficacy of African pharmacopeia. Indeed, many non-Africans conceptualize traditional medicine largely in terms of a narrowly defined cluster of practices involving the collection, preparation, and administration of "herbal remedies." Healing is thus perceived as a process of providing herbal "cures" for specific diseases. This kind of reductionism oversimplifies African therapeutics and leads to inaccurate interpretations about the role of plant medicines, in particular.

Few TMPs in Mathare or elsewhere in Africa discussed their patients' problems or shared their technical knowledge of diseases and dosages with other TMPs. With regard to plant medicines, most Kamba TMPs exhibited great concern about the freshness and related potency of their *miti shamba*. Most also made frequent safaris home to Kitui, one of the reasons being to collect plants from the bush for their use in Mathare. In actual practice, however, there is little evidence that TMPs consistently used identical plant preparations for the same malady, although different plants may have common active properties. Moreover, dosages are also imprecise, which reflects Okwu's (1979, p. 23) argument that "traditional medication in Africa has no measurement . . . and even the medicine itself is ritualized. The point is that the African does not expect to be healed by medication alone." To be effective, medication must be "associated with divinity and with sacrifice to or invocation of the deity or ancestors" (Okwu, 1979, p. 23). In essence "medical stuffs are all 'placebos' because they are regarded as incapable of producing any cures by themselves" (Okwu, 1979, p. 23).

The use of herbal medicines is thus inseparable from the *art* of human healing. For example, Kamba TMPs used their customary God-given power to transform certain plant species into *ng'ondu*, a term which signifies magical healing power. One form of *ng'ondu* was used for ritual "washing" and "cleaning" of the body. In infertility therapy, for example, TMPs tied pieces of *ng'ondu* "roots" around the women's waist to promote conception. In neither case did the patient ingest the preparation.

It is worth emphasizing that above all, the TMP's most important task is to restore social order. As Maclean (1971) notes, traditional medicine "is bound up with the concept of magical power or influence and its possible effects are not limited to what we would regard primarily as pharmacological cures."

The present study was not designed to assess the biochemical and therapeutic effects of medicinal herb products (teas, powders, and so on) ingested by the patients of Kamba TMPs. Although our understanding of traditional medicines remains impoverished without this knowledge, the design complexities and high costs of such work largely explain the limited progress in this area. Botanical identification and verification alone presents an imposing task. However, the TMPs did provide the vernacular names and specified the applications of many plants in common use in Mathare. A list of 51 plants used in Mathare, including their botanical names and uses, is found in Appendix B.

The condition of many patients who sought out Mathare TMPs was complicated by intercurrent diseases and symptoms. Thus in a particular case where the primary focal zone of discomfort in a physical sense was "abdominal pains," for example, the patient also reported dizziness, headache, and joint pains. Similarly, the child treated for gastroenteritis also had a fever and suffered from worms and a middle ear infection. Indeed, the burden of illness in the African population was and is large. Although it is tidy and convenient to refer to patients as having a singularly classified disease, such cases are probably exceptional. These factors should be remembered in connection with the following descriptions of therapies for specific diseases.

Madness (Nduuka)

Half of all *waganga* in the Mathare survey identified "madness" as one of the three illnesses they perceived themselves as best qualified to treat (Table 6-6). Katz and Katz (1981) also worked with TMPs in Nairobi and found that 60% felt especially qualified to treat "madness" as well as other symptoms and disorders of the central nervous system (CNS). Since it is difficult to compare traditional and biomedical diagnoses, they included convulsions, "epilepsy," headache, spirit possession, "mental confusion," and nervousness in the CNS group—"all terms used by the healers themselves" (Katz & Katz, 1981, p. 7). In the present study several of the patients observed in treatment for "madness" had previously been inpatients at the adjacent Mathari Mental Hospital operated by the Ministry of Health.

As the following capsule summaries demonstrate, Mathare TMPs employed a variety of methods in treating *nduuka*, or madness. A patient often stayed with a TMP for 3 days or longer for intensive treatment and observation.

> A Kikuyu man, held and restrained by four other adults, was brought to a Kamba female healer. Those accompanying the patient said he had been sick for 6 months, including a hospital stay of 1 month, and that he sometimes slept without waking for 3 days. The TMP promptly slapped the patient in the face, which put him to sleep. After he awakened several hours later, she crushed the bark of *mwalika* (*Tinnea aethiopica, Litoralis vollensem?*) into powder form and roasted it in a tin over a fire. She then held the back of the patient's head so that he could inhale the *mwalika* smoke. This caused the patient to sneeze and eject "the worm" responsible for the madness.

Complex ritual was the featured therapy in another case of "madness," also treated by a Kamba female TMP. K., the patient, told us he had already received one treatment from this TMP and, having experienced improvement, had returned for a second session. His illness persisted for 2 years prior to consulting this healer, during which time he had been involved in several forms of anti-social behavior, including going around naked. K. had also been an inpatient at the Aga Khan Hospital, Kenyatta National Hospital (KNH), and at Mathari Mental Hospital.

> When it was his turn for treatment K. was placed on the floor in a sitting position with his legs straight out. The TMP looped a string with seven knots in it around his neck and tied the ends around the third finger on each hand. After balancing impala horns on K.'s shoulders, the TMP rubbed a powdered medicine on top of his head, spat on it, and circled around him whistling, chanting, and touching his head, body, and legs with another set of horns. This circling ritual was repeated several times, and the knots in the string were rapidly undone with quick wrist snaps. The horns were removed from K.'s shoulders, replaced, and removed once again. The TMP next held an impala horn on top of the patient's head, chanted a song, and spat ritually on top of his head. K. then had to "taste" (lick) the horn twice, and this was followed by a clinical examination of his eyes and tongue. A powdered medicine was then wrapped inside a small piece of white cloth which was placed in an impala horn filled with water. After firm orders from the TMP, K. swallowed the entire contents of the horn. Thereafter, the TMP made small cuts on K.'s breastbone with a razor, into which powdered medicine (*muthea*) was rubbed. An impala horn was held over this position briefly and then removed, symbolizing extraction of the power causing the illness. The TMP explained that the medicine K. drank would move to and remain in the place where she made the cuts on his chest.

Next, another powdered medicine was mixed into a glass of water and K. was instructed to drink it. Finally, the TMP gave K. some additional medicine to take home and a small red amulet which she ordered him to carry at all times.

A third therapy for "madness" illustrates a test for efficiency based on the patient's ability to carry out practical tasks.

In this instance the TMP was a Kamba male who informed us that when he encounters an actively disturbed patient he initially says prayers to calm him down. The TMP also noted that some mad people like food very much, but food given on demand might worsen the patient's condition. Moreover, the behavior of a mad person usually gets worse at noon and 5:00 P.M., depending on "how the sun is moving."

Therapy begins with divination. Afterward the TMP shows the patient a fire burning on the end of his medicine stick. He then takes a blunt chisel, grips it tightly, touches it against the patient's head, and beats the end of it with a hammer (this procedure is also followed in treatments for epilepsy and dizziness). Later, the TMP ritually washes the patient with a mixture of *mwaiitha* (*Dalbergia lactea*) in hot water and kerosine. The patient also ingests medicine prepared from the crushed tuber of *muvunda* (*Moringa* sp. ?) in a soup containing the brain of a goat. Treatment is continued over a 3-day period. During this time the TMP will try to send the patient to a shop to buy some item. If the patient succeeds and also returns with correct change, the TMP "knows the patient has recovered." He can then be discharged.

The effectiveness of traditional therapy for mental disorders is difficult to validate in the absence of rigorous studies specifically designed to follow patients through an entire course of therapy, and beyond. Even then, understanding the relationship between the therapy process and outcome is problematic because psychotherapeutic techniques (in contrast to routine mechanical procedures of curative and preventive medicine) may not readily transfer, if at all, between cultures. As many researchers have observed, the methods of psychotherapy have evolved from sociocultural beliefs, practices, and the world view that are shared by patients and healers (Katz & Katz, 1981; Lambo, 1978). Evaluation is thus confounded by the paradigm conflict between psychiatry in the biomedical tradition, in which "value-free technique" is viewed as primary, and therapy in the African tradition which stresses the creation of expectancy and relies upon the healer's "supreme self-confidence" (Lambo, 1978) and charisma to achieve results (Rappoport & Rappoport, 1981). Furthermore, as Luborsky, Singer, and Luborsky (1975) note, regardless of the type of therapy about 80% of all psychiatric patients will improve if the practitioners exhibit empathy and if the patients have hope and believe they will be helped (cited in Katz & Katz, 1981). Ideally, the most favorable circumstances for successful treatment of mental disorders may spring from cooperative cross-cultural efforts such as Dr. T. A. Lambo's (1978) family- and community-based "network" therapy program in Aro, Nigeria (see Chapter 10).

Epilepsy (Kizala *or* Mung'uthuko)

Epilepsy, 90% in the form of grand mal, reportedly has a much greater incidence in Africa than in either North America or Europe. Different studies reveal rates that range from 1.4% to 14.7% of the African population (Nordberg, 1983). Because modern medication is inaccessible to the great majority of epileptics in Africa, attacks are also much more frequent than in technologically advanced societies. Many African epileptics also suffer greatly from social stigmatization by their communities. Even where health centers with supplies of appropriate drugs are theoretically accessible, distance, travel time, and poor comprehension of treatment regimes by patients often interfere with the control of epilepsy's symptoms. A recent study at KNH in Nairobi found that literacy increases compliance and possibly seizure control. It points to the need for community education and recommends ways that biomedical practitioners can improve the transfer of information about treatment to patients (Lisk, 1984).

Kamba TMPs in Mathare identified several causes of "epilepsy," including bewitchment, "brought by God," a curse (e.g., from a barren stepmother/cowife), and hereditary factors. Consequently, they believed epilepsy can be transmitted from one generation to the next. It was widely believed that seizures signify when "the worm in the head turns," causing the sufferer to fall down. Kamba also distinguished between the "common" epileptic and one who had been severely burned and scarred from a fall into the hearth fire. It was believed that the epileptic burn victim can never be cured.

Draat's (1981) survey, cited by van Luijk (1982, p. 107), revealed the painful social stigmatization that accompanies epilepsy among the Kamba.

> Thus the epileptic patient is hidden away by his family and shunned by the society. The mental retardation of the epileptic that was often noticed further isolates the patient from social contact. The stigma of an epileptic also strongly reduces the possibility for marriage. The majority of the respondents stated that it is impossible and inconceivable that epileptic patients would marry each other. The fact that people believe that epilepsy is sexually transmitted makes it very difficult for an epileptic woman to have children (Draat, 1981).

In the Mathare TMP survey, 16% of all healers ($N = 45$), especially women, and 20% of the Kamba identified "epilepsy" as one of the three diseases they felt best qualified to treat. Another 16% of the TMPs (19% of Kamba) believed that "epilepsy" is incurable. When asked to identify which diseases are best treated by biomedicine and traditional medicine, only 4% of the Kamba thought to mention cure of "epilepsy" as a strength of biomedicine, and only 6% credited traditional medicine.

Treatment of epilepsy involves divination to confirm the cause (Table 6-4) of the disease, elaborate rituals, plus one or more of several other procedures such as ingestion of plant medicines, inhalation of medicinal smoke, the mixing of saliva, and clinical examination. Two of several cases we observed are described here to illustrate common and contrasting approaches to therapy.

Rose, a Kikuyu teenager, had a history of convulsions going back several years. She had attended primary school for 3 years, but her parents did not send her thereafter because the seizures caused embarrassment. Rose arrived at the Mathare clinic of a female Kamba healer accompanied by her mother, who said that her daughter had received hospital treatment but had not improved. This TMP generally charges Shs. 300/ (U.S. $40.) for treating epilepsy, and the fee is paid in two installments of Shs. 150/. After asking them questions about the case, the TMP dabbed a white chalklike powder (*ila*), said to come from sacred Mt. Kenya, in the center of her own forehead and on the outside ends of her eyebrows. This was done to ensure the help of her spirits in revealing the disease (this TMP does not use herbal medicines in treating epilepsy).

Following divination, which involved the use of her magic bow, shaking "marbles" from her calabash, and interpreting images that appeared in a hand-held mirror, the TMP proceeded to smear a powder of undetermined origin on Rose's head—having first tasted it herself. The TMP then held up a pair of impala horns and, chanting, circled around the patient several times, touching almost every part of her own and Rose's body with the horns. A live, reddish-brown chicken, purchased by Rose's mother, was thereafter made to stand motionless on top of Rose's head while the TMP danced around her, holding the horns, singing, whistling, and ritually spitting (a form of blessing) on the patient's face and head. The TMP next got the chicken to jump off Rose's head by biting its tail feathers! The patient was then instructed to lie down on the floor, and a blue powder was smeared on the chicken in preparation for its sacrifice. A liquid was sprinkled over the patient's head, and ash was placed under her head. After reciting a prayer, the TMP cut the chicken's head off with a large knife. Some of the chicken's blood was directed to flow onto Rose's head, and the chicken's head was placed under her head on the ashes. The TMP prayed for the patient's recovery, examined her head, mouth, face, fingernails and eyes, and took her pulse at the wrists. Rose was instructed to sit up, whereupon the practitioner sprayed a mouthful of water on her head—saying that the chicken now "goes away" with her epilepsy problem. Following another brief clinical examination, Rose was told to stand at the door facing the outside. The TMP proceeded to spray more water on her head, face, shoulders, and toes. Finally, while water was being poured on her head, the patient was instructed to walk out the door and not look back.

No herbal medicines were ingested or otherwise used in treating Rose for epilepsy. Treatment was based on integrated ritual and psychotherapy. Specific procedures included divination and the use of *ila*, the white earth that transmits power to the healer; the use of impala horns, which contain a powerful *kithitu* (fetish); the sacrifice of a chicken; ritual spitting (a form of blessing) and cleaning; and clinical examinations.

Nyegera, a teenage Embu girl, was a second epileptic we observed. She was first treated "without improvement" at Nkubu Mission Hospital in Meru District. When we met her she had returned for fourth round of therapy as a continuing patient of one of Mathare's most popular and charismatic female TMPs. She was accompanied by her mother.

> Following divination, Nyegera was told to remove her clothes. A white sheet was then tied around her waist. She was instructed to sit on the floor and a long necklace of beads was draped over her right shoulder. The TMP's male assistant tied two separate strings around Nyegera's upper arms. Next he extended the strings to the patient's feet, and the end of each string was tied around her corresponding big toes. A red substance (*thiriga* in Kikuyu) was smeared on the patient's back, head, and legs. The TMP next took a piece of chalk, drew a rectangle on the floor, and instructed the patient to sit inside the demarcated area. He then placed a medicine bag (*ngusu*), several ram and cow horns, large sea shells, several dried seed pods of raffia palm (*mwale*), and small calabashes around the patient. *Muti muiu*, an herb mixed with fat from a male goat, was dabbed on the floor around the rectangle. The patient was ritually "washed" with a green bough of the plant *njou ya iya*, and several of these were also set around her on the floor. The TMP then used a match to light the *muti muiu*, creating in effect several burning "candles" at the patient's sides. Nyagera was instructed to lie on her back and close her eyes. A glass of water containing *muti muio* and another medicine (*muti wa wendo*) was placed on the patient's forehead while the "candles" burned.

> With the above scene set, the TMP prayed to "all of the spirits of Kikuyuland, Embu, and Ukambani" and poured libations of water on the floor. She then removed the sea shells from the floor and dabbed *ila* near the outside corner of Nyegera's eyes. The TMP's assistant brought in a young chicken, cut a hole in its chest, and removed its heart. The heart was placed on the patient's chest where it continued to beat for about 90 s. This signified that the chicken (which was thrown away, not to be eaten) would "carry the disease away."

> Following this phase of the treatment the TMP mixed a powder made from the *muguuka* herb with *muti muiu* and water. She placed drops of this concoction on the patient's tongue from the end of a knife. The practitioner then prayed while her assistant ritually "washed" Nyegera with a mixture of *thiriga*, *ng'ondu*, and *njou ya iya* in water.

> To conclude the therapy, the chicken's heart was roasted and crushed into a powder. This was mixed with *muguuka* and *muti muiu*, to be taken orally by the patient after she returned home. Nyegera's mother paid half of the TMP's Shs. 200/ fee (U.S. $25.65) before they left.

As in Rose's case, Nyegera also received prior hospital treatment for her epilepsy. It is not known whether the failure to control her condition was attributable to the patient's ignorance and noncompliance with drug therapy, laxity or improper procedures by hospital staff, or other circumstances. Although there are areas of overlap in the therapeutic procedures of the TMP's who treated Rose and Nyegera, including divination, the use of *ila*, and a sacrificial chicken, the differences in ritual style and substance are also striking. (Two other cases of epilepsy handled by different TMPs featured use of the patient's saliva in the therapy process.) In contrast to Rose's treatment, Nyegera was administered concoctions containing medical plants, in which case there must be allowance for some potential but unknown pharmacological action and physiological response. Both cases, however, illustrate the scope of the systemic differences in concepts of disease and therapeutic practices that often separate biomedicine from African traditional healing. These cases also underscore the formidable problems that confront scientific investigators who may wish to evaluate the efficacy of traditional therapy for epilepsy.

Infertility (Ngungu)

As in other parts of the world, human infertility is a cause of major social, psychological, and medical problems in Africa. High overall rates of population growth notwithstanding, in tropical Africa it is estimated that up to 30% of marriages are involuntarily childless (Mbizvo, Chimbira, & Mkwanzi, 1984; Sherris & Fox, 1983). Over 20 years ago Bennett (1962, 1965) estimated that up to a third of all women in the large urban centers of East Africa had become sterile by age 30. Nordberg (1983, p. 450) recently asserted that "it can be assumed that 10–20% of women over 30 . . . are sterile and that few of these have already born the desired number of children." Gebbie (1974) argued that infertility is "the chief reason" that Kenyan women seek biomedical attention.

Virtually everywhere in Africa, childlessness is perceived as a "pitiable condition"—a source of misery, emotional stress, and stigma that can result in divorce (Caldwell & Caldwell, 1983; Imperato, 1977; Mbizvo et al., 1984; Molnos, 1973). Unfortunately, the social problems associated with barrenness are compounded for African women because by tradition the fertility of African men is rarely questioned unless successive wives fail to conceive (Imperato, 1977; Mbizvo et al., 1984). Such attitudes, of course, have no basis in fact because, according to andrology studies conducted in Africa and elsewhere in the world, male causative factors play a role in as many as 40% of childless marriages (Guest, 1978; Mbizvo et al., 1984).

In Kenya, Uganda, and other areas of sub-Saharan Africa pelvic inflammatory disease (PID) is the most frequent gynaecological disorder (Arya, Taber, & Nsanze, 1980; Nsanze, 1980; St. John & Brown, 1980). No longer an "urban" problem, PID has spread into rural and remote places. Gebbie (1974, p. 493), citing Carty, Nzioki, and Verhagen (1972), asserted that "there can be little doubt that the prime factor in the etiology of the disease [PID] is gonorrhea." A study by laparoscopy in young Kenyan women with primary infertility found that over 80% had chronic PID. Defects of male spermatogenesis formed the second commonest finding (Gebbie, 1974 citing Mati *et al.*, 1973). Each episode of PID carries a 30% risk of infertility (Nordberg, 1983). Thus the key to eliminating PID, and reducing infertility, is the eradication of gonorrhea. Unfortunately, in many African countries the incidence of gonococcal infections has now reached epidemic levels, and sequelae such as infertility and ectopic pregnancies are also more common (Owili, 1983).

In the Mathare survey, the TMPs identified a variety of causes of infertility ranging from natural causes to witchcraft and ritual error. Significantly, some TMPs also cited venereal disease and "diseases from prostitution" as a cause of (female) infertility (Table 6-5). Overall, 22% of the Mathare *waganga* (and over twice as many males as females) identified infertility as one of the three illnesses they are "best qualified" to treat (Table 6-6). Infertility and gynecological disorders ranked third among illnesses reported on the Mathare client record forms (Table 6-8).

As practiced by the Kamba TMPs in Mathare, traditionally based therapy for infertility includes extensive ritual and psychotherapy as well as the use of herbal medicines which are ingested. We observed and recorded numerous cases of treatment for infertility, some of which stretched out over 3 days. The following case is illustrative.

The attending TMP in this instance was a Kamba woman, 35 years old and herself biologically childless, whose apparent successes in promoting conception had gained her a large following maintained by word-of-mouth. "Mary," the patient, was a Meru woman in her late 20s. She had two living children and desperately wanted a third, but she had now gone several years without conceiving. We met her shortly after her arrival at the TMP's clinic in the company of her sister, who was 6½ months pregnant. Before her own treatment began Mary asked if the TMP could help the sister abort her fetus (*kutambaika*). The TMP scolded both of them, saying that she did not undertake such activities. Speaking to Mary, the TMP said "tell your sister to buy herself a coffin!"

Mary's treatment for infertility began on July 11, 1979, the day she arrived, and continued on July 12 and 13. On July 11 and 12 the TMP concentrated on *kuvungua*—a ritual, cathartic-like process of "opening" the patient. A brief divination on the first day confirmed that Mary had been bewitched by her jealous co-wife. The latter had obtained some of Mary's menstrual blood and mixed it with other substances in order to harm her. Mary was given two types of herbal medicines. *Mulenda*, the first of these, was to be drunk in the morning and evening in order to treat a problem such as gonorrhea that may exist somewhere in her reproductive tract. Mary was instructed to drink *mulivindi*, the second liquid, in the evening to "warm up her eggs." During the *kuvungua* ritual on July 12 Mary was required to step over various piles of horns, shells, and amulets. During part of this ceremony she remained seated on a traditional three-legged stool. Meanwhile, the TMP kept a flame burning inside a shallow pan. She also placed a long blue string with seven knots in it around the back of Mary's neck and attached the ends to the patient's right big toe. Subsequently, this string was removed and the knots were snapped open over different locations on Mary's body.

July 13, Mary's third day as a patient, featured the main treatment—described by the TMP as *uima theatre biu*, or "the real operating room." The procedure was as follows:

> After Mary arrived the TMP took chalk and drew a rectangle on the floor. At the back of the rectangle the TMP printed the name of her own village in Kitui District, and at the front she wrote "Kiui"—"a big river." She also drew three sides of a much smaller rectangle inside the large one, drew an oval inside it, and placed crushed green pieces of an herb called *mwalula* inside the oval.
>
> Next, the TMP took small, short sections of plants known as *mulali* and *ndatakivumbu* (*ngondu*) [*Talinum portulacifolium*?], crushed them on the concrete floor with a stone, and combined the product with *ila* (white powdered rock) and soil. Mary was instructed to take some soil from a container handed to her and add it to this mixture. The TMP then walked to the doorway and scattered some tobacco (*mbaki*) outside, asking for God's blessing. Mary was handed an empty Baby Cham drink bottle and asked to bring a specimen of her urine in it. The patient then departed for the nearby public toilets. On return, Mary handed the urine sample to the TMP, who wrapped the bottle in pink toilet paper and set it aside.
>
> With Mary standing, the TMP next unrolled a piece of black crocheting thread from a ball in her medicine bag (*ngusu*) and measured this to fit the circumference of the patient's hips. (The TMP mentioned that neither she nor her male assistant should have sexual relations for 24 hr in order to give the patient an opportunity to have intercourse with her husband.) Then she took a small (5 cm × 5 cm) piece of yellow fabric cut from Mary's handkerchief. A portion of the mixture of *mulali, ndatakivumbu, ila*, and soil, noted above, was placed on the piece of cloth. The cloth was then folded over and bound tightly with thread to the black string that the patient would eventually wear around her hips until she became pregnant. This charm was then dipped into a half-calabash containing a liquid medicine.
>
> At this point the TMP blindfolded Mary and instructed her to undress completely, which she did. Curtains had been drawn so that the several other patients waiting for treatment could hear but not see this part of the ceremony. [In the meantime, a woman accompanied by her mother-in-law stopped at the clinic to thank and praise the TMP for the successful result of her treatment for infertility. To the great pleasure of the practitioner, this woman displayed a color photograph of her new baby and paid the practitioner the remaining Shs. 150/ (U.S. $20) due on her fee. I observed similar testimonies on several other occasions.] The TMP next dabbed the white powder, *ila*, on Mary's upper arms, both sides of her thighs, on her pubis, and between her breasts. The magical charm prepared earlier was then tied around the patient's hips, and a red sheet was wrapped around her from the waist down. A chicken was then brought into the room by the TMP's assistant. Its head was cut off, and its body

was held so that the blood dripped onto the herbs that had been deposited on the floor inside the small quasirectangle that had earlier been drawn with chalk.

The practitioner next laid a blue, knotted string lengthwise across the inner and outer rectangles. The patient, who was in the third day of her menstrual cycle and still blindfolded, was now directed to sit on the mixture of chicken blood and herbs inside the rectangle. Her private parts had to rest directly on top of this mixture so that the chicken and menstrual bloods could mix. (The cultural interpretation of this mixing of bloods was that the process neutralized the *thahu*—a Kikuyu term for an internalized evil power causing the infertility.)

Mary remained seated in this position for about 15 min. Meanwhile, the TMP lit a strong-smelling smoky fire on a plate and gave the patient a glass of a slightly golden, viscous liquid to drink which was made from parts of the plant called *ndatakivumbu*. This plant is the main source of magic in the powerful ritual healing medicine known as *ngondu*. As Mary drank the medicine the TMP chanted "have baby, male or female" several times.

After Mary had spent several minutes sitting on the blood and herbs, the TMP poured the urine specimen on the nape of the patient's neck so that it would run down her back and mix with two bloods and medicine around her private parts. She was then ritually "washed" with a plant swisher dipped in the mixture of *mulali, ndatakivumbu, ila*, and soil. In the final phase of therapy the TMP, using her finger, collected a sample of the mixture of two bloods and herbs and deposited it on a piece of toilet paper. Mary was told to insert this into her vagina and then to put her clothes on and leave immediately. She was further advised not to sit on a stool in anyone's house, but to go straight home and sit on her bed. She was to remove the medicine from her vagina as soon as her husband got home in the evening, and not fail to have intercourse with him that night. The TMP stated quite emphatically that Mary would be pregnant before she had another period. Mary said she had no money to pay an installment on the TMP's Shs. 315/ (U.S. $42) fee for treatment, but agreed to return to the clinic on July 16 with a first payment of about Shs. 200/ (U.S. $27).

Mary's treatment provides a particularly vivid illustration of the extensive ritual and symbolism used in infertility therapy. It also highlights the strongly individualized features in a TMP's repertoire, as in the intriguing use of the "map" that embodied the patient and, apparently, joined her to the original geographic sources of the practitioner's healing powers.

Biomedical specialists and others certainly want to know if and how traditionally based treatment of infertility leads to conception. Once again, no studies exist that have been conducted according to biomedically acceptable standards. Unfortunately, despite its high profile as a set of interrelated organic and emotional problems that affects a large segment of the population, infertility and its root causes are also evidently low-priority policy concerns among Kenyan health authorities.

During the present study we witnessed numerous testimonials from women who believed they had been successfully treated by a TMP. Some wrote letters and sent or brought photographs of their child. Others carried the "evidence" of effective treatment in their arms. One can certainly agree that "success" in these cases is more than a woman *thinking* she's been helped (Katz & Katz, 1981). It is reasonable to adopt the position that:

> Whatever the biologic reasons—relief of muscular spasm of the fallopian tubes, change in the chemical environment in the vagina and uterus, improved ovulation, any or all mediated by the central nervous system—the resultant baby is concrete proof of success. Each such baby enhances the healer's reputation and puts prospective patients in that psychological state which makes them most amenable to treatment (Katz & Katz, 1981, p. 5).

Based on observations in Mathare, and until such time as there are studies that prove otherwise, it seems reasonable to believe that substantial numbers of African women experience increased capacity for conception under the care of selected TMPs.

Abdominal Problems (Kwalwa Ni Ivu, And So Forth)

Abdominal complaints are probably the most numerous of all conditions treated by traditional healers in Africa (Chavunduka, 1978; International Development Research Center, 1980). Janzen (1978), for example, emphasizes the centrality of *vumu*, the abdomen, among the Kongo people in Zaire. *Vumu* is likened to Kongo society's lineage house, and thus incorporates both physical and social symbolism. It refers to reproductive anatomy and processes, including vulva, uterus, pregnancy, breast, and lactation, as well as the abdomen, "food substance," and those who are "descendants of the same mother." According to Laman (1936) in the *Dictionnaire Kikongo-Française*, as interpreted by Janzen (1978, p. 170), *vumu* "is a verbal category of ideas, both physical and social, drawn from a dominant idea about subsistence, identity, and well-being."

Abdominal disorders ranked at or near the top of complaints presented to TMPs in Mathare (Tables 6-7 and 6-8). Nearly 40% of local *waganga* (and over twice as many women as men) identified abdominal problems as one of the three illnesses they are best qualified to treat. Most patients presented chronic forms of illness and many had intercurrent symptoms including headache, joint pains, dizziness, loss of appetite, general malaise, constipation, gonorrhea, menstrual pains, or the presence of foreign objects (*kindu nda*) that had entered the abdomen through witchcraft. A large proportion of the patients we encountered who had abdominal disorders had previously attended a clinic or hospital for the same illness but did not experience the anticipated "cure" due to misdiagnosis, inadequate adherence to or foreshortened medication regimes, or alienation occasioned by contact with impersonal approaches to biomedical treatment. Some continued to seek concurrent help at biomedical outpatient clinics in Nairobi and elsewhere.

Through divination, most TMPs confirmed that witchcraft was the dominant causal agent in abdominal complaints. All such therapy included divination and ritual procedures, and most treatments included the use of herbal medicines. Powdered medicines may be rubbed into small subcutaneous cuts made with a razor at strategic locations on the body. Traditional medical practitioners also extracted foreign objects (*kumya kindu nda*) from the abdominal area without surgery.

In one case a teenage school girl presented with chronic, generalized abdominal pains.

> The TMP diagnosed the cause as witchcraft perpetrated by a neighbor, and proceeded to massage the girl's stomach. Before going further the TMP asked her about her monthly period and told her to be sure to return in a month because she might be pregnant. The practitioner then made 12 tiny incisions on the girl's body: 4 each on the abdominal surface above the ovaries and 4 just above the pubis. Then she rubbed a medicine into the cuts. Next, the TMP poured a blackish liquid into a half-calabash, deftly applied and reapplied an antelope horn (*kithitu*) to the girl's abdomen, and emptied its "contents" into the half-calabash container. The black liquid was drained from the half-calabash and, to the girl's amazement and relief, maize kernels, beans, and pieces of glass which had been "removed" from her body to the *kithitu* now remained in the bottom of the container. Following this powerful demonstration the girl was instructed to dress, go home, and to report back the next day for a checkup.

Another case involved a rural Kikuya woman, age 52, from a village in Kiambu District.

> Her abdominal pains were apparently related to an old and chronic case of elephantiasis. Her past medical history included periods of dizziness and loss of appetite. She had been treated for these symptoms as an inpatient at Kiambu District Hospital near Nairobi, but did not show improvement. She said she then turned to TMPs for assistance because she was disappointed with hospitals. Thus a year ago she started coming to "M.", a Kamba male TMP in Mathare who succeeded in "curing" her illness. Recently she had experienced abdominal pains and had now come back so that she could be treated by M. The TMP diagnosed the cause of her illness as witchcraft, which someone had perpetrated "because of progress she has made in her home." As treatment, M. first advised her to "eat a lot of vegetables." He then gave her a glass of a medicinal tea prepared by boiling the barks of *mukanu, muthunji, mukinga, mutote,* and *mutongatongwe.* M. next poured the equivalent of four glassfuls of this medicine into a large beer bottle and instructed the patient to drink one glassful on each of the next 4 days. She was also told to return for a checkup in 3 days. M.'s fee for this treatment was Shs. 17/ (U.S. $2.27).

Mumbi, a Kikuyu woman of 42 years who complained of abdominal pains, joint pains, and headaches, also came to M. for treatment from rural Kiambu District.

Through divination M. concluded, as he had in the previous case, that witch-craft was the cause of Mumbi's problems. M. encouraged us to ask Mumbi questions to elicit details of her efforts to find a cure. She had previously been treated at several biomedical units but experienced no improvement. However, when she came to M. he treated her and her condition improved. She stayed well for 3 months, but returned because the pains had shifted to different parts of her body.

In treating Mumbi, M. first applied cow's cream to her shoulder, elbow, and knee joints, and massaged these locations until the pains subsided. He then waved a burning firestick several times near the skin surface at these joints. Next he made small razor cuts on top of Mumbi's head and applied a powdered herb known as *muguuka* on these cuts "to calm the nervous system." The patient also inhaled some of the *muguuka*. Finally, Mumbi was given a large bottle of an herbal medicine prepared by boiling parts of the following six plants together: roots of *mutote* (*Carissa edulis*), leaves of *mukiliulu* (Harrisonia abyssinica), and the barks of *mukanu, mutongatongwe, muteta* (*Stryehhnos henningsei?*), and *mukinyai* (*Euclea divinorum*) (Appendix B). M. instructed the patient to drink one glass of this medicine daily until she finished it, and advised her to come back if she did not improve. He agreed that she could pay his fee of Shs. 30/ (U.S. $2.67) later.

Edema (Mwimbo)

The Kamba term *mwimbo* refers to a generalized swelling of the body or extremities. Biomedically the condition constitutes an excess of water in the tissue fluid. Kamba TMPs in Mathare attributed witchcraft as the primary cause of *mwimbo*, although they also recognized several other causes (Table 6-5). Adults are the principal group at risk, although *mwimbo* also denotes the swelling associated with the malignant malnutrition (kwashiorkor) common in young children. In discussing *mwimbo* Kamba TMPs usually also mentioned "pitting edema" (*kiathi* or *kithakame*), which occurs when the flesh is pressed.

Mwimbo ranked third among illnesses Mathare TMPs said they are best qualified to treat. It was the second most common problem the TMPs could recall treating during the week prior to their interview (Table 6-7), and ranked fourth [with sexually transmitted diseases (STD)] on the client record forms (*N* = 468) maintained by selected TMPs (Table 6-8).

I suspect that the *mwimbo* syndrome is widespread in Africa, although it receives little discussion in the literature. Neither Lindblom (1969) nor Thomas (1971) mentioned it in their respective studies of Kamba life and health care. Van Luijk (1982), writing about the Kamba in northern Macha-kos District, devoted a paragraph to *mwimbo* but did not indicate its frequency. Imperato (1977) mentioned edema only as a complication of pregnancy among the Bambara of Mali. The Canadian International Development Research Center (1980) study in Zaire referred to "swelling" as one of the recognized categories or principal disorders. Chavunduka (1978) identified several cases of edema in his Zimbabwe study. *Nganga* featured prominently in these individuals' courses of therapy.

Since *mwimbo* is most often attributed to witchcraft its treatment requires divination to confirm the source, as well as other forms of ritual and herbal treatment. According to the TMPs, the average time required to treat such edema is about 6 days, with a range from 2 days to 2 weeks. All gave their patients liquid medicine to drink. Some also massaged the affected area, bathed the patient with a medicine, or rubbed plant preparations into small incisions on the skin. The following descriptions allow a comparison of medications used by six Kamba TMPs who treated *mwimbo*.

1. Liquid from boiled roots and bark of *mwaiitha* (*Dalberigia lactea*) and leaves, stems, and roots of *muteta* (*Strychnos henningsei*) is drunk by the patient. The medicine causes diarrhea and improves the blood.

2. *Mwaiitha* roots are crushed and boiled in water. The resulting decoction resembles soapy water, and the patient drinks one cup of it. The TMP noted three effects of this medicine: it relieves constipation, combats the "germs" of *mwimbo*, and causes the patient to urinate (red urine) frequently.

3. Patient drinks a tea made from boiled root of *mutiu*. This was said to cause sweating and to increase urination, thus reducing the swelling. Also, a powder made from crushed and dried root of *mutiu* is added to hot water. The patient is washed with this preparation.

4. A sheep is slaughtered and the sheep's fat is boiled in water containing the herbs *mutote* (*Carissa edulis*) and *muguuka*. The patient drinks this "soup." "*Tatha*" (?) is also boiled in water and patient bathes with this liquid.

5. After the TMP goes over the patient's body with a fire stick "to open the pores," the patient is massaged with a medication made from *nzou ya yia*, *mwaiitha* (*Dalbergia lactea*), *mwenu* (*Cassia* sp.), and *muteta* (*Strychnos henningsei*) steeped in hot water. This massage "improves the blood circulation by mixing the diluted and thick blood together." A special diet is prescribed, and the patient is given a supply of a liquid medicine made from *mukandu* (*Ocimum suave*?) bark and roots of *mwalula*, *mutongatongwe*, and *mutula*. At home, this medicine is mixed with soapy water or milk and taken in the amount of one glassful three times a day for 3 days. The patient returns to the TMP after 3 days and the whole process is repeated—and again after 3 more days.

6. Roots of *muvunda* (*Moringa* sp.) together with the barks of *mwaiitha* (*Dalbergia lactea*), *mulandathe*, *mukanu*, *mwashiumui* and *kyusha* are boiled in water. This medicine is taken orally and also massaged into the skin. It relieves heart failure (said to be a likely symptom of *mwimbo* together with nosebleeds and swelling) and constipation, dilates blood vessels, and clears the veins.

Infant Gastroenteritis (Nyunyi)

Diarrhea and vomiting in infants and very young children was generally seen by TMPs and the community at large as a naturally occurring disorder. Greenish diarrhea ("teething" diarrhea) and depressed fontanelle are the commonest symptoms of *nyunyi*. Overall, 16% of the Mathare TMPs (mostly females) identified *nyunyi* as one of the three illnesses they are best qualified to treat (Table 6-6). The extent to which this level of involvement was indicative of the actual urban demand for TMPs to treat *nyunyi*, as opposed to its representing a specialty they use(d) mainly in the rural areas, is not known. One would suspect that a majority of mothers in Nairobi would take a child with *nyunyi* to a clinic or hospital since it is an acute illness whose symptoms can be usually quickly relieved with oral rehydration therapy and other biomedical procedures. In rural Kilungu I also observed five cases where mothers brought infants with *nyunyi* symptoms to TMPs for treatment and return checkups. Although one would anticipate a greater reliance on rural TMPs for acute conditions in view of the poor access to biomedical services there, two of the children in Kilungu (Kalongo Sub-location) were treated by a TMP whose house was within 600 meters of the Kikoko Mission Hospital and outpatient clinic.

Although Mathare TMPs exhibited considerable variation in their treatment of *nyunyi*, most said that divination is not a necessary part of the process because the symptoms are clearly seen. Rituals, particularly those that use a chicken, are a frequent part of therapy together with prayers, herbal medicine, and massage. The following five capsule summaries of *nyunyi* therapy, obtained through discussions with TMPs, illustrate the areas of common and dissimilar practice among practitioners. Such information about the nature of and variation in therapeutic procedures is essential to any effort by biomedical investigators to assess the efficacy of the traditionally based treatment for gastroenteritis in infants and children.

1. Boil parts of *munatha* (*Courbonia comporum*?) and *mwaiitha* (*Dalbergia lactea*) in water called *maji ya mganga*. This colloquial Swahili term means "water from the patient's home area." When this liquid is warm, bathe the child with it to reduce the edema and relieve discomfort. The above medicine is also taken orally: one teaspoonful and one tablespoonful three times a day for 2 days for infants and larger children, respectively. The child is washed with the medicine for at least 2 days.

2. Stems of a plant called *kamama* are boiled together with the foot of a chicken. This decoction is rubbed on the infant's gums and administered orally in a dose of one teaspoonful three times daily. *Kamama*, *muripidi*, *kibutua butui*, and *nguku* are boiled together. The patient is covered with a sheet and inhales the steam from this liquid once each day for 3 days. Steam inhalation "heals the illness in the blood," while the oral medicine "treats the affected organs and makes the stools return to their normal color and form."

3. A chick is first placed on the head of the child. Next, one of the chick's toes is cut off and the chick is allowed to run away. *Ikoka,* a creeping plant, is boiled in water. The liquid is given to the child for 2–3 days. This medicine stops the diarrhea and changes the stools from green to normal.

4. First the child is laid down. Cuts are made on the crest of a chicken into which *muthea,* a ritually prepared black powder "made from the leaves of *katolekya,*" is rubbed. The chicken is held by the TMP and circled over the child seven times. Roots of *muvunda (Moringa* sp.), obtained from Kitui District, are boiled in water and this mixture is given to the child over 3–4 days. The medicine causes the child to "start breathing normally, stops vomiting and diarrhea, and changes the stools to normal."

5. A chicken is necessary so that it can symbolically fly off with the problem. The TMP also boils a bird nest *(kathonjo)* in water and sponges the child with this when the liquid is tepid. *Mutaa (Ocimum americanum)* is boiled in water and the resulting medicine is dosed to the child for 2–3 days.

Gonorrhea (Kisonono)

Sexually transmitted diseases, and particularly gonococcal infections, have become epidemic in Africa during the past two decades. This is attributed to changes in sexual behavior, including earlier first coitus and increased promiscuity of young people. Treatment and control of gonorrhea is also increasingly adversely affected by extensive use of antibiotics, which has fostered the development of new patterns of antibiotic resistance. For example, an alarming increase in the incidence of penicillin producing *Neisseria gonococci* strains has been reported for West and East Africa (Owili, 1983).

For Kenya, Verhagen (1974) estimated an incidence rate of 20 (diagnosed) cases of gonorrhea per 1,000 adults per year. However, probably not more than 10% of actual infections are treated at a biomedical facility and many of these treatments are inadequate. Consequently, it can be assumed that most cases of gonorrhea are never diagnosed or recorded. Repeated episodes are linked with a risk of infertility that exceeds 50% (Nordberg, 1983).

Mathare TMPs universally recognized *kisonono*—the Swahili term for gonorrhea—as a STD whose primary symptoms (at least in males) include painful urination and pus discharge. As is true of biomedical diagnoses, the extent to which the TMPs interpreted these symptoms correctly cannot be certified without laboratory confirmation.

Overall, only 11% of all Mathare TMPs, and 14% of the Kamba, identified gonorrhea as one of the three diseases they are best qualified to treat (Table 6-6). This compares well with the 12% of TMPs who identified gonorrhea, syphilis, or both as one of their specialities in the Nairobi survey conducted by Katz and Katz (1981). Sexually transmitted diseases ranked fourth as a category of illnesses recorded on TMP's client record forms in Mathare (Table 6-8), and sixth among illnesses presented to Kamba TMPs during the week before their survey interview (Table 6-7).

The therapeutic regimen for gonorrhea used by four Mathare TMPs is recorded below. All use herbal preparations as the primary treatment. None reported the use of a plant mentioned by one of the others.

1. *Mukaya* roots are boiled and the patient drinks a half-glassful of the decoction. The patient is given a bottle of this medicine and instructed to take the same dose three times daily. When the medicine is gone the patient is expected to return for more. The disease is "cured" in 5 days.

2. This TMP does employ divination first to determine both the specific cause and extent of the infection—for example, "to see if it is beyond a cure." If the patient has "open wounds" (confusion with chancroid or syphilis?), the TMP washes and rubs a powdered plant medicine on the affected genital areas. (This TMP mentioned that a woman will scream when he does this.) Medicine is prepared from parts of seven plants: roots of *mwaliki, muvavai,* and *muguuka,* roots and leaves of *chonge,* bark and roots of *mukolekia,* paw paw (papaya) fruit; and an entire plant called *munyoloko upesi* (the TMP's nickname for the plant.) One effect of this medicine is to produce diarrhea, "which relieves lower abdominal pains." "The time required to treat gonorrhea depends on the length of time the patient has had the disease. If the patient's attack is recent (within the last few days), one treatment is enough. If the patient has had the disease for 3 years, 1–2 weeks of treatment and lots of herbal medicine will be needed."

3. Roots of *mwaiitha* (*Dalbergia lactea*), *munatha* (*Courbonica* sp.), and *muva* are boiled in water and the patient drinks this decoction. The effect of the medicine is to diminish pain when urinating and decrease pus discharge. Two to 3 weeks are needed to treat gonorrhea.

4. Roots of *kivavai* and *mwelengwa* (*Cissus aphyliantha*) are crushed, allowed to dry, and crushed again to form a powder which is mixed with water. The patient drinks half a glassful about four times. Four days are required to cure the patient.

Uvulectomy (Kitila Kilimi)

Surgical removal by traditional practitioners of the uvula, the tissue that hangs down from the back of the soft palate, is evidently a long-established and widespread custom in Africa. In addition to Kenya, the practice is reported in Tanzania (Haddock & Chiduo, 1965), Sierra Leone and Nigeria (Fleischer, 1975; Ijaduola, 1982; Imperato, 1977; Maclean, 1971), and Mali (Imperato, 1977). It is generally believed that an elongated uvula causes or contributes to several throat and digestive disorders, the symptoms of which are relieved by uvulectomy. The operation is usually performed with a sickle-shaped knife or scissors. Biomedical opinion generally views uvulectomy as "undesirable and sometimes dangerous" (Haddock & Chiduo, 1965, p. 334) and an "unnecessary and useless operation" (Maclean, 1971, p. 66).

In northern Nigeria, Hausa barber surgeons excise the child's uvula with a sickle-shaped knife on the third day after birth (Imperato, 1977). The belief underlying this custom is that the uvula will swell and rupture if it is left intact, causing the child to suffocate (Fleischer, 1975). In a hospital-based study in Jos, northern Nigeria, Fleischer (1975) found that 96.2% of Hausa and Fulani children had had uvulectomies.

An investigation in coastal Tanzania found that uvulectomies had been performed on people from 27 different ethnic groups. Of 314 inpatients examined at Dar es Salaam's Muhimbili Hospital, 48.1% had had the operation (Haddock & Chiduo, 1965). Coughing is the indication these patients mentioned most frequently (73%) as cause for uvulectomy, followed by general weakness/anemia (18%), sore throat (5%), and no illness/prophylaxis (4%).

A hospital-based study of 541 patients in Nairobi found that 37% had experienced uvulectomy. Among Kikuyu patients, who formed the majority (44%), 56% had had the operation—most of these before age 11 (Jarvis & Mwathi, 1959). This age-incidence pattern for uvulectomy was corroborated by our evidence from Mathare, but it differed sharply from the age distribution reported in a 2-year study by Ijaduola (1982), an otolaryngologist in Lagos, Nigeria. Ijaduola's subjects comprised 180 people (more females than males) who had come to Lagos University Teaching Hospital because of complications following uvulectomy (38%) or, without attributing any complications to the surgery, because of persistence of symptoms for which the operation was first performed. Uvulectomy had been done most often (78%) in the age group above 21 years. Only 8% were under 11 years old at the time of their uvulectomy (Ijaduola, 1982).

Although the medical literature curiously neglects age as a factor in postoperative complications from uvulectomy, Ijaduola's data strongly suggest that adverse aftereffects increased with age at the time of surgery. Of 68 patients who presented with complications, 56% had postoperative bleeding, 24% acute otitis media, 13% damaged palate, and 4% epiglottis. There was one severe case of tetanus in a man who had an uvulectomy 6 weeks earlier. Examination revealed a deep puncture wound in the posterior pharyngeal wall caused by a knife.

In contrast, in areas where uvulectomy is performed mainly on younger children, as in Kenya (Jarvis & Mwathi, 1959) and Tanzania (Haddock & Chiduo, 1965), the majority of patients appear not to have suffered serious aftereffects. In the only Kenyan study available, only one of 199 people who had had uvulectomies experienced "definite harmful effects." This single case was a man who had his uvula cut at age 28 and thereafter developed persistent nasal speech (Jarvis & Mwathi, 1959, p. 437). In the Tanzanian study, delayed diagnosis of tuberculosis was considered the most adverse consequence of uvulectomy (Haddock & Chiduo, 1965).

If the incidence of complications is actually much lower in Kenya than in southwestern Nigeria, the lower risk in the former country may be due to the youthful age at which the operation is performed and the fact in a majority of cases the surgery "is far from radical and a stump of varying length remains" (Jarvis & Mwathi, 1959, p. 437). Given the almost universal practice of uvulectomy on newborns in northern Nigeria, a low incidence of postoperative complications in that population would strengthen the hypothesis of youthfulness and lowered risk.

The reasons for uvulectomy are generally similar in the areas of Africa surveyed. In the Nairobi study by Jarvis and Mwathi (1959, p. 437), for example, the most common explanations included irritative cough (33%), pain in throat (19%), elongated uvula (15%), anorexia or vomiting (9%), and general weakness (6%). Other reasons included hoarse voice and "custom" Ijaduola (1982, pp. 772–773) developed a revised diagnosis of previously uvulectomised patients with persistent symptoms ($N = 112$). He found that "the commonest lesion for which uvulectomy was wrongly performed is chronic tonsillitis (44.6%)." This is followed by chronic laryngitis (20.6%), chronic pharyngitis (15.2%), chronic sinusitis (11.6%), nasopharyngeal carcinoma (3.6%), and carcinoma of the larynx (1.8%). There was one case each of chronic myeloid leukemia, carcinoma of the tonsil, and Hodgkin's lymphoma.

In Kenya, uvulectomy is apparently most common among the Kikuyu cluster of peoples, including the Embu and Meru. It is also practiced with somewhat less frequency among the Luo, Kamba, and certain other groups. Among the Kamba, uvulectomy is the only traditional surgery apart from male and female circumcision and the practice of making incisions on a patient's body in order to rub medicine directly into the bloodstream.

Two Kikuyu people (close relatives) provided the uvulectomy (*kitila kilimi*) services available in the Mathare study area. One was 66 years old and illiterate; the other was 35 had completed 7 years of school. Known as "*dakatali wa kilimi*" (Swahili), they performed the operation on "anyone who comes with uvula problems—from 2 weeks of age to as old as grandparents." The groups they treated most frequently included, in rank order: (1) Kikuyu, (2) Luo and Luhya, (3) Kamba, and (4) others (including about four "Asians" per year). Handwritten records contained clients' names, the date of surgery, and the fee charged. These revealed that the fee (cash only) rose from Shs. 15/ (U.S. $2.00) in early 1976 to Shs. 20/ in May 1977 and Shs. 25/ (U.S. $3.33) in October 1977. This 67% fee increase in just 30 months may have been a response to the boom in coffee prices that injected considerable sums of money into the hands of Kenya's growers and other sectors of the economy during the late 1970s.

According to the Mathare *kilimi* surgeons, the operation "prevents coughing, tonsillitis, hoarseness, and general weakness." Those conditions occur when "the uvula grows too long and tickles the back of the tongue, causing coughing and throat irritation. So it has to be cut off." In adults the uvula is only partially removed (to avoid hemorrhage), whereas in children it is "cut to the roots because it will grow again. A child doesn't have strong vessels to bleed badly."

To stop bleeding, the patient gargles with hot salty water. After the operation it is advisable to:

> eat something crusty so it will scrape off the clotting blood and speed up healing. Soft foods will only stick to the clotting blood and cause infection. After a few days the patient should continue eating his normal diet.

I observed two *kilimi* operations. The first was an 11-year-old Kikuyu boy who was brought by his mother because he would vomit after eating. The younger of the two *kilimi* specialists performed the operation with a hand-fashioned metal instrument, about 18 cm long, that works on the cigar cutter principle. The tool was placed into the child's mouth and the uvula was positioned into a rectangular opening at the end of a metal strip. This strip fits into a metal housing which serves as its guide and also has a razor-sharp edge at the end near where the uvula is placed into the metal strip. A quick forward snap of the housing severed and caught the uvula inside the opening in the metal strip. The entire procedure took about 3 s. Immediately thereafter the practitioner wrapped the uvula in a scrap of paper and threw it out the door into a drainage ditch. The boy was allowed to spit out a little blood and was given a drink of water, followed by a cupful of a brown bouillonlike liquid. Notably, the surgical instrument used on this child was not washed prior to the operation and even had traces of blood from a previous operation caked on it. The practitioner said that the wound would take about 4 days to heal.

The second case witnessed was that of an infant under 3 months old who was brought to the elder *kilimi* specialist by its Kikuyu mother. The manner in which this old woman approached the child is remarkable. She moved almost rhythmically—slowly, calmly, and with obvious empathy for the child. She used an instrument identical to that used by the younger practitioner, also with no thought of cleaning it prior to inserting it into the infant's mouth. A quick stroke of the tool produced an excised uvula about 0.65 cm long. The infant cried hoarsely for a few seconds, and then with a reflexlike action the mother began to suckle it. Meanwhile, the practitioner stirred a powder (a "shop medicine" the name of which was overlooked) into a small glass of water which turned white and translucent. This was spoon-fed to the infant. Shortly thereafter the mother paid the Shs. 25/ fee, tied the baby on her back, and left.[3]

Malignant Neoplasms ("Kansa")

Overall, the total incidence of cancer in Africa may be only one-quarter of the rate in North American and Europe. In part this reflects the comparatively low incidence in Africa of cancer of the lung, breast, colon, and rectum, and the fact that the African population is more youthful. Nevertheless, certain types of malignancies are much more frequently observed in Africa than in North America and Europe, including those of the esophagus, liver, nasopharynx, and cervix as well as Kaposi's sarcoma, scar epithelioma, and Burkitt's lymphoma. Cervical cancer is the most common site in females in Kenya, Uganda, and Tanzania, while cancer of the esophagus has a high incidence among males. Many of these cancers exhibit significant spatial variations over relatively short distances (Alexander, 1985; Linsell, 1974).

Studies of cancer as a problem of interest in traditional medicine are rare. A recent exception is a retrospective hospital-based study of 119 Tanzanians with cancer (76% of whom were residents of rural villages) who were interviewed about their use of traditional therapy for cancer preliminary to evaluation for radiation treatment (Alexander, 1985). Nearly half ($N = 53$) of this group had been treated with traditional medicine, and of these 74% had consulted a TMP *before* going to a biomedical doctor. Treatments received included ingestion of herbal medicine, topical applications, subcutaneous injections, and combinations of these procedures. According to the patients' ($N = 53$) own impressions of the outcome of their traditional therapy, 2% ($N = 1$) improved, in 53% the disease progressed, and 43% reported no change.

The problems of case-finding, diagnosis, and treatment of cancer in Africa are greatly influenced by indigenous perceptions, labeling, and other behavioral aspects of the patient and the disease process itself—particularly its capacity to imitate many other illnesses. Alexander (1985) observed that the exceptional individuality and variability of the disease process in cancer victims (e.g., in site, growth rate, stage, and degree of malignancy) make it difficult to predict how they will react to it. Since many years are usually necessary for a malignancy to manifest itself, the early symptoms may be overlooked or played down until the cancer is well advanced and hospitalization is imperative. Admission to the hospital is a unique and often traumatic event for patients and their families because cancer is often associated with disfigurement or death. Moreover, cancer therapy in the hospital usually requires victims to leave their home villages for days or months at a time—

causing alienation, family disruption, and loss of productive activity. Alexander (1985, p. 58) concluded that:

> The significance of a diagnosis of cancer in Tanzania can be understood only by knowing the individual who has the disease. Its nature is unknown to the vast majority of people. The society as a whole does not equate cancer with pain, disfigurement and death as customarily is done in Western cultures.

Traditional medical practitioners in Mathare made occasional references to "*kansa*," but none claimed it as an area of expertise. For most people, patients and healers alike, "*kansa*" is probably at best only dimly understood as a disease process. Moreover, as Katz and Katz (1981) observed elsewhere in Nairobi, the only direct experience a TMP has with cancer *qua* cancer is when a patient who has been labeled by a hospital as a cancer victim arrives at the TMP's clinic seeking help. When the hospital diagnosis is "inoperable cancer," the TMP is virtually the last hope for the sufferer and his family.

Maclean (1971) is undoubtedly correct in asserting that the curative potential of traditional magical-herbal remedies is "totally inadequate" in the face of a terminal diagnosis. However, there is another significant and largely overlooked dimension to the care that TMPs can provide such cancer patients. This is their capacity to create a supportive, mentally healthful environment for a dying person. Hospitals generally discharge terminally ill patients so that their beds can be made available to someone else. Although it can appear cold and insensitive, this "practical" *modus operandi* does allow the sufferer to spend his or her remaining period of life in the much preferred setting of home, or at least in a far less alien environment where family and friends, and TMPs, can be the caretakers. Traditional medical practitioners are thus involved in creating a hospice for the terminally ill. As such, they are positioned to perform an essential function that biomedicine is not prepared to render. This is yet another area of complementarity between traditional medicine and biomedicine in that cooperation between the two systems could yield important health and humanitarian benefits.

We had an opportunity to follow the case of a terminally ill person who had been discharged from KNH in Nairobi. Those "coopted" into the therapy managing group included the patient's son, a primary TMP and two "consulting" TMPs, my research associate, a Kenyan doctor from the University of Nairobi Medical School, and myself. The usual family support system was not operative in this case due to a broken home and the strong, vengeful animosity directed toward the patient by certain close family members.

The patient, a literate, gray-haired Kamba man of about 60 whom I shall refer to as "J.," came from Machakos District and had been employed as a medical assistant for many years. He was admitted to Machakos Provincial General Hospital during the latter part of October 1977. He was then referred to KNH where he was admitted on October 25 and discharged on October 31, 1977. His case summary prepared by the consultant surgeon at KNH included the diagnosis "cancer bronchus, invading the oesophagus with secondaries in the spine," and concluded that "it's clear that nothing surgical can be done. Discharged to the nearest district hospital (Machakos Hospital)."

Details of J.'s whereabouts immediately after his discharge from KNH are not clear. Presumably his son, an articulate Kenya Army corporal, arranged for him to be transferred back to Machakos Provincial General Hospital, where he was once again discharged because of his terminal illness. We first encountered J. in Mathare on November 11, 1971. By this time J.'s son had already contacted a young, female Kamba TMP in Mathare II and she had agreed the day before to try to help his father. She had arranged to have J., accompanied by his son, transported by taxi from an undisclosed location to a place on Juja Road (Figure 6-4) only 30 meters from her residence/clinic. When the taxi arrived the son, driver, and another man on the spot carried J.—a catheter dangling from his body—from the roadside to the TMP's room. They set out a blanket on the concrete floor and laid J. on top of it, placing a pillow under his head. J. was alert but extremely weak and hollow-eyed. He appeared ashen and gaunt, had difficulty breathing, had a wracking, painful cough of long standing, and could swallow only with great difficulty. He had been incontinent and constipated for "some days" up to his first contact with the TMP, who gave him an herbal medicine that enabled him to void his bowel.

J. looked like a man close to death, and it was evident that his son was anxious to try almost any strategy that might save his father. He recognized the seriousness of his father's condition and later confided that he expected that traditional medicine would also be ineffective against the already advanced stage of the malignancy. Yet he was not ready to believe that the case was hopeless, which reassured J. and most certainly contributed to a brief but remarkable rally against the disease.

We eventually assisted the son to obtain additional "compassionate" leave from his Army duties by contacting his commanding officer (CO). In the end, however, this process was nearly sabotaged by an angry family member from Machakos who anonymously telephoned the CO's adjutant at the Army base and provided contradictory and false information concerning the father's health and the son's motives for requesting special leave![4]

J. was to remain in the care of the young Kamba TMP for 46 days. On this first day of our contact with the patient therapy began with a divination session, conducted while J. remained stretched out on a blanket on the floor. I was asked to contribute Shs. 8/ ($1.07) for the divination and was subsequently seated on the floor in front of and facing the TMP who was also seated, her legs outstretched. She played the magic bow and sang several refrains of a by then familiar, high-pitched song. The small room was soon crowded with ten people. The divination rituals served to quickly establish a quiet, watchful, and solemn (but not mournful) atmosphere for the proceedings. After about 5 min the TMP's male assistant motioned to me to get up so that J.'s son could participate in the main divination ceremony. I handed the son Shs. 8/ which he placed beside a small pouch next to the TMP's knee. He also placed five Sportsman cigarettes in an ashtray as an offering to the spirits (*aimu*). The divination now proceeded in Kikamba, and the TMP's questions permitted the son to respond to points raised about his father, himself, and their family. On more than one occasion the room filled with laughter, and J. experienced a noticeable lifting of his spirits—even smiling one or two times.

After the divination the TMP announced that she would attempt to nurse J. back to health. She had already rented a room in Mathare II for him to stay in while he recuperated. Following a detailed physical examination, J. was transported on a stretcher to this room. Our research group was invited to return in 3 days to check on J.'s progress. The son hoped to return from his army post within a week in order to remain at his father's bedside indefinitely. It was at this point that we intervened at the son's request and wrote to his CO requesting "compassionate" leave—a request that was soon honored.

During the first 2 weeks under the TMP's care J. began a slow but visible improvement. The TMP insisted that his room remain well lighted and ventilated to provide "fresh air," and instructed J. to keep moving his joints while lying in bed in order to "help the blood circulation." Within 6 days he had enough strength to hold a cup of millet porridge (*uki*). He was given several herbal medicines to "stop the pain and coughing." Three herbs—*mwaiitha, muvunda,* and *muteta*—were ground with a stone and boiled in water. J. drank one cup of this mixed with fresh milk at intervals. He also received warm sponge baths with water in which *mwaiitha* and *muvunda* stems, and a leafy green vegetable known as *sukuma wiki* had been boiled. This herbal mixture was also added to milk, to be drunk by the cupful. The TMP made small cuts with a razor blade on J.'s chest. This enabled her to check the color and "circulation" of the blood. A powdered medicine was then rubbed into (subcutaneous injection) these incisions with leaves of *sukuma wiki*.

Initially J. was nursed in a sitting position on his bed because this lessened his breathing problem and chest pains. An empty Kimbo tin and a bed pan were kept under the bed, and he was supplied with chewing gum to "prevent mouth odor." He was made to stand up with the support of others several times daily, and the TMP made certain that he could open his bowels and urinate several times each day. His nourishment included milk, rice, potatoes, tomatoes, and other soft food.

By the beginning of December J.'s physical and mental condition had dramatically improved. He was able to sit up on his own, his cough had diminished considerably, and his appetite and attitude were both good. In addition to the TMP's care, J.'s son also deserved considerable credit for his father's apparent rebound. Once he returned with army leave he remained at his father's side to feed him and ensure daily exercises. Indeed, the son left no stone unturned. In late November, for example, he became anxious and visited a prominent commercial astrologer-herbalist (a charlatan) who advertised in the daily press and worked out of a hotel suite in downtown Nairobi. Hopeful of gaining some added confidence and a potent medicine for his father, he spent Shs. 400/ (U.S. $53) on this fruitless venture.

During the week before Christmas 1977, J. was able to stand up and walk around without assistance. He had gained weight, maintained a sense of humor, and appeared to be in the midst of a remarkable recovery. J. died in his room at Mathare on December 26. His son took his body home to Machakos for burial. Under ideal circumstances he would have died at home, surrounded by family members. However, under the care of an urban TMP and loyal son, J. had lived through the terminal stage of his illness in a relatively humane and even health-producing environment. His death was dignified.

The preceding discussion illustrates the diversity of illnesses presented to and the range of treatments employed by TMPs in Nairobi's Mathare Valley. There can be little doubt that TMPs had become a significant force in urban health. The management of J.'s terminal cancer offers a striking example of the positive contributions to human welfare that some TMPs provided.

NOTES

1. This is an estimated range based on observation in Mathare and Murang'a, an up-country town. For several reasons, including the practice of installment payments on assessed fees, and privacy, it is extremely difficult to obtain accurate income data. Analysis of 153 client record forms kept by one of the most reliable and successful TMPs in the survey between September 28 and November 12, 1977, indicated an average charge of Shs. 28/ (U.S. $3.73) and a total income of Shs. 4,292/ (U.S. $572.) Prorated for a year, this level of activity yields Shs. 38,628/ (U.S. $5,150). I consider this to be far below the actual income earned by this TMP because her average patient load is much higher than the level she was able to record during the period noted.

2. This report, published in a *New York Times Magazine* article, does not cite a source of evidence.

3. Both of the Kikuyu research assistants working with me on this day recoiled when they saw the operation, but remarked that they too had had their uvulas removed—one as a young child, the other as a teenager.

4. Documentation for this case includes the following: (1) case summary, Kenyatta National Hospital; (2) letter from regimental survey officer to the dean, University of Nairobi Medical School; (3) letter from son to C. M. Good, January 23, 1978; and field notes.

9

Interpreting the Case Studies and Comparative Evidence

Traditional medicine is a persistent, pervasive, and popular enterprise in contemporary Kenya. It coexists with sophisticated biomedical technology, global satellite telecommunications, rapid urban growth, internationally qualified Kenyan physicians, and an infant mortality rate that is eight to ten times greater than levels in the United States or Western Europe. To grasp the role of traditional medicine in African health care one is compelled to ask where, when, how, and why it is practiced and connected to the other parallel therapies that combine (depending on location and other accessibility factors) to form Kenya's pluralistic ethnomedical systems. Furthermore, as I have attempted to illustrate throughout this study, traditional medicine in Kenya is at once an integral part of a much broader social tapestry that includes the peoples' individual and collective consciousness and the geographical pattern of the economic and cultural systems they have fashioned to support rural and urban life. One cannot understand the true place or significance of traditional medical systems without holistic and spatial approaches to the subject. Such a perspective is necessary to avoid the pitfalls of narrow "sectoral" analyses that lead only to partial, distorted, and stereotyped popular views of traditional medicine. Examples of such notions include (1) traditional medicine is important primarily because TMPs can lead scientists to discover indigenous plants with pharmaco-active properties that may result in the mass production of new and potent drugs[1]; (2) traditional medicine is synonymous with quaint herbal cures, superstition, and ignorance; (3) traditional medicine is largely confined to rural backwaters and will soon be "dead as the dodo" (Vuori, 1982, p. 129); (4) traditional medicine is a static, anachronistic, and unsavory element of indigenous culture that interferes with the effectiveness of biomedical services and retards the modernization of society; and (5) most TMPs are charlatans who have little if any positive contribution to make to health care.

In this final chapter I review the key findings of the Kenya case studies in relationship to the five main themes (Chapter 1) and the theoretical perspectives around which the study has been organized. One of those themes concerns the means by and extent to which traditional medicine and biomedicine can be linked to promote and extend primary health care. I will conclude with an extended focus on this cooperative dimension of health care policy in Chapter 10.

Despite their fundamentally distinct paradigms regarding disease, illness, and health and different systems of etiology, diagnosis, and treatment, it is evident that traditionally based therapies and biomedicine are not perceived by the general population of Kenya as *a priori* mutually exclusive, incompatible choices of health care. The evidence from Kilungu and Mathare shows that people of all socioeconomic circumstances and classes are pragmatic and often remarkably eclectic in their search for effective treatment of illness. Therapy selection is a complex process involving many "inputs" such as distance to a particular provider, transportation capabilities, expectancy of a successful outcome based on direct or vicarious prior experience, the attitude of therapy managers toward a particular medical culture or treatment strategy, stage of the illness, condition of the patient, and probable cost. In short, efforts to construct a theory of therapeutic choice and to find consistent evidence of a hierarchy of resort beyond specific cases continues to be frustrated by the diversity and complexity of individual behavior.

As case after case demonstrates, the form of therapy considered appropriate can vary quite dramatically as a disease progresses and as treatment experience is evaluated during the course of an illness. Although accessibility factors such as distance, travel time, and cost of treatment and transportation may predetermine the choice of therapy (e.g., the nearest dispensary may be more than a day's walk from home), the perceived value of a given treatment or therapist is always a key influence on actual behavior. In Kenya as in Zimbabwe (Chavunduka, 1978), the redefinition of an illness from "normal" to "abnormal"—and a corresponding shift from biomedicine to a TMP— often occurred when the sufferer's symptoms had progressed to a chronic stage. Such an interpretation signaled a need to pay greater attention to ultimate causation and thus to the psychosocial domain of the TMP.

The choice of one therapy over another, including shifts among therapies during successive episodes of the same illness, is not contingent upon the patient's or the patient's therapy managers knowledge of the disease, or of how or why an alternate form of treatment actually works. This is true whether such treatment is based on a supernatural or biochemical process, or both. In other words, in Kenya as in the United States, many people are licensed to drive motor vehicles although only a minority of them understand the mechanics of the internal combustion engine; similarly, few individuals in either society ignore the possible health benefits of vaccines because they have not mastered the science of immunology. Such pragmatic behavior cuts across ethnic lines and class differences. Thus educated and sophisticated members of Kenyan families may consult a TMP when an illness that threatens life or family welfare becomes chronic and fails to respond to biomedical therapy. Of course, such consultations are not restricted to crises or illnesses that require a "social consensus" (Janzen, 1978) in matters of therapy. Similar to what Maclean (1971) discovered concerning sickness behavior among elite Nigerian families in Ibadan, I found through evidence gathered informally that elite members of some class-differentiated extended families in Nairobi also made occasional use of TMPs' remedies for common *acute* illnesses such as colds, chest congestion, coughs, and headaches.

The dramatic expansion of treatment strategies within ethnomedical systems in Kenya during this century surely has important consequences for the health of both rural and urban populations that have yet to be adequately documented in any study. Pharmacies, "bush doctors," astrologers, and spiritual healing in the independent churches are but a few of the options people may be positioned to consider alongside TMPs and biomedical workers (Figure 2-4). Levels of hygiene and sanitation have apparently improved in many areas as well. In the latter example such change occurs gradually as an incremental social process. It begins as "self-treatment" through the adoption of promotive-preventive behavior by individuals, and it signifies and facilitates the practice of more effective self-care.

Circumstantial evidence from the case studies, for example, the propensity of many individuals to shift from one form of treatment to another during an illness, suggests that the availability of plural therapies does exert an important influence on (1) the construction of patients' explanatory models (EMs) of disease and illness and thus on (2) their geographical mobility and on (3) the nature and relative success of treatments received. Explanatory models are formulated from cultural attitudes, collective experience, and individual cognition. They are operationalized as part of a process that involves the perception, labeling, communication, and management of illness among patients, their families and wider social networks, and therapists—the latter inserting their own EMs into the transaction. How to monitor and translate patients' EMs into improved diagnosis and treatment of their health problems remains a crucial issue for ethnomedical analysis and the biomedical profession. The challenge requires a joint effort from medical anthropologists, cross-cultural psychiatrists, and the biomedical community. In this connection the World Health Organization's patient-oriented Reason-for-Encounter Classification (RFE-C) could become a valuable aid toward understanding lay EMs. Rather than following the standard practice of having the health care provider interpret a patient's illness, disease, or injury, the RFE-C is based on obtaining the patient's reasons for the encounter before any judgment concerning their validity or accuracy is made by a health worker (Mead, 1983).

Variations on the themes of witchcraft and sorcery continue to figure prominently in the perception of illness causation and in behavioral responses to actual or possible misfortune in Kenya and the rest of Africa. Indeed, these institutions are "mentifacts" of daily life. Together with the concept of spirit-related illness, witchcraft and sorcery beliefs are important sources of dissonance between the biomedical and traditional African approaches to healing and they could interfere with efforts to promote cooperation between practitioners of the respective systems. Ironically, in biomedical circles there has been a propensity to underestimate, and thus to overlook or misinterpret, the authority and power of witchcraft, magic, and superstition in disease and illness in Africa. As Kato (1970) observes, those who say that witches and witchcraft don't exist

> confuse the language of one area with the facts of another. For those who believe in it, witchcraft does exist. One can draw the analogy of belief in God. For the atheist God does not exist, but for the believer God does exist as a powerful force.

The present study and other investigations of traditional healing in contemporary Africa suggest two dominant perspectives of witchcraft and sorcery beliefs. The first of these perspectives can be called the culture-relative, or functional, interpretation and is most clearly articulated by Chavunduka for the Shona of Zimbabwe (1978). In this view witchcraft provides an acceptable explanation of why an event such as abnormal illness (sudden, persistent, dysfunctional) or other misfortune occurs. In contrast, people in a Western culture will generally attribute the event to chance and not try to explain why a particular individual is victimized (Gluckman, 1944). Witchcraft also functions as a force that supports the moral code of society "since it particularly acts against the moral code[s] which are likely to bring about such acts of sorcery or accusations of witchcraft. In Shona society a man who commits incest, for example, is regarded as a witch" (Chavunduka, 1978, p. 78).

Tensions in social relations are also brought into the open through witchcraft accusations because a sorcerer is typically denounced in public and expelled from the community. Traditional medical practitioners' (TMPs') perceived abilities to counter witchcraft, and in particular their corresponding powers to detect would-be thieves (a power vividly demonstrated in the clinics of some Mathare TMPs), is believed to discourage many people from undertaking antisocial acts (Chavunduka, 1978).

The second perspective on witchcraft and sorcery is a more "external" interpretation that views the phenomena of witchcraft and sorcery as a brake on human social betterment in Africa in the 1980s. Surely there is validity and wisdom in both perspectives. However, this second perspective views witchcraft as a fear system whose persistence—indeed expansion—poses a threat to the social order in Africa's cities and rural areas. It is perceived as a negative social force with little redeeming social value. It feeds on peoples' insecurity and is a dominant source of what Janzen (1978, p. 205) terms "other-inflicted illness." More than twenty years ago Parrinder (1963, p. 9) observed that "still in modern Africa belief in witchcraft is a great tyranny spreading panic and death." This statement remains valid today.

As Mutungi (1977), a Kenyan legal scholar, points out the belief in and practice of witchcraft, whether intentional or unconscious, knows no educational, occupational, or social boundaries. An African theological student in Malawi recently observed that "from early life the existence and fear of witchcraft is hammered into our minds" (Hopkins, 1980, p. 56). A survey of 25 students in the same Christian theological college in Malawi provided detailed evidence of the pervasiveness of the witchcraft syndrome and its effects on human attitudes and behavior (Hopkins, 1980).

In Chapter 3, I presented evidence from Kenya that TMPs are frequently scapegoated as culprits who manipulate peoples' fears and maintain the witchcraft dynamic of magical and harmful reciprocity. However, I think Janzen (1978) is essentially correct in arguing that the best diviners do not consciously use their analyses of interpersonal conflict to fan the flames of conflict. I will not debate this here except to argue that witchcraft is a major institution of social control, and that maintenance of witchcraft beliefs requires participation of the public at large as well as the Kamba *andu awe* and TMPs from other ethnic groups. These specialists are simultaneously respected and feared for their (putative) inherent powers to identify and foil witches and sorcerers (as well as thieves and liars), and their antidotes for treachery are readily available for a fee. In effect, there is a kind of obligatory symbiosis that maintains the witchcraft phenomenon.

Several important geographical aspects of traditional medicine emerge from this study. The shift of focus from rural to urban settings has, of course, been underway for some decades. Urbanization in Kenya has increasingly concentrated large numbers of people in Nairobi and other cities and towns who do not, and in many respects cannot, rely on the biomedical services to meet all their needs for health care. The in-migration of TMPs from Kitui and other rural areas is in response to these needs and to the resulting opportunities for potentially lucrative rewards of traditionally based medicine as "free enterprise" (Chapter 8). The out-migration of TMPs from rural areas, together with the diminishing number of people who become TMPs today in rural areas such as the Kilungu Hills (the "social reproduction" problem), means that urban places, not rural villages, are the main arenas for the growth and adaptation of traditionally-*based* medicine. This expansion can be expected to occur regardless of the degree of availability of conventional biomedical services in a place. As such this urban-oriented growth process does not conform to classical modernization theory, which holds that traditional institutions are comparatively static and inflexible and thus give way to more advanced institutions and practices under the influence of urbanization, education, science, and technology (Van Etten, 1976).

In contrast, those rural areas that are not producing new TMPs who stay and practice in the villages are experiencing a net loss of health resources. I have documented this in the case of Kilungu, but Kwale District in southeast Kenya and many other rural areas may be experiencing similar effects.[2] This is emphatically true where government health services have not maintained pace with population growth.

I have emphasized the term "traditionally based" medicine in connection with the urban scene because the core practices of TMPs and the expectations and behavior of their patients remain rooted in long-established cultural patterns, including witchcraft beliefs, that will not disappear from the corporate psyche for generations to come. Nevertheless, as the Mathare evidence demonstrates, accommodation of the changed circumstances of urban life and the informal, demonstration effects of biomedicine and alternative therapies (Figure 2-4) give traditional medical practices a strong dynamic character and an emerging syncretistic ambience. Borrowed language (e.g., disease labeling), and new ideas (e.g., hygiene and contagion) and techniques (e.g., administration of antibiotics, use of stethoscope) come primarily from biomedicine; yet effective application of such knowledge is hampered because it is acquired in a fragmented, *ad hoc* fashion and because the TMPs are undereducated. On the other hand, as I show in Chapter 10, TMPs expressed a desire to "upgrade" their biomedical knowledge and to cooperate with biomedical professionals. If this were realized it would enhance, not endanger, the soundness of health practices in urban (and rural) communities.

In geographical terms traditional medical systems have dual characteristics because patients and TMPs interact functionally and spatially in both the rural and urban areas. They are at once part of a single system operating at different scales that sometimes involves people in movements that span hundreds of miles. In this sense the interaction of TMPs with their patients, with each other, and with their respective places of residence and work, and the movement of patients from rural to urban and, to a lesser extent, urban to rural areas to consult with TMPs, is a microcosm of Kenya's evolving economic and social systems. The city is a safety-valve and a powerful magnet that pulls heavily on the resources of the villages—in this case attracting able-bodied men and women with special skills in traditional medicine away from the rural areas. For Kamba TMPs in Mathare, their home areas in Kitui District also served as the major source of supply of essential medicinal plants for their urban practices. In addition, the urban TMPs depended on their rural families for occasional supplies of food and as an ultimate haven in the event of their economic failure in Nairobi (the "success" rate of Mathare TMPs remains a question mark), serious illness, or retirement.

One of the salient contrasts between urban and rural TMPs stems from the fact that the former are to a much greater extent commercial and entrepreneurial—they operate almost entirely on a fee-for-service basis. Certainly the selfish and fraudulent TMPs are too numerous, as is the case in any poorly regulated profession. Nevertheless, in spite of the inevitable competition, cheating, and medical and business "failures" that this traditionally based urban occupation produces, there was also much evidence among TMPs of professionalism, integrity, and compassion in their dealings with patients.

It is tempting but inaccurate to view urban/rural relations of the kind exemplified by the locational behavior of participants in ethnomedical systems as a one-way, "parasitic" process. Despite the economic leverage exerted by Nairobi, TMPs did not perceive the city as the ideal place in which to spend their lives or to raise a family. Nairobi was and is a means, not an end. Consequently, most TMPs had few if any children or adult family members living with them in Mathare. Instead, families remained in the rural areas and TMPs carried or sent home whatever portion of their earnings was available after expenses to help support them. Some TMPs also invested in material improvements to their homes and farms as well as in retail shops and other enterprises that could create income. A systematic study of urban income generation and rural investment strategies of TMPs would yield valuable insights concerning the economic impacts of core/periphery relationships at different geographic scales. In short, just as the health care behavior of the African population cannot be divorced from its many connections to the wider political economy of the peasantry (Chapter 7; see also Feierman, 1984), the economic and related spatial behavior of TMPs represent a valid and fruitful focus for the study of trends in regional and national economic systems.

Generalizing from the Mathare survey, Nairobi's TMPs tended to cluster in the lowest income areas outside the central business district and elite suburbs. This pattern reflects the city's extreme shortage of affordable better quality housing and the apparent desire of TMPs not to project a noticeably affluent life-style. Only one of the TMP's who operated a clinic in Mathare had a city residence outside Mathare proper. In this case the TMP rented an "Asian" type house of comparatively high quality that was conveniently located only a few hundred yards away from Mathare.

While charisma and a reputation for effective therapy were a *sine qua non* of the successful urban TMP, survival depended ultimately on their functioning as both diviners and herbalists, that is, as generalists.[3] Many also sought to boost their reputations and provide mutual support services in joining one or more quasiprofessional organizations such as the Waganga wa Miti Shamba Society created by and for Kamba TMPs. Although a few of these societies showed evidence of potential to stimulate the development of professional standards and create a stable system of mutual aid, the lack of effective leadership and organizational goal-setting meant that TMPs' confidence in these societies was never high and was subject to rapid erosion—particularly in connection with the management of funds contributed by the membership.

Rural TMPs also included some generalists, although most practiced only part time or as needed and remained primarily dependent on agriculture for their livelihoods. However, a majority of Kilungu TMPs practiced some speciality such as herbal medicine, ritual treatment of infertility and protective medicine for newborns, divination, exorcism, and midwifery (Table 5-5). The erosion of valuable skills in traditional medicine and the apparent decline in the numbers of new "recruits" to the ranks of rural TMPs in Kilungu (longitudinal studies are needed to test this hypothesis) is likely to continue, in part because the rural communities cannot routinely pay fees for services that are high enough to support some of the younger people who might become TMPs if they could earn an adequate living from it. This is not to say that TMPs are about to disappear from the local scene. However, coupled with the limited availability and marginal quality of biomedical services, the attrition within the ranks of TMPs in rural areas such as Kilungu suggested a net loss of health resources in rural communities and highlighted the inadequacy of the ethnomedical system as a whole. The extent to which this gap was partially compensated for by positive contributions from "informal" biomedicine, that is, injections, antibotics, and other treatments made available by persons who operate in the shadows outside "accredited" health service points, remains unknown. In this regard it also remains to be seen whether Kenya's new national system for the supply and management of some 40 essential drugs will limit losses through pilferage and black market activities to any significant degree (G. D. Moore, 1982).

The Mathare findings indicated that biomedicine was often generally the first resort for people in the city, and especially so for acute and incapacitating illnesses and conditions that require surgery, repair of fractures, and trauma. Traditional medicine complemented biomedicine, and not only in the sense of providing a hospice for a terminally ill patient. Traditional medical practitioners did see cases of gastroenteritis, sexually transmitted diseases, and other infectious diseases. However, most of their patient load included people with chronic nonincapacitating illnesses—including psychological and social disorders. Many patients presented as people desperately striving to be more successful and secure—socially and economically—in the competitive and materialistic urban environment. Some TMPs revealed an uncanny ability to identify with these insecurities and to comfort their patients. Such health care needs are not addressed as effectively by any other form of therapy, and thus TMPs continue to be relevant at a time of Kenya's rapid incorporation into the world system.

In contrast, biomedicine was not and is not oriented toward nor otherwise equipped to manage the expanding caseload of mental and sociopathic disorders which increasingly approach the epidemiological status of infectious and communicable diseases. Moreover, there are no dramatic "penicillin-type" cures for the chronic diseases of the epidemiological transition such as cancer, diabetes, hypertension, "stress," and other degenerative conditions (Nchinda, 1976) that are often referred to in Africa today as "executive diseases." Such problems are emerging as urbanization and life expectancy both increase, and they can be expected to cause new anxieties and therapeutic quandries. The Nairobi evidence suggests that under these rapidly changing circumstances more people with chronic ailments can be expected to turn the TMPs for assistance. Once again there are important theoretical and practical implications of medical pluralism for organizing health care services.

The effectiveness of African traditional medicine in relieving symptoms and curing disease is not impossible to measure in acceptable scientific terms, but such evaluation will require sustained, multidisciplinary collaboration. The persistence and, in urban areas, growing prominence of traditionally based medicine is evidence that it is valued because TMPs do treat patients with sufficient regularity and success to ensure their viability. In essence, traditional medicine survives today because it is a culturally understood approach to managing a wide variety of physical and social ills. It is an integrated system of social control and curative procedures with strong individual and corporate health effects.

The main characteristics and current trends in traditional medicine and medical pluralism that have been identified in the case studies of urban and rural Kenya are not unique; rather, they reflect a general pattern in Africa south of the Sahara within which important local variations occur. Evidence from other studies that have been cited in this book supports this interpretation.

Medical pluralism is an underdeveloped resource in Kenya and the rest of Africa at a time when most countries are struggling to maintain the status quo of very limited coverage by the officially recognized health services. In the final chapter I examine opportunities for and approaches to cooperation between TMPs and the biomedical sector.

NOTES

1. Vuori (1982) argues that the vision of a "green revolution" in medicine based on phytotherapy with traditional medicinal plants has not materialized in part because the original use of the plant by the TMP, and the TMP's own knowledge of a drug, have been ignored in the rush to "enrich the pharmaceutical arsenal of modern medicine" (p. 137). Moreover, there are few common African plants that have not been studied either in Europe or North America. Citing Ekong (1979), Vuori noted that (p. 136) "most of the plants investigated have yielded substances with no significant activity or with activity inferior to that of other well-known products used in modern medical practice."

2. According to one observer who recently spent 2 years working and living with the Duruma people of interior Kwale District, younger people are not joining the ranks of established TMPs. Thus the ratio of competent traditional specialists to population is declining. Personal communication, Joel Walukos, March 6, 1986.

3. Similar findings were reported from Zaire. The proportion of generalists to specialists in Kinshasha (86%) and the towns (78%) compared with 68% in the rural areas of Zaire (International Development Research Center, 1980).

10

Ethnomedical Systems and Community Health in Africa: Policy Considerations, Needs, and Approaches to Cooperation

One of the primary goals of the present study has been to examine the prospects and potential value of closer cooperation between traditional medical practitioners (TMPs) and biomedical professionals. I began fieldwork with the assumption that no one in the biomedical sector would enter a serious discussion of possible avenues of cooperation in the absence of organized empirical information. Based on the Kenyan and comparative evidence I have concluded that health policy makers have much to gain for the benefit of national health by supporting systematic experiments designed to evaluate the merits of limited cooperation between TMPs and the biomedical services. African health authorities also have the backing of the World Health Organization (W.H.O.) for such initiatives (Akerele, 1984). Carefully designed, community-supported pilot projects need not be expensive. They can also be a litmus test to guide planners who direct national health programs to choose wisely among the well-known range of national policy options (Green, 1980) (Table 10-1). Although Kenya's health care system is the ultimate focus of discussion here, my argument is partly based on the evidence from other African countries.

DIVIDES AND "BRIDGES" TO EFFECTIVE COOPERATION[1]

In theory, the idea of further developing the linkages of traditional medicine and biomedicine has inherent appeal. However, health policymakers may find it useful to review some of the factors that can inhibit a cooperative approach and to consider them in relation to others that link the traditional and biomedical systems.

Several divisive features thwart cooperation, but the problems are not insurmountable. First and most general is the paradigm conflict discussed earlier (Chapter 1). Second, the restrictive attitudes held by some biomedical practitioners toward TMPs does not augur well for strengthening ties through such means as "peer" observation, information exchange, mutual

Table 10-1. Policy Options Concerning the Role of Traditional Medical
Practitioners in National Health Services

Kikhela, Bibeau, and Corin (1981)	Pillsbury (1982)
Option 1: *Making traditional medicine illegal*	1. Illegalization/severe restriction of TMPs (no known evidence of effective implementation).
Option 2: *Informal recognition of traditional medicine*	2. Nonformal recognition and occasional cooperation with TMPs.
Option 3: *Simple legislation regulating traditional medicine*	3. Ignore TMPs (officially).
Option 4: *Gradual cooperation with TMPs,* oriented toward modification of whole health care structure	4. Formal recognition of TMPs and their institutions.
	5. Recruitment and training of TMP's for utilization in the biomedical primary health care program. Strategies for "upgrading" of skills and recruitment to program are key variables.
	6. Licensing/registration of TMPs, possibly combined with (4) and (5), above.
	7. Integration of traditional and biomedical systems with mutual cooperation and referral.

Abbreviations: TMP, traditional medical practitioner.

referrals, and cooperative therapy (Pearce, 1982).[2] However, if it is determined that such activities are in the interest of national health, part of the problem is already solved because, as evidence from Kenya presented later on shows, most TMPs favor closer working relationships with the biomedical profession. A third, related barrier is inertia: even where health policies call for developing parts of the two systems in tandem, to date little progress has taken place regarding TMPs. Pillsbury (1982) cites several reasons for this inertia of health ministries, including the shortage of evaluative findings on TMPs therapeutic effectiveness; the poor performance generally of national health care programs in raising health standards of rural populations, even where other types of community health workers (presumably more acceptable to biomedical practitioners than TMPs) have been employed; and the unspoken realization that an effective TMP component will depend on the success of the overall health care program. Pillsbury (1982, pp. 1829–1830) stresses an additional related factor: that too often the "priorities of key decision-makers and implementers take precedence over improving rural health." These "other agendas" include the use of rural outreach programs to create or extend a regime's political influence at the district or township level through patronage; bureaucratic "turf" interests among central government program managers who wish to maintain or expand their authority over personnel and budgets; donor manipulation of international assistance programs in primary health care for political ends; and the frequent preoccupation of government doctors and administrators at the local level with advancing their personal careers and their families' socioeconomic advantages.

At present, low levels of literacy and formal education pose inescapable constraints on what most TMPs can be expected to achieve through a systematic relationship with biomedical workers. However, reasonable and carefully considered objectives and procedures with regard to TMPs' acquisition and utilization of new skills and medical knowledge can lead to positive effects on community health that would not otherwise be realized. Supporting evidence for this view includes the initial results of Ghana's Techiman project, discussed below.

What "bridges" currently provide positive connections between biomedicine and traditional medicine in African countries, and consequently also support closer working relationships between them? One factor is that each system has its characteristic strengths, some of which are potentially transferable to the other. In the case of biomedicine, the strengths that have "borrowing" or "goodness of fit" potential include the values of sanitation and hygienic procedures, the emphasis on standardization of therapeutic procedures, drugs, and dosages; simple but usually efficacious cures for many acute conditions (e.g., oral rehydration therapy for infant diarrhea); and the practice of communicating and consulting with other practitioners about specific medical case histories in order to share the knowledge base, develop protocols, and improve the results of therapy.

Strengths of traditional medicine that have some transferability include an assortment of "indigenous technical knowledge." For example, their patients would benefit if biomedical workers can be encouraged, through both formal training and role modeling, to value and understand the psychosocial context of illness, patient–healer communication, and treatment of the whole person. Such training would include practical knowledge of patients' explanatory models (EMs) incorporating cultural assumptions, labels, and expectations regarding disease causation, diagnosis, therapy and therapists, and outcomes of treatment. Skeptics may view this argument as glib or naive because professional orientations and styles are as much the links to status and power as they are necessary to appropriate and effective therapy. Nevertheless, the fact that biomedical courses generally do not provide the requisite training and thus rarely instill the necessary values in students or practitioners is not evidence that such changes are impossible.

Several other "bridges" connect the traditional and biomedical systems. First, there is the support of international agencies, the W.H.O. in particular, for national policies that aim to find effective means of incorporating TMPs with various kinds of expertise (and not just midwifery) into community health care programs. The recent publication of *Traditional medicine and health care coverage; a reader for health administrators and practitioners* by Bannerman, Burton, & Wen-Chieh (1983) for the W.H.O. is illustrative.

Second, there is a small cadre of biomedical workers, including physicians, in every African country who do have the interest, insight, and motivation to work with TMPs in order to improve the quality of health care (McEvoy & McEvoy, 1976). On the other side are the TMPs, a substantial *majority* of whom desire to increase their medical knowledge and skills through cooperation and exchange with biomedical workers (Good & Kimani, 1980; Warren *et al.,* 1982). Again, for those skeptics who argue that TMPs are interested in training and cooperation because it promises economic rewards, it remains true that self-interest can also be a positive and powerful motivator.

A third linkage is that the two systems are already in place. Where they overlap geographically people see them as complementary and often use them concurrently or serially. In other areas the systems are spatially discrete and biomedicine is accessible only with great effort due to the distance between people and a health care facility. Regardless of their locational pattern, the fact that traditional medicine and biomedicine are already informally linked in many areas points to a valuable resource. Innovative programs of cooperation could begin without requiring large new expenditures on health care providers and facilities because an adaptable infrastructure already exists. Thus the problem is not so much a shortage of resources as it is one of imaginative ideas and policy initiatives!

WHERE IN AFRICA IS THERE EVIDENCE THAT COLLABORATION IS FEASIBLE AND EFFECTIVE?

A comprehensive census of specific projects is beyond the scope of the present discussion. Several small-scale programs, most with limited objectives and involving the use of traditional midwives, have been implemented and some continue to function. Published information about active projects in Africa where traditional healers, including herbalists and religious-medical specialists, are actually utilized as a part of health care programs sanctioned at the national level appears limited to Nigeria and Ghana. In this connection *The professionalisation of African medicine,* a collection of new papers by Last and Chavunduka (1986), will be of great interest. Also, a new experiment under the ministry of health in Swaziland should be instructive. They have recently initiated a 5-year program to establish a dialogue and cooperation between biomedical workers and TMPs that is focused on "priority areas of health care, viz. diarrheal diseases, childhood immunization, and maternal/ child health" (Green & Makhubu, 1984, p. 1077). While there are frequent references to policy initiatives in several other countries, few details are provided. For example, Akerele (1984) reported that the W.H.O., with support from the United Nations Development Program, recently "collaborated with Ethiopia in the integration of traditional medicine into its national health system; a national training program for traditional practitioners was also initiated." Unfortunately, no confirming evidence accompanies this report.

References to Zimbabwe's utilization of *banganga* in its national system can be found, but again without specific documentation of the program's character or results. The 15,000-member Zimbabwe national Traditional Healers Association—which *The Wall Street Journal's* Harare correspondent recently referred to as "the AMA of witchdoctors" (Kronholz, 1982)—has evidently won official recognition since the establishment of a black African government. However, documentation of anything substantive beyond this apparent gain in TMPs' status was not available for Zimbabwe at the time of writing.

Tanzania provided official support for the study of its indigenous health care systems with the establishment of the Traditional Medicine Research Unit in Dar es Salaam in 1974 (Beck, 1981). Although information about the current status of this organization was not accessible, during the 1970s it reportedly analyzed 2,000 commonly used herbs, conducted a survey of TMPs in Bagamoyo (1974) and Kisarawe Districts (1976) and in the capital region (1976–1977), held seminars, coordinated some activities with the Department of Chemistry and Microbiology, and created a chair of traditional medicine at the University of Dar es Salaam (Miller, 1980; Traditional Medicine Research Unit, 1978). For the Tanzanians, finding a means to standardize the traditional and biomedical systems was said to be crucial. During the 1970s discussion focused on the local feasibility of adapting the Han Chinese system of codifying traditional medicine (Miller, 1980). As in Zimbabwe, however, there was little evidence of TMPs being used in support of biomedical health services. The prospect for cooperation may be brighter in the Usambara region in northeast Tanzania. There, planners of the Ministry of Health are reportedly "anxious for the *waganga* to accept village health posts, because they are interested, trusted, and much less likely to leave for paying jobs" (Feierman, 1981, p. 404).

In Kenya, a traditional medicine unit was recently established in the Kenya Medical Research Institute to identify and coordinate projects of potential interest to the Ministry of Health. Permission to conduct a small experimental project in which TMPs received brief exposure to biomedical practice was given by the Medical Board to the University of Nairobi's Department of Community Health. In this project, 24 urban *waganga* received lectures on first aid, heart and pediatric problems, and elementary anatomy and physiology (Kimani, 1981). Although the practical results of this experiment are not known, the fact that it was run with government sanction suggests that the official stance toward TMPs is flexible and subject to reevaluation.

The Nairobi-based African Medical and Research Foundation (AMREF) has also taken an interest in investigating the practices of TMPs among Kenya's pastoral Maasai, the Kamba, Rabai and Kisii, and Luhya peoples through its Health Behavior and Education Department (Githagui, September, 1985). Another AMREF program aims to upgrade the knowledge, attitudes, and practices (KAP) of traditional birth attendants (TBAs) in part of Machakos District.[3]

Although a full account of officially sanctioned research on Kenya's traditional medical systems is not possible at this time, available evidence suggests that progress remains slow—particularly in obtaining the kind of systematic evaluations of traditional medicine and TMPs that are essential to formulating workable policies for effective collaboration. The major emphasis in values and actions continues to be on the acquisition of medical technology, the development of secondary and tertiary biomedical services, and specialization and professional achievement (cf. Bennett & Maneno, 1986).

In Mathare my approach to the "theoretical" issue of cooperation between the two medical systems was to focus the TMPs' attention on a few categories of KAP that seemed most appropriate to the topic. The tables included below reflect TMPs' spontaneous responses to specific questions. No attempt was made to encourage them to give exhaustive or definitive accounts. Among the findings was that Mathare *waganga* are, as a rule, sensitive to many of their limitations as therapists. They perceived asthma, epilepsy, fractures, and tuberculosis as particularly intractable conditions (Table 10-2). That tuberculosis accounts for 10% of the illnesses the TMPs

Table 10-2. Illnesses Mathare Traditional Medical Practitioners Perceive as Incurable

Illness	Percentage of all responses	
	All TMPs ($N = 45$)[a]	Kamba TMPs ($N = 35$)[b]
Asthma	17	12
Epilepsy (if victim has been burned)	16	19
Tuberculosis	14	14
Chronic madness	5	4
Curse	3	3
Chronic heart disease	3	4
Diabetes	2	—
Leprosy	2	1
Cancer	2	3
Witchcraft	2	3
Tropical ulcer	2	1
Chronic edema	2	1

Note: From field survey, 1977–1978.
Abbreviations: As in Table 10-1.
[a]All TMPs accounted for 89 responses.
[b]Kamba TMPs accounted for 70 responses.

Table 10-3. Illnesses/Conditions Mathare Traditional Medical Practitioners Report They Are Unable to Treat

Illness/condition	Percentage of all responses	
	All TMPs $(N = 45)^a$	Kamba TMPs $(N = 35)^b$
Fractures	13	12
Epilepsy (if victim has been burned)	11	13
Tuberculosis	10	11
Asthma	7	9
Obstetric-gynecological	4	5
Severe cut/wound	4	5
Infusion-transfusions	4	4
Cancer	3	2
Accidents	3	2
Anemia	3	2
STDs	3	4
Curse	3	4
Witchcraft	3	1
Surgical conditions	2	1

Note: From field survey, 1977–1978.
Abbreviations: STD, sexually transmitted disease; TMP, traditional medical practitioner.
[a] All TMPs accounted for 103 responses.
[b] Kamba TMPs accounted for 82 responses.

said they cannot treat (Table 10-3) can be interpreted as a positive factor from a biomedical perspective. On the other hand, the fact that epilepsy and tuberculosis account for 16% and 14%, respectively, of the illnesses perceived to be "incurable" (or presumably "uncontrollable") suggests areas of ignorance that may affect community health.

Experience in Mathare and elsewhere in Nairobi (Katz & Katz, 1981) confirms that TMPs freely used such labels as "TB," "pneumonia," "gonorrhea," and "*kansa*" without understanding their pathology. This will surprise no one. However, through such ignorance TMPs can also cause a patient to delay receiving appropriate curative therapy—which is especially critical in cases of communicable disease (Katz & Katz, 1981). Since they are so numerous and can significantly influence lay therapeutic actions it will be advantageous to train as many *bona fide* TMPs as possible to recognize the symptoms and signs of common diseases. This will help to improve case-findings by TMPs, and the referral practices they currently engage in would then be more rational and effective.

When asked about specific actions they take when faced with a condition they cannot treat, 56% of the Mathare TMPs said they referred patients to the hospital. I witnessed such hospital referrals on several occasions. Another

42% indicated they would refer the patient to a hospital or another TMP. Thus only 2% excluded hospital referral in their response. Mathare TMPs also resorted to biomedicine for some of their own health problems, as confirmed by personal observation, including our occasional involvement in their own therapy management. In contrast to Mathare, Chavunduka (1978) found that only 37% of the Shona healers ($N = 145$) in his Zimbabwe (Rhodesia) study referred patients to hospitals or to other TMPs, while 57% said they did not refer patients to either alternative.

Mathare TMPs perceived that biomedicine and traditional medicine each treat quite distinctive sets of illnesses more successfully than the other (Table 10-4). The findings indicate the TMPs appreciated biomedicine's strengths in antibiotic therapy and in the treatment of acute and surgical conditions. Furthermore, chronic and psychosocial illnesses were perceived as the main strengths of traditional medicine, which again underscores the complementarity of the two systems.

When asked, "Are there areas of medical practice that require collaboration between TMPs and biomedical practitioners?" 91% of the Mathare practitioners responded affirmatively. Ninety-three percent of the TMPs also expressed interest in participating in a government-sponsored course aimed at increasing their medical knowledge and skills. They also suggested a variety of general and specific content areas which they believed would be beneficial in such a course (Table 10-5).

There can be little doubt that the majority of Mathare's TMPs favor, even advocate, cooperation with the biomedical sector. They are also inter-

Table 10-4. Illnesses/Conditions Treated Most Successfully by Traditional Medical Practitioners and Biomedical Practitioners (Perceptions of Mathare *Waganga*)

A. TMPs		B. biomedial practitioners	
Illness/condition	Rank[a]	Illness/condition	Rank[a]
Witchcraft	1	Tuberculosis	1
Madness	2	Surgical problems	2
Abdominal pains	3	Malaria	3
Edemas	3	Transfusions/infusions	4
Gastroenteritis	4	Fractures	5
Infertility	4	STDs	6
Epilepsy	4	Pneumonia	7
Curse	4	Anemia	7
STDs	5	Epilepsy	8
		Orthopedic surgery	8

Note: From field survey, 1977–1978.
Abbreviations: As in Table 10-3.
[a]Based on percentage of total responses by 45 TMPs (A = 101; B = 105).

Table 10-5. Medical Knowledge and Skills Traditional Medical Practitioners Desire from a Government-Sponsored Course

Knowledge/skills	Rank[a]	
"Diagnosis and Treatment of diseases"	1	(27%)
"Additional medical knowledge"	2	(17%)
Preservation of herbal medicines to prevent spoilage	3	(16%)
Learn how to assess potency, longevity, and proper dosage of drugs	4	(11%)
Use of medical instruments in diagnosis, e.g., thermometer, stethoscope	5	
Midwifery	6	
Surgery	7	
Injections	8	
Infusions/transfusions	9	
Orthopedics	9	
Treatment of gonorrhea	9	
Laboratory techniques	9	
Literacy	9	

Note: From field survey, 1977–1978.
[a]Based on percentage of total responses ($N = 82$) by 45 TMPs.

ested in improving their basic skills to enhance the quality of services they can offer their patients.

NIGERIA: A SUCCESSFUL, REPLICABLE DESIGN?

Dr. T. A. Lambo's family- and community-based "network" therapy program began in 1954 in Aro, Nigeria, a Yoruba settlement of farmers, fisherfolk, and artisans comprised of four villages in a rural suburb of Abeokuta. The Aro program reportedly integrates the "best practices of traditional and contemporary psychology" (Lambo, 1978, p. 37). According to Lambo, who has also been deputy director general of the W.H.O., Aro has been guided by the premise that utilization of "the therapeutic practices that already existed in indigenous culture," including "the power of the group in healing," could, when paired with modern psychiatry, form an unorthodox but effective hybrid system for treating a wide range of mental disorders. Key features of the system are family participation (required), a day hospital and a "boarding-out" village care program in which ill people have daily, unrehearsed and voluntary contact with "settled, tolerant, healthy people" (Lambo, 1978, p. 38). In recent years all kinds of patients have reportedly been accepted into the program, including "violent persons, catatonics, schizophrenics, and others whose symptoms make them socially unacceptable or withdrawn. The system is particularly effective with emotionally disturbed and psychotic children" (Lambo, 1978, p. 38).

The average length of stay and recovery at Aro was about 6 months. This is said to be about one-third of the average stay at other hospitals, particularly for the various forms of schizophrenia. Neurotic patients are said to have the most rapid response to therapy. Lambo (1978, p. 39) contended that "because of its effectiveness, the Aro system has been extended to four states in Nigeria and to . . . Kenya, Ghana, and Zambia."

Although Lambo has not provided case histories, recovery and recidivism rates, or other information bearing on the efficacy of the Aro system, the *prima facie* evidence suggests that the planned physical setting and social geography are factors that are crucial to its apparent accomplishments. In this connection the Rappoports (1981, p. 779) observe that

> the location of the program in a rural village context has been instrumental to its success. In such an agrarian community, the shaman is undoubtedly able to function effectively through the presence of familiar and essential cultural "props." However, future planning will have to confront the growing African urban communities where tribal structures are dispersed and close-knit groupings harder to find. Under such conditions, the integration of systems may be more difficult. In the Western context where the traditional healer is almost always invited into a medical setting, the problem is even more acute.

GHANA AND THE PRIMARY HEALTH TRAINING FOR INDIGENOUS HEALERS PROGRAM: COLLABORATION AND ACHIEVEMENT

One of the most interesting and possibly the most successful of collaborative experiments in Sub-Saharan Africa is the Primary Health Training for Indigenous Healers (PRHETIH) program in Techiman, Ghana. Formally inaugurated with national media coverage in June 1979, PRHETIH is the brainchild of Dennis M. Warren, an anthropologist based at Iowa State University whose interest grew from research he conducted in Techiman District on Bono disease concepts and therapy strategies during 1969–1971 (Warren, 1974). During this period he introduced the administrator and other staff members of Techiman's Holy Family Hospital to many of the local healers. Warren et al. (1982, p. 1876) have since recalled that "these relationships were nurtured over the years as staff members attended festivals of the healers and healers came to the hospital for treatment of conditions they could not manage themselves."

PRHETIH was the newest component of primary care supported by the Techiman Hospital in the early 1980s. The project was designed to provide better health care to the population in Techiman District by augmenting the

biomedical knowledge and skills of local TMPs and fostering closer cooperation and understanding between the TMPs and biomedical practitioners.
Because of its focus on TMPs (both herbalists and religious-medical specialists), as opposed to being limited to TBAs, the health outcomes, lessons, and
longevity of this program are likely to provide significant guidance for
structuring similar projects elsewhere in Africa in the future.

Prior to designing the PRHETIH program detailed information was
gathered from the local TMPs regarding their techniques, beliefs, perceived
needs for training, and desire to participate in such a program. It was also
necessary to select and train biomedical health workers (Ghanaians as much
as possible) as trainers of the TMPs. A 6-month survey conducted in Techiman Township (pop. 20,000) included 45 TMPs (69% herbalists and 31%
priests/priestesses representing 12 ethnic groups). The TMPs

> unanimously supported the program from its inception and by the time the
> initial survey had been completed healers from outlying villages were asking to
> be included in future training programs. The common assumption that tradi-
> tional healers by and large are unwilling to cooperate in such ventures was not
> the case in the Techiman area (Warren *et al.*, 1982, pp. 1876–1877).

Information concerning the public's utilization of TMPs and biomedical
services was also gathered from the community. Some 69% of the population
said they use the services of traditional healers, while 94% said they go to a
health center or the hospital. These proportions thus underscore the overlap
and joint use of the two systems.

Only the main features and results of the PRHETIH program are
presented here. Further details are available in Warren *et al.* (1982). In its
essentials, the initial program consisted of two 6-week cycles of weekly classes
(Table 10-6), plus review sessions, conducted by specially trained persons
from the Holy Family Hospital staff, district-level Ministry of Health
workers, and field coordinators. The teaching model featured a question–
answer format, the demonstration/return method, visual aids, and drama
followed by a summary. The training syllabus also contains topics of special
interest to TMPs, such as methods for storage and preservation of medicinal
herbs. Plans to instruct the TMPs in the use of several basic drugs were not
adopted because at the time pharmaceuticals were too scarce in Ghana to
assure supplies to healers trained in their use. Traditional medical practitioners responded most favorably to those parts of the syllabus that complement
the practice of herbalism, such as improved storage and preservation of
medicinal herbs. Family planning received the lowest rating, "probably because contraception is antithetical to the beliefs and practices of this highly
natalistic society" (Warren *et al.*, 1982, p. 1878).

The PRHETIH program in Ghana has continued. By mid-1983 over 80
TMPs had been trained, and there were plans to extend the program into
surrounding rural communities. In 1985 the Dormaa Presbyterian Primary

Table 10-6. Syllabus, Primary Health Training for Indigenous Healers Program (PRHETIH), Techiman, Ghana[a]

Part I
1. Hygienic preparation and storage of medicinal herbs (1)
2. Houseflies and the spread of disease (4)
3. Diarrhea/dehydration; preparation of oral rehydration fluid (3)
4. Work of the medical field unit; traditional vaccinations; typhoid fever (6)
5. Basic nutrition (5)
6. Convulsions (2)
7. Review session

Part II
8. Preservation of herbal medicine in liquid form (1)
9. Weaning foods (3)
10. Family planning; venereal disease (6)
11. First aid (2)
12. Measles (5)
13. Leprosy (4)
14. Review session

Note: From "Ghanaian national policy toward indigenous healers-The case of the Primary Health Training for Indigenous Healers (PRHETIH) Program" by D. M. Warren, G. S. Bova, M. A. Tregoning, and M. Kliener, 1982. *Social Science and Medicine*, 16, p. 1873. Reprinted by permission, Pergamon Journals Ltd.
Abbreviations: As in Table 10-1.
[a]Numbers in parentheses indicate the descending order of popularity of the sessions among the TMPs involved.

Health Care Programme, located in the same region of Ghana, adopted the Techiman model and successfully trained their first group of TMPs.[4] The positive results evident from the initial pilot programs should be of great interest to health authorities in other areas of tropical Africa who are searching for ways to make appropriate health care more accessible to their populations. For example,

> Follow-up visits to the healers used a standardized questionnaire which revealed that the trainees retain over 60% of the basic material they were taught. A 6-month follow-up survey on the original trainees is currently in process. It focuses primarily on behavioral changes, and the changes so far observed or reported by the trainees are that they now store their herbs in clean plastic bags, as opposed to keeping them in a box or on the floor. Those who treat children with febrile convulsions are now using cold water sponging instead of herbal enemas or herbal fumigation. Oral rehydration fluid is now used to treat diarrhea, and high protein diets are prescribed for children with kwashiorkor. Some healers have been instrumental in improving environmental sanitation by establishing village refuse dumps (Warren *et al.*, 1982, p. 1877).

EFFECTS OF SPATIAL ORGANIZATION ON
THE RESULTS OF COLLABORATION

If available and accessible, Africans commonly select biomedicine first for acute and life-threatening diseases. However, both systems are perceived as acceptable and viable choices by a large proportion of the population. Each has unique styles and limited capacities for healing.

The widespread need for expanded access to more appropriate, acceptable, and effective therapy at low cost is widely recognized today. In practice, however, there is inadequate appreciation of the fact that effective health care must be integrated with the actual spatial and social behavior patterns of the people it is intended to serve (Stock, 1980). This principle assumes added importance where plans to promote cooperation between two or more medical systems are being developed.

Minimizing physical distance and travel time to a health service point is probably desirable in a majority of circumstances. However, as a rule the bridging of social and psychological distance may be equally or even more important. In the case of Lambo's community-based psychiatric therapy in Nigeria, for example, the program's physical situation in a village is apparently crucial to its reported achievements. In other words, for traditional healers to be effective the "place" of therapy cannot be merely "anywhere." This is probably most important in the case of religious-medical specialists such as diviners and priests; but it may also require consideration in relation to some herbalists and other specialists who are less (or not at all) reliant on magical-supernatural kinds of interventions and interpretations. Familiar settings away from a clinic or hospital, such as the healer's own residence or the back of his or her shop, may be the most appropriate locations and should be carefully considered if a local community wishes to support formal collaboration between their TMPs and biomedical workers.

Rappoport and Rappoport (1981) propose a two-stage method for linking biomedicine and traditional healing which is inspired by a process currently used in Western psychosomatic medicine. For example, a patient with a peptic ulcer receives medication, diet, or surgery for his or her organic disturbance. Thereafter, the physician, having also identified underlying emotional problems, frequently refers the patient to a psychotherapist to find the "nonphysical" cause of the ulcer. This two-stage process

> is close to what the modern African is doing with a broad spectrum of disorders . . . and there is every reason to believe that such a model could work in settings in which a traditional healer is indicated. The system would entail letting the scientific specialist treat the symptom [e.g., using chemotherapy and/or hospitalization] and then referring the patient to a shaman who functions under his or her own auspices in another physical setting (Rappoport & Rappoport, 1981, p. 78).

The Rappoports suggest that systematic refinement of this process would continue an already established pattern in Africa. It would also promote and strengthen the utilization of the best of both systems in their separate physical locations.

How can a community's health resources be organized spatially so that the advantages of formal collaboration between TMPs and biomedical workers may be optimized? There is little experience or theory to rely on for guidance beyond incidental observations, ideas such as the satellite villages model under development in the rural areas outside of Techiman Township in Ghana (Warren *et al.*, 1982), and the Rappoports' (1981) two-stage/two-place model based on psychosomatic medical practice.

Community patterning in Africa varies widely in terms of scale, density, and the locational expression of social organization. Ideally, a system designed to upgrade the practices of TMPs and strengthen cooperation with the biomedical services will incorporate as much as possible the existing community structure rather than imposing a new layer of bureaucracy. In principle, community social boundaries, defined as informal boundaries within which individuals (including TMPs) interact with each other on a more or less daily or periodic basis, should be a sound basis for defining primary health service areas. The health implications of such a model of cooperation find expression in Chaffe's (n.d.) idealistic definition of "community" as "a group of people who have committed themselves to comprehensive and responsible care for each other's future." Viewed from this approach, the boundaries of areas in which one would find biomedical workers cooperating with TMPs to improve the quality of health care might then coincide with nucleated or dispersed villages, weekly market regions, herding groups, divisions of urban shanty towns, or other localized, interdependent household functional units in which people are in face-to-face contact and customarily provide mutual assistance.[5]

An example of how a decentralized system of cooperation might be organized in agrarian communities is shown in Figure 10-1. This generalized arrangement assumes that TMPs have already been selected by their communities for the program and that they are receiving ongoing training and evaluation in a "village health worker" (VHW) type of role (the VHW label may not be necessary or desired) of the kind provided in Ghana's Techiman Project and recently proposed in Swaziland (Green & Makhubu, 1984). The model suggests the need to emphasize community-based preventive and curative care in the hierarchy of local health care needs. In this scenario, community-selected TMPs (e.g., well-regarded herbalists and midwives) have acquired the knowledge and skills to make several routine diagnoses (e.g., dehydration, trachoma, kwashiorkor), screen some patients, treat simple ailments and wounds [e.g., diarrhea using UNICEF packets of oral rehydration salts (ORS), minor burns, tropical ulcer, and trachoma], administer safe dosages of effective herbal remedies and a few common pharmaceuticals and

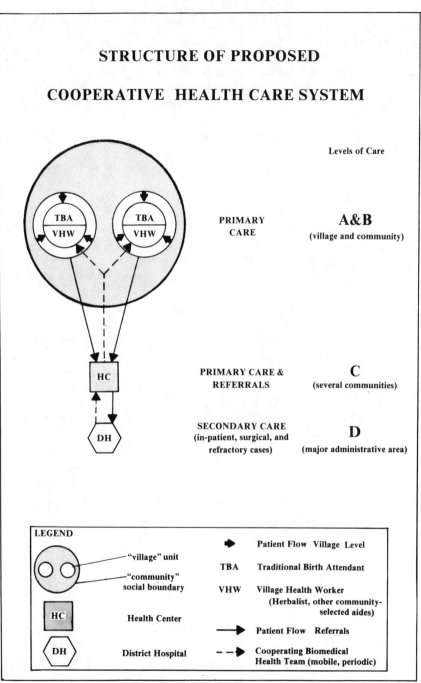

Figure 10-1

ointments, and offer guidance on hygiene and maternal and child health practices. Traditional birth attendants in this setting are able to provide improved delivery services as well as prenatal and postnatal counseling. Respected TMPs with special skill in diagnosing and treating psychosomatic and mental disorders may also form part of a "health team" and function as "clinical psychiatric associates" at each level (not shown). Referrals are made through a nested hierarchy formed of the local (A and B level) communities, a health center or dispensary (C level), and a district or mission hospital (D level). Several nested villages comprising a larger B level community receive visits by a health center team (C level) on a regular rotation cycle at a place designated by the community's health council. In addition to offering special services such as immunization, maternal and child health services, screening, and health education (including child spacing), the health center team would review, evaluate, and reinforce the biomedical skills and knowledge of local TMPs and do joint consultations with TMPs on new and continuing cases that are particularly difficult or otherwise important. The health team would possibly restock the participating TMPs with a few common drugs and ointments appropriate to local conditions, and help to arrange for the evacuation of patients needing hospital care at a district, provincial, or national location (D level).

Assuming it is acceptable in principle, some *locally adapted* version of this decentralized model has the potential to:

1. Expand access to a better quality of the most elemental health care, including sanitation and hygiene.

2. Facilitate simple curative treatment, first aid, and improved midwifery practices on a more or less *continuous* basis even at times when the health center team is prevented from holding its periodic satellite clinic due, for example, to nonfunctioning vehicles, lack of fuel, or impassable roads.

3. Make more effective use of respected TMPs, TBAs, and beneficial indigenous medical practices already available in a community.

4. *Reduce* costly, difficult, and often unnecessary travel to health centers and hospitals through improved primary care within the local community; thereby also alleviating some of the patient load at these higher-level facilities. As Shaffer (1984) observes, "in most dispensaries, a clogging of 30–40% of attendances are community-preventable problems brought in for therapeutic recycling." (p. 10).

5. Encourage communities to take greater initiative in determining solutions to their own health needs.

6. Combine the best and eliminate the least desirable features of traditional medicine and biomedicine.

Creating and adapting this and other kinds of models for use in similar and different environments (e.g., African towns and cities) requires the reinterpretation and reorganization of existing health resources. The problems encountered and the solutions required are multidisciplinary in nature. Success with pilot projects and later programs will depend on close communication and cooperation among health authorities, local communities, and social scientists.

In Kenya a few well-designed pilot projects—including one in an urban area—would test the feasibility and validity of these ideas without committing the Ministry of Health to a new philosophy or operational structure. The question for health policymakers is, why not do this experiment? While the information obtained from such experiments may demonstrate that cooperation between TMPs and the biomedical service is at best marginal and therefore will not lead to improvements in community health care, the probability of achieving positive results is much greater. The findings could then be evaluated in the light of the current national commitment to (1) increase the coverage and accessibility of health services in rural areas; (2) strengthen the rural to urban referral system; (3) emphasize services that reduce the country's morbidity, mortality, and fertility rates; and (4) increase alternative financing mechanisms through such means as community-based health care (Republic of Kenya, 1983).

Kenya has many settings in which pilot projects could be initiated. Rural Kilungu Location in Machakos District is but one example. There are numerous active TMPs in Kilungu who are respected leaders in their communities and who also see large numbers of the local population as patients. These *akimi wa miti* (herbalists), *isikya* (TBAs), and other TMPs regularly see patients who need, for example, first aid; whose babies are suffering from diarrhea and dehydration; and who have fevers, respiratory and gastrointestinal ailments, skin diseases, broken bones, helminthic infections, depression, or other acute or chronic disorders (Chapter 5). Untrained TBAs currently perform at least eight of ten deliveries in the location and offer local women whatever their experience allows in the way of ante- and postnatal advice. Traditional birth attendants thus complement the activities of local TMPs, who attempt to cope with a heavy burden of ill-health without any external assistance or recognition from government health authorities. Together with TBAs, Kilungu's TMPs represent an underdeveloped resource whose potential to improve the quality of health care in their communities has been neglected. In Kilungu and the rest of Kenya it is important to begin now to study and assess the factors that would predispose, enable, and reinforce cooperation between TMPs and biomedical workers.

A recent experiment aimed at improving midwifery practices, including family planning education, among a group of 24 illiterate Kamba TBAs in another area of Machakos District (Kibwezi) is indicative of the potential gains from pilot studies involving TMPs (Mwalali, 1983). These TBAs, who had assisted over 70% of all deliveries in the study area, reportedly demonstrated a strong capability to learn new skills and to cooperate with biomedical personnel on referrals. Evaluation and follow-up of the TBA training (instruction lasted 2 weeks) reportedly produced "astonishing" results. Over-all midwifery skills of the trained TBAs proved superior to those of untrained TBAs (control subjects) in every category. Still birth rates had improved for both domiciliary (75.4% of all deliveries) and hospital deliveries, dropping from 24 to 20 and 44 to 28 per 1,000 births, respectively. Furthermore, "no malpractices were found" (Mwalali, 1983).

In principle, formal collaboration between the two dominant medical systems is feasible in a variety of rural and urban settings in Kenya and other African countries. A sensible, unromanticized strategy for extracting greater health benefits from *existing* resources is required. Research and training programs must be targeted on a few areas (e.g., see Table 10-5) that have been determined, through sound testing and evaluation, to be the most promising. Successful experiments and follow-through programs will be characterized by genuine community participation from the start, will incorporate the patient's perspective (Bloom, 1985), and will be actively supported by the Ministry of Health—including key biomedical personnel such as nurses (Barbee, 1986).

Initiatives in cooperation should certainly not be undertaken with the goal of achieving "integration," or a synthesis of the two systems. As Warren *et al.* (1982, p. 1873) emphasize in the case of the innovative PRHETIH experiment in Ghana,

> no responsible or realistic person advocates melding the disparate systems. The challenge is to identify areas where the traditions can best complement each other and to establish a working, dynamic contact in these areas.

The importance of careful planning and monitoring of pilot projects cannot be emphasized too often. The process demands flexibility, sensitivity, humility, and cooperation from everyone involved, that is, from the univer-sity-trained doctors, clinical officers, nurses, and TMPs to local community leaders, participating social scientists, and decision-makers and planners in the Ministry of Health. Some of the major tasks and concerns associated with pilot projects are as follows:

1. Find the most *appropriate locations* for the projects.
2. Create a strategy that will help to identify and mobilize *appro-priate community resources* (political, administrative, economic, human [e.g., highly motivated TMPs and TBAs]).
3. Definition of *goals and objectives*.
4. How best to enhance the *compatibility and transferability of essential ideas and practices* that are rooted in divergent medical con-cepts and procedures.
5. Target *specific health needs* to be addressed through collabora-tion (strong community input in identifying and ranking needs is essen-tial).

6. Determine the kinds of *joint training* needed for health team members, and the community and professional settings in which to conduct it.

7. Create the *appropriate spatial organization* of people and resources that will optimize the intended advantages of the cooperative system.

8. Establish criteria and allocate resources for *project evaluation*.

CONCLUSION

Cooperation between biomedicine and traditional medicine is a controversial issue. Inertia, vested interests, and an overrepresentation of self-interest versus public service in the bureaucracies can quickly undermine efforts for change. The politics of personal and class values permeate virtually all *de jure* and *de facto* policy-making by government authorities and insure that possible innovations in health care cannot be successfully addressed primarily in terms of therapeutic, preventive, and structural goals. In Africa, as in most countries around the world, the biomedical establishment is powerfully represented in the health ministries. It often disassociates itself from popular and radical initiatives, and is always powerful enought to determine what changes and programs are acceptable. Furthermore, if a proposal to promote cooperation between biomedicine and traditional medicine comes from a foreign (non-African) source it can easily be dismissed on the grounds that "they don't know our problems," or "they want us to be satisfied with second-class health care." Against these arguments is the indisputable evidence that unless innovative, radical, and low-cost alternatives to most of the current national systems are developed, the majority of people in Africa will have to accept the status quo of the mid-1980s as the best health care they can expect. This is unnecessary and indefensible. Meanwhile, the quantitative and qualitative deficiencies of all health services are exacerbated by the high fertility rates found throughout most of the region.

Practically all of the physical and human resources required to build cooperative programs between biomedicine and traditional medicine are currently available in every African country. Medical pluralism is a resource, not a barrier to better community health. Imaginative, practical ideas, and political responsiveness to the needs of average citizens are in greatest shortage.

NOTES

1. The remainder of this chapter is an adaptation (by permission of the publisher) of part of C. M. Good, "Community health in tropical Africa: Geographical and Medical Viewpoints." In R. Akhtar (Ed.), *Environment, health, and health care*

planning in tropical Africa. London: Gordon and Breach Science Publishers, Ltd., 1987, pp. 13–50.

2. Such attitudes are reinforced by sensational reporting in the mass media of unconfirmed reports accusing TMPs of committing atrocities. Recently, for example, a leading article in *The Standard* (Nairobi) entitled "When witchery spells murder," noted that "according to Councillor Erastus Nyanga, people are being killed and beheaded in Mt. Elgon sub-district. The heads are then being sold to witchdoctors for as much as KSh. 5,000 [U.S. $313] each." Cited in R. Stanbridge, "When traditional and Western medicine meet," *New African*, July 1985, p. 6. Such stories are rarely investigated further and TMPs have no effective means to respond to the charges.

3. Conference correspondence, African Medical and Research Foundation (AMREF) to National Council for International Health, Washington, D.C., 1983.

4. Personal communications, Sr. Mary Ann Tregoning, June 13, 1983, Washington, D.C., and May 26, 1986, as Diocesan PHC Coordinator, Sunyani, Ghana.

5. For example, in western Kenya the *lusomo* is an ancient division of the Luhya people composed of 100–400 households under a headman known as the *liguru*. In the late 1970s Dr. Miriam Were of the Department of Community Health at the University of Nairobi encouraged several thousand Luhya to achieve greater self-reliance through an innovative community-based health care project. The people themselves decided that the *lusomo* is the most appropriate definition of "community" for such a program, and proceeded to establish a health committee in each *liguru's* area and to appoint a local health worker to be trained and thereafter paid by *lusomo* members (Black, 1978).

Appendix A

ILLNESSES AND REMEDIES RECOMMENDED BY
TWO KILUNGU HERBALISTS

The following lists contain information on the uses of specific plant medicines (*miti shamba*) prescribed by Nzeki and Kutenga. The latter are the fictive names of two persons (introduced in Chapter 5) who were among the most professionally active herbalists (*akima wa miti*) practicing in the Kilungu Hills in 1978. These lists represent a portion of the material acquired through several hours and multiple sessions of systematic questioning and discussion with each practitioner. Both men were generous with their time and patient in accommodating us with thoughtful answers to the numerous questions put to them. The lists contain only a portion of the illnesses that TMPs in Kilungu collectively treat. However, they are fairly representative and include most of the commonest conditions presented to local herbalists. In Kutenga's case a few therapeutic tools and first aid procedures are also included.

Extensive, additional research is necessary to determine the degree to which these combinations of drug plants are typical among Kilungu TMPs or unique to specific individuals. The value of these lists is that they provide scholars, and those who make decisions about national and local health programs, with otherwise scarce empirical information about diseases, illness syndromes, and therapeutics at the community level.

Kutenga's Remedies

    1. Headache (*kwalwa ni mutwe*).
    2. "Peptic ucler" (*kavoo/kavaso*).
    3. Fever (*ndetema*).
    4. "Pneumonia" (*kyambo*).   See Chapter 5, Table 5-8
    5. Edema (*mwimbo*).
    6. "Syphilis" (*teko*).
    7. Gonorrhea (*kisonono*).
    8. Continuous menstruation (*kuua nthakame kuma ukani*). Juices in the stem of *mwany'a nthenge*, a climbing plant, are squeezed into

patient's mouth. Patient also chews bark of *muthika* (*Warburgia ugandensis?*) and swallows the juices.

9. Infertility (*ngungu*). Juices of *mwany'a nthenge* plant are swallowed. Barks of *mwaitha* (*Dalbergia lactea?* or *Entada leptostachya?*) and *mukolekya* are crushed and soaked in water, which is thereafter drunk.

10. Epilepsy (*mung'athuko*). Leaves of *musanduku* (*Eucalyptus* sp.) are used.

11. Stomach ache (*kwalwa ni ivu*). Leaves of *muteta* (*Strychnos henningsii?*), *mukenia* (*Polygala sphenoptera?*), *mutula* (*Ximenia caffra?*), and *muthaa* are squeezed and the juices swallowed.

12. Chest pain (*kwalwa ni kithui*). Root of *mwany'a nthenge* plant are dried and crushed and mixed with water.

13. Infant gastroenteritis (*nyunyi*). Barks of *mukenea* (*Fagara chalybea?*) and *mukolekya* are crushed and mixed with water. *Mathimoti* (oil-like medicine available in shops) is applied on depressed fontanelle and chest.

14. Swelling in abdomen associated with a nonspecific disease of the pancreas (*ndumo*). Root of *mwany'a nthenge* is dried, crushed, and added to water. Patient drinks this.

15. Pain in joints (*mutambuko*). Root of *mulingula* is dried, crushed, and added to water. Patient drinks this.

16. Diarrhea (*kwituua*). Root of *mulingula* is dried, crushed, and added to water. Juices squeezed from roots of *kyuvi* and *mutula* are swallowed.

17. Whooping cough (*mutitino*). Barks of *mukoleyka* and *muthika* (Warburgia ugandensis?) are crushed and added to soup or water.

18. Excessive bleeding. Juice of sugar cane is squeezed on the wound to stop bleeding.

19. Eliminate numbness. Stem of *mukungula*, a climbing plant, is dried and crushed to a powder, mixed with milk and drunk. "It helps the blood to circulate and removes numbness from a part of the body."

20. Local anesthesia. Leaves and stems of *isongosya* and *isoambumbu* tree are warmed near a fire; white latexlike substance is then squeezed out and externally applied to affected area for about 5 min. Numbness lasts up to 1 hr.

21. Repair of wound—for example, in case of cut on lower leg. Roots of *mukulila* are dried and crushed into powder form, which is applied to wound and covered with gauze. If this does not work well, add stems of *mukoiwa*.

22. Snake bite (*kuumwa ni nzoka*). First spits on wound. A powder made from crushed and dried stem of *ndonga* tree is then rubbed into wound. If fang marks are not large, small cuts are made at the site and the powder is rubbed into them.

Kutenga mentions 28 different plants and one shop preparation (*mathi-moti*) in the list above. Only seven of these plants are used for more than one illness or treatment. They are *mwany'a nthenge* (six illnesses or treatments), *mukolekya* (four), *mulingula* (three), *mukenia* (two), *muthika* (two), *mwaitha* (two), and *mikoi wa ulenge*.

Nzeki's Remedies

1. Headache (*kwalwa ni mutwe*).
2. "Peptic ucler" (*kavoo/kavaso*).
3. Fever (*ndetema*).
4. "Pneumonia" (*kyambo*).   See Chapter 5, Table 5-8
5. Edema (*mwimbo*).
6. "Syphilis" (*teko*).
7. Gonorrhea (*kisonono*).

8. Worms (*nzoka sya nda*). For adults, roots of *mukolekya* and *muuka* (*Microglossa densiflora?*) and stems of *musoka* are boiled. Patient drinks the decoction. For children, leaves and bark of *muteta* (*Strychnos henningsii?*) and stems of *muuku* are boiled and patient drinks decoction.

9. Pancreatic (nonspecific) disease (*wasyungu*). First, the herbalist lifts up the sternum bone (*kavoo*). A powdered medicine prepared from the crushed and burnt root of *mulingula* is then added to water, tea, or porridge and given to patient.

10. Eye disease/conjunctivitis (*uwau wa metho*). Leaves of *mukengeka* (*Cassia singueana?*) are squeezed and juice is dropped into patient's eyes. Juice of *muti* (*Aspilia pluriseta* or *A. mossambiensis*) is commonly used in treating children.

11. Teething problems (*uwau wa ini*). Applies soda ash (*iati*) on child's gums.

12. Stiff neck/transient deviation of neck (*mukiki*). Applies saliva on patient's neck and massages the area.

13. Upper abdominal-rib cage swelling (*iwethyu*). Associated with *kavoo/kavaso*. Patient is given a mixture of *mulingula, mutheketha, kauthilu,* and *kiema uvunyie*.

14. Kidney disease (nonspecific) (*mbio*). Herbalist (1) applies oil on his hands and lifts up small ribs which are pushing into the flesh covering the kidneys; or (2) puts a stone in cold water and then places it on the affected region. It then heals.

15. Liver disease (nonspecific) (*itema*). Leaves of *muthiati* are boiled and patient drinks the decoction.

16. Nose bleed (*kuua na manyu*). Leaves of *muthia* (*Acacia mellifera?* or *Cordia ovalis?*) are squeezed and juice is added to a half-glass of

water which patient drinks. Thereafter herbalist rubs bridge of patient's nose and nape of his or her neck.

17. Hydrocele/swollen testicles (*kwimba malee*). *Muvou* roots and *ng'ondu* leaves are mixed and rubbed on the "blood vessels."

18. Skin discolorations (*malanga*). Yellowish spots itch. Cause is "in the blood." Mix kerosine and *iati* (soda ash from Lake Magadi) together. Soak a cloth in it and then wash affected skin. Also apply juice from mashed *kakumbatu* roots directly on spots.

19. Ringworm (*tinea capitus*) on scalp (*kinguu*) and on face, back, and neck (*kiea*). Soda ash (*iati*) and juice from *kakumbatu* roots are mixed with butterfat from cow's milk and applied to affected areas. In the old days *kinguu* was treated by encircling the affected areas with the needle that is used to repair broken calabashes (presumably to isolate the infection).

20. Blockade (perceived) of throat (*ngome*). Sometimes described as tight "ring"/constriction of throat and wound inside throat that renders swallowing almost impossible. To treat, a certain stone called *uana* is heated over a fire until it becomes red hot. It is then put in cool water to warm the water. Patient then drinks this heated water (mineral exchange?) If this is not effective a piece of *mukukuma* (*Terenna graveolens?*), a creeper or climbing plant, is cut and burnt and ground into powder. Patient licks powder off his or her palm.

21. Poisoning (*mwikue sumu*). "The best treatment for poisoning is milk."

22. Genitourinary system infection (*muluo*). Symptoms are thick whitish substance in urine and painful micturation. Said to affect both adults and children. Cause unknown, but Nzeki insisted it is not gonorrhea. A Mkamba physician at Kenyatta National Hospital called it "a venereal disease, especially gonorrhea." *Muluo* is treated with two plants: the leaves of *mwenye* and bark of *mwisiya*. The two are boiled together and patient drinks decoction.

23. Incontinence of urine (*mumao*). Excessive urination at short intervals (e.g., after 5 min). Affects adults and children. Urination act stops suddenly, as if urethral passage is blocked. Treated with *mwenye* and *mwisiya*, as above for *muluo*.

24. Severe chronic headache (*kitau kya mutwe*). Pain is on top of head instead of temples. Accompanied in the head area (sinuses) by mucus that contains pus. Inside of ears are dry and whitish. Leaves of *mulavuta* (*Helichrysum odoratissimum?*) and *muemaaka* are squeezed and added to water. Several applications of this mixture are put in patient's ears.

25. Encephalitis (*mumama*). Described by a Mkamba physician at Kenyatta National Hospital as "encephalitis-mental retardation accompanied by some degree of dullness." Nzeki uses leaves of five plants in remedy: *wamama* and *kinyinywa*, which are both creepers and climbers;

muvila; *muthaa*; *muvindavindi* (*Fagaropsis hildebrandtii*?), and *mwi-siya*. All leaves are boiled together in a sufuria which is placed between the patient's legs. The patient sits under a blanket, which becomes a steam tent, and sweats profusely. Procedure is repeated once every 3 days, three times.

    26. Pain in foot (*kiato*). Heel of foot swells and becomes painful; possible pus discharge. Cause unknown. Mkamba physician at Kenyatta National Hospital described *kiato* as "a pain deep in sole of foot, said to be caused by stepping on a certain type of caterpillar which has spines." Nzeki treats by pressing his foot on the patient's foot for a short time.

Nzeki mentions 33 different plants, soda ash (*iati*), milk, and a special stone in the list above. Thirteen plants are used for more than one illness or treatment, but only *mulingula* (three illnesses or treatments) and *mwisiya* (three) are used more than twice.

    Of the total of 61 different plants used by Nzeki and Kutenga, only 5 appear on both lists. These are *mulingula*, *mukolekya*, *muteta*, *muthaa*, and *muvindavindi*.

Appendix B

REGISTRY OF SELECTED MEDICINAL PLANTS USED BY MATHARE TMPs[a]

Vernacular name[b]	Botanical name	Reported Uses
1. *Kalaku*	a. Becium sp. (*Labiatae*);	epilepsy
	b. also *Fuerstia africana* (*Labiatae*)	
2. *Kikau*	*Melia volkensii* Guerke (*Meliaceae*)	abdominal pains
3. *Kiluma*, or *Kiruma* (Kikuyu)	*Aloe secundiflora* Engl. (*Liliaceae*)	impotency, pneumonia
4. *Kinyonywe*	*Rumex abyssinicus*	bloody diarrhea, stomach ulcers
5. *Kithea* (also *Muthiia*)	*Acacia mellifera* Benth.	?
6. *Kithia*	*Dobera glabra* (Forsk) (*Salvadara*)	abdominal pain
7. *Kitungu*	a. *Launea alata* (Engl) (*Acanthaceae*)	abdominal pain, fever
	b. *Commiphora africana*	

8.	*Kiva*	*Papea ugandensis*	?
9.	*Kutukumwe (Rukiga)*	*Centella asiaticaa* L urban *(Chubelliferae)*	?
10.	*Mugaa* (Kikuyu)	a. *Acacia sublata valthe (Mimosaceae)*	infant gastroenteritis, worms, anti-witchcraft, abdominal problems, madness
		b. *Acacia seyal Del. var. multijuga (Mimosaceae)*	
11.	*Mukaa*	*Dombeya Mastersii*	edema, heart disease
12.	*Mukandu*	*Ocimum suave*	Ref. Kokwaro (1976: 341)
13.	*Mukenea*	a. *Carissa edulis*	abdominal pains, epilepsy, peptic ulcer, love affairs, pneumonia, continuous menstruation
		b. *E. Burtii*	
		c. *Fagara chalbea*	
14.	*Mukiliulu*	*Harrisonia abyssinica*	?
15.	*Mukinyai*, or *Mukinyei*	*Euclea divinorum*	constipation, pneumonia, abdominal pains
16.	*Mukula*	*Combretum splendens*	?
17.	*Mukuswi*	a. *Acacia breyispica*	convulsions
		b. *Acacia pennata Willd.*	
		c. *Acacia atascacantha*	
18.	*Mulawa*	*Triumphera rhomboides Jacq. tiliac*	infant gastroenteritis
19.	*Mukawa*	*Carissa edulis*	?
20.	*Munatha*	a. *Courbonia comporum*	infant gastroenteritis, gonorrhea, rheumatism
		b. *Courbonia glauca (Capparidaceae)*	
21.	*Mungendiathenge*	*Senecio genus*	?
22.	*Muoa/mwoa*	*Albizia authelmintica (Mimosaceae)*	abdominal pains, pneumonia, malaria

325

23.	*Musemei*	*Acacia sublata valke* (*Mimosaaceae*)	abdominal pains, impotency
24.	*Musemei*	a. *Acacia nilotica*	?
		b. *Acacia gerrardii*	
25.	*Mutaa*	a. *Ocimum americanum* (*Labiatae*)	birth pains, infertility, spirit possession, infant gastroenteritis
		b. *Ocimum basilicum*	
26.	*Mutanda*	*Gymnosporia cuteolo* (*Celastraceae*)	?
27.	*Mutandambogo*	a. *Pterolobium stellatum* (*Leguminosae*)	epilepsy, anti-witchcraft
		b. *Scutia buscifolia*	
		c. *Scutia indica*	
28.	*Muteta*	a. *Strychnos henningsei* (*Loganiac*)	abdominal problems, infertility, edema, infant gastroenteritis, generalized pains (from curse/bewitchment), joint pains, continuous menstruation
		b. *Strychnos reticulata*	
29.	*Mutheu*	a. *Rhus aff. flexicaulis* (*Anacardiaceae*)	generalized pains (from curse/bewitchment)
		b. *Rhus vulgaris Miekle*	
30.	*Muthiia*	*Acacia mellifera*	infant gastroenteritis, abdominal pain
31.	*Muthulu*	*C. megalocarpus*	abdominal pains, pneumonia, edema, urinating blood, madness, continuous menstruation
32.	*Muthunzi*	*Maytenus putterlickiodes*	abdominal pains
33.	*Mutote* (also *Mukenea*)	*Carissa edulis*	induces relaxation, abdominal pain, infertility, edema

34.	*Mutula*	a. Commiphora sp. b. *Ximemia caffra Sond. (Olocaceae)*	bloody urine, stomach ulcers, headache, malaria, fever, abdominal pain, madness, infertility
35.	*Muua*	*sclerocarya birrea (Anacord)*	worms, abdominal pain
36.	*Muuka*	*Maerua* sp.	constipation, madness, hypertension, infertility, pneumonia, TB, abdominal pain
37.	*Muvatha*	*Vernonia lasiopus*	infertility, palpitations
38.	*Muvindavindi*	*Fagarasis hildebrandtii (Rutaceae)*	diarrhea, abdominal pain
39.	*Muvingo*	*Dalbergia melanoxylon (Papilianaceae)*	joint pains, heavy cold
40.	*Muva*	*Pappea capensis*	gonorrhea
41.	*Muvu*	*Grewia villosa*	?
42.	*Muvunda*	*Moringa* sp.	worms, gastroenteritis, edema, madness, infertility, impotence
43.	*Mwaiitha* *Mwaritha* (Kikuyu)	a. *Dalbergia lactea* b. *Eutada leptostachya (Mimosaceae)*	infant gastroenteritis, joint pain, abdominal pains, edema, backache, anemia, hemorrhage, dizziness, high fever, malaria, love-making, tapeworm, common cold, helps produce breast milk, madness, epilepsy, infertility, gonorrhea, pneumonia
44.	*Mwale*	a. *Pennisetum* b. *Rhodognaphalon Schumannianum (Bombaceae)*	?
45.	*Mwalika*	*Tinnea aethiopica (Labiatea)*	epilepsy, impotence, madness, eliminate poverty, infertility

46. *Mwelengwa*	*Cissus aphyliantha (Vitaceae)*	infant gastroenteritis, infertility (ritual use), gonorrhea
47. *Mwenu*	a. *Cassia longiraceucosa Vatke (Coesalpiniaceae)* b. *Cassia didymobotrya* c. *Cassia goratensis Fresen*	convulsions
48. *Ndatakivumbu* (also "*Ng'ondu*")	a. *Ceuchrus ciliciris* b. *Talinum portulacifolium*	infertility
49. *Ng'ondu*	a. *Jasminium portifolium* b. *Sphaeranthus gomphrenoides (Compositae)* c. *Portulaca oleracea (Portulacaceae)*	edema
50. *Nzo*	*Cajanus indicus*	?
51. *Uthuko*	*Evolvulus alsinoides*	love medicine, infertility, abdominal pain

[a]Practically all entries represented by at least one specimen in the East African Herbarium.
[b]Vernacular names in Kikamba unless noted otherwise.

References

UNPUBLISHED MATERIALS

Oral Records

Set of 55 tapes of interviews with 32 TMPs in Mathare/Nairobi (24); Kilungu Location, Machakos District (5); and Murang'a District (3). Also three tapes (270 min) of *gukunura mundu mugo* ceremony in Murang'a District, including interview with Mr. Stanley Kiama Gathigira of Tumu Tumu. All tapes recorded by the author in 1977–1978 and left in Kenya to be placed on public deposit by Violet N. Kimani, Research Associate.

Official: Kenya National Archives, Nairobi

Lists of District Officers who served in Machakos District (1933–1952) and Kitui District Central Province. (1920–1963).
    Annual Report, 1911–1912. PC/CP. (4/2/1)
    Annual Report, 1916–1917. PC/CP. (4/2/2).
    Annual Report of Provincial Commissioner, 1936. PC/CP. (4/3/1)
Kikoko Mission Hospital, Kalongo Sublocation
    Annual statistical returns, January 1–December 31, 1977. Personal file, field notes, 1978.
Nunguni outpatient returns
    May 1977–February 1978.
Kitui District
    Political Record Book, 1898–1912. PC/CP. (1/2/1)
    Provincial Commissioner's Inspection Report, 1910. DC/KTI/8/1.
    Annual Report, 1912–1913. DC/KTI. (1/1/1)
    Annual Report, 1914–March 31, 1915. DC/KTI. (1/1/1)
    Annual Report, 1915–March 31, 1916. DC/KTI. (1/1/1)
    Provincial Commissioner's Inspection Report, 1915. DC/KTI, 8/1.
    Annual Report, 1936. DC/KTI. (1/1/4)
    Annual Report, 1956. DC/KTI. (1/1/13)
Machakos District
    Annual Report, 1980. Office of the district commissioner.
    Political Record Book. Up to 1910. (Pt. 2). P. 12.
    Political Record Book, Vol. 2. Up to 1910. DC/MKS. 4/10.
    Annual Report, 1910–1911. DC/MKS. (1/5/7)
    Annual Report, 1912. DC/MKS. (1/5/10)

Annual and Quarterly Reports, 1918–1919. DC/MKS. (1/10)
Annual Report, 1921. DC/MKS.
Annual and Quarterly Reports, 1922. DC/MKS. (1/1/10)
Annual Report, 1924. DC/MKS. (1/1/2)
Annual Report, 1926. Appendix No. 17. Medical Officer's Report. DC/MKS. (1/1/22)
Annual Report, 1927. DC/MKS.
Annual Report, 1928. Appendix No. 17. Medical Officer's Report. DC/MKS. (1/1/22)
Annual Report, 1929. Appendix 29, Native Tribal Customs. DC/MKS. (1/1/22)
Annual Report, 1930. DC/MKS. (1/1/22)
Annual Medical Department Report, 1931. DC/MKS. (1/7/1)
Annual Report, 1932. DC/MKS.
Annual Report, 1933. DC/MKS. (1/1/25)
Annual Report, 1934. DC/MKS. (1/1/25)
Annual Report, 1935. DC/MKS. (1/1/25)
Annual Report, 1936. DC/MKS. (1/1/4)
Annual Report, 1948. DC/MKS. (1/1/30)
Annual Report, 1950. DC/MKS. (1/1/2)
Annual Report, 1951. DC/MKS. (1/1/30)
Annual Report, 1954. DC/MKS. (1/1/32)
Annual Report, 1955. DC/MKS. (1/1/33)
Annual Report, 1956. DC/MKS. (1/1/33)
Annual Report, 1957. DC/MKS. (1/1/34)
Letter to chairman, Machakos African District Council, February 18, 1963. DC/MKS. (10/3)
Nyeri District
Annual Report, 1915–1916. DC/NYI. (1/2)
Ukamba Province
Annual Report, 1908–1909. PC/CP. (4/2/1)
Quarterly Report, October–December 1909. DC/MKS. (1/5/1)
Provincial Commissioner's Inspection Reports, 1910. DC/KTI/8/1.
Annual Report, 1917–1918. DC/MKS. (1/5/11)
Ulu District
Annual Report, 1913–1914. DC/MKS. (1/1/2)
Annual Report, 1914–1915. PC/CP. (4/3/1)

PUBLISHED MATERIALS

Acuda, S. W. Mental health problems in Kenya today—A review of research. *East African Medical Journal,* 1983, 60, 11–14.
Ademuwagun, Z. A., *et al. African therapeutic systems.* Waltham, MA: Crossroads Press, 1979.
Adjala, K.V. The development of the medical services in Kenya. *East African Medical Journal,* 1962, 39, 105–114.
African Business. Call to ease Kenya's urban cramming. October 1985, p. 21.
African Medical and Research Foundation. *AMREF in action,* Annual report, 1985. Nairobi: AMREF, 1985.
Ahmed, P. I. & Kolker, A. The role of indigenous medicine in W.H.O.'s definition of health. In P. I. Ahmed & G. V. Coelho (Eds.), *Toward a new definition of health.* New York: Plenum Press, 1979.

Aikman, L. *Nature's healing arts*. Washington, DC: National Geography Society, 1977.

Akerele, O. WHO's traditional medicine programme: Progress and perspectives. *WHO Chronicle*, 1984, 38, 76–81.

de Albuquerque, K. Non-institutional medicine on the Sea Islands. In M. S. Varner & A. M. McCandless (Eds.), *Proceedings of a symposium on culture and health: Implications for health policy in rural South Carolina*. Center for Metropolitan Affairs and Public Policy, College of Charleston, Charleston, October 1979.

Alexander, G. A. A survey of traditional medical practices used for treatment of malignant tumours in an East African population. *Social Science and Medicine*, 1985, 20, 53–59.

Alland, A., Jr. *Adaptation and cultural evolution: An approach to medical anthropology*. New York: Columbia University Press, 1970.

Amis, P. Squatters or tenants: The commercialization of unauthorized housing in Nairobi. *World Development*, 1984, 12, 87–96.

Arya, O. P., Taber, S. R., & Nsanze, H. Gonorrhea and female infertility in Uganda. *American Journal of Obstetrics and Gynecology*, 1980, 138, No. 7, Pt. 2, 929–932.

Atkinson, P. From honey to vinegar: Levi-Strauss in Vermont. In P. Morley & Wallis, R. (Eds.), *Culture and curing*. Pittsburgh: University of Pittsburgh Press, 1979.

Audy, J. R. Measurement and diagnosis of health. In P. Shepard & D. McKinley (Eds.), *Environ/mental: Essays on the planet as a home*. Boston: Houghton Mifflin, 1971.

Balcomb, J. Statistics shed new light on Kenya. *UNICEF News*, 1978, 96, 3–7.

Bannerman, R. H., Burton, J., & Wen-Chieh, C. (Eds.), *Traditional medicine and health care coverage*. Geneva: World Health Organization, 1983.

Barbee, E. L. Biomedical resistance to ethnomedicine in Botswana. *Social Science and Medicine*, 1986, 22, 75–80.

Beck, A. History of medicine and health services in Kenya (1900–1950). In L. C. Vogel, A. S. Muller, R. S. Odingo, Z. Onyango, & A. DeGeus (Eds.), *Health and disease in Kenya*. Nairobi: East African Literature Bureau, 1974.

Beck, A. *Medicine, tradition, and development in Kenya and Tanzania*. Waltham, MA: Crossroads Press, 1981.

Bennett, F. J. The social determinants of gonorrhea in an East African town. *East African Medical Journal*, 1962, 39, 332–334.

Bennett, F. J. The social, cultural and emotional aspects of sterility in women in Buganda. *Fertility and Sterility*, 1965, 16, 243–251.

Bennett, F. J., & Maneno, J. (Eds.). *National Guidelines for the Implementation of Primary Health Care in Kenya*. Nairobi: Ministry of Health, Government of Kenya, 1986.

Beresford-Stooke, G. Ceremonies designed to influence the fertility of women. *Man*, 1928, Nos. 128–130, 177.

Bernard, F. E. Planning and environmental risk in Kenyan drylands. *Geographical Review*, 1985, 75, 58–70.

Bernard, F. E. & Thom, D. J. Population pressure and human carrying capacity in selected locations of Machakos and Kitui Districts. *Journal of Developing Areas*, 1981, 15, 381–406.

Bhardwaj, S. Attitude toward different systems of medicine: A survey of four villages in the Punjab-India. *Social Science and Medicine*, 1975, 9, 603.

Bhardwaj, S. Medical pluralism and homeopathy: A geographic perspective. *Social Science and Medicine*, 1980, 14B, 209–216.

Bhardwaj, S. & Paul, B. K. Medical pluralism and infant mortality in a rural area of Bangladesh. *Social Science and Medicine*, 1986, 10, 1003–1010.

p'Bitek, O. *Africa's cultural revolution*. Nairobi: Macmillan, 1973.

Black, M. Deciding in the daylight: Community care in western Kenya. *UNICEF News*, 1978, 98, 4–9.

Bledsoe, C. The reinterpretation of "Western" pharmaceuticals among the Mende of Sierre Leone. Paper presented at International Health Conference on Traditional Healing and

Contemporary Medicine, National Council for International Health, June 13–15, 1983, Washington, DC.

Bloom, A. Introduction: The client's perspective in primary health care. *Medical Anthropology*, 1985, 9, 7–9.

Bradley, D. Water supplies—the consequences of change. In K. Elliott & J. Knight (Eds.), *Human rights in health*. Amsterdam: ASP North Holland, 1974.

Caldwell, J. C. & Caldwell, P. The demographic evidence for the incidence and cause of abnormally low fertility in tropical Africa. *World Health Statistics*, 1983, 36 (1), 2–34.

Campbell, D. J. Land use competition at the margins of the rangelands: An issue in development strategies for semi-arid areas. In G. Nordcliffe & T. Pinfold (Eds.), *Planning African development* (p. 39–61). Boulder, CO: Westview Press, 1981.

Carty, M. J., Nzioki, J. M., & Verhagen, A. R. The role of gonococcus in acute pelvic inflammatory disease in Nairobi. *East African Medical Journal*, 1972, 49, 376.

Chaffe, J. Testimony on HR3580: the "Rural Development Policy Act of 1979." United States Congress, House Committee on Agriculture, Subcommittee on Family Farms and Rural Development, Washington, DC.

Chambers, R. A., Longhurst, R. & Pacey, A. (Eds.), *Seasonal dimensions to rural poverty*. Totowa, NJ: Allanheld, Osmun and Co., 1981.

Chandler, J. A., Highton, R. B., & Hill, M. N. Mosquitoes of the Kano Plain, Kenya. II. Results of outdoor collections in irrigated and non-irrigated areas using human and animal bait and light traps. *Journal of Medical Entomology*, 1976, 13, 202–207.

Chavunduka, G. L. *Traditional healers and the Shona patient*. Gwelo, Rhodesia: Mambo Press, 1978.

Collier, P. & Lal, D. Why poor people get rich: Kenya, 1960–79. *World Development*, 1984, 12, 1007–1018.

Colson, A. The differential use of medical resources in developing countries. *Journal of Health and Social Behavior*, 1971, 12, 226–237.

Conco, W. Z. The African urban Bantu traditional practice of medicine. *Social Science and Medicine*, 1972, 6, 283–322.

Daily Nation (Nairobi). Witchcraft warning. . . , January 27, 1978.

Daily Nation (Nairobi). Witchcraft no cure for cholera, March 3, 1978.

Darnton, N. & Corbett, M. The magic force of witchdoctors. *New York Times Magazine*, October 19, 1980, 74.

Davis-Roberts, C. *Kutambuwa ugonjuwa*: Concepts of illness and transformation among the Tabwe of Zaire. *Social Science and Medicine*, 1981, 15B, 309–316.

Dawson, M. Smallpox in Kenya, 1880–1920. *Social Science and Medicine*, 1979, 13B, 245–250.

Desowitz, R. S. Epidemiological-ecological interactions in savanna environments. In D. R. Harris (Ed.), *Human ecology in savanna environments*, 1980 (pp. 457–477). New York: Academic Press.

Diesfeld, H. J. & Hecklau, H. K. *Kenya— a geomedical monograph*. Berlin: Springer-Verlag, 1978.

Dingwall, R. *Aspects of illness*. London: Martin Robertson, 1976.

Dorozynski, A. *Doctors and healers*. IDRC-043e. Ottawa: International Development Research Center, 1975.

Draat, P. J. E. M. Knowledge, attitude and practice survey towards epilepsy in a rural population in Kenya. March, 1981. (Mimeographed)

Duke, J. A. *CRC Handbook of medicinal herbs*. Boca Raton, FL: CRC Press, 1985.

Dundas, C. History of Kitui. *Journal of the Royal Anthropological Institute*, 1913, 43, 480–549.

Dunn, F. L. Traditional Asian medicine and cosmopolitan medicine as adaptive systems. In C. Leslie (Ed.), *Asian medical systems*. Berkeley: University of California Press, 1976.

Edgerton, R. B. Traditional treatment for mental illness in Africa: A review. *Culture, Medicine and Psychiatry*, 1980, 4, 167–189.

Eisenberg, L. Disease and illness: Distinctions between professional and popular ideas of sickness. *Culture, Medicine and Psychiatry*, 1977, 1, 9–23.

Ekong, D. E. U. African medicinal plants under the microscope. *UNESCO Courier*, 1979, 32.

Elling, R. H. The capitalist world-system and international health. *International Journal of Health Services*, 1981, 11, 21–51.

Evans-Pritchard, E. E. *Witchcraft, oracles, and magic among the Azande*. London: Oxford University Press, 1937.

Fabrega, H. Jr. *Disease and social behavior*. Cambridge, MA: Massachusetts Institute of Technology Press, 1974.

Fabrega, H. The scope of ethnomedical science. *Culture, Medicine and Psychiatry*, 1977, 1, 9–23.

Family Health Institute. *A working paper on health services development in Kenya: Issues, analyses, and recommendations*. For USAID, Technical Assistance Program to the Government of Kenya. Washington, DC, May 16, 1978.

Feierman, E. K. Alternative medical services in rural Tanzania: A physician's view. *Social Science and Medicine*, 1981, 15B, 339–404.

Feierman, S. Change in African therapeutic systems. *Social Science and Medicine*, 1979, 13B, 277–284.

Feierman, S. *The social origins of health and healing in Africa*. Commissioned by American Council of Learned Societies/Social Science Research Council Joint Committee on African Studies. Presented at 27th Annual Meeting of the African Studies Association, October 25–28, 1984, Los Angeles.

Fendall, N. R. E. Health centres in Kenya. *East African Medical Journal*, 1960, 37, 171–185.

Fendall, N. R. E. & Grounds, J. G. Incidence and epidemiology of disease in Kenya. *Journal of Tropical Medicine and Hygiene*, 1965; Part I: Some diseases of social significance (April), 77–84; Part II: Some important communicable diseases (May), 113–120; Part III: Insect-borne diseases (June), 134–141.

Ferguson, A. Women's health in a marginal area of Kenya. *Social Science and Medicine*, 1986, 23, 17–30.

Ferraro, G. P. Nairobi: Overview of an East African city. *African Urban Studies*, 1978–1979, 3, 1–14.

Field, M. J. *Religion and medicine of the Ga people*. London: Oxford University Press, 1937.

Fleischer, N. K. F. A study of traditional practices and early childhood anemia in northern Nigeria. *Transactions, Royal Society of Tropical Medicine and Hygiene*, 1975, 69, 198–200.

Fonaroff, L. S. Man and malaria in Trinidad: Ecological perspectives of a changing health hazard. *Annals, Association of American Geographers*, 1968, 58, 526–556.

Foster, G. M. Applied anthropology and international health: Retrospect and prospect. *Human Organization*, 1982a, 41, 189–197.

Foster, G. M. Community development and primary health care: Their conceptual similarities. *Medical Anthropology*, 1982b, 6, 183–195.

Foster, G. M. & Anderson, B. G. *Medical anthropology*. New York: Wiley, 1978.

Foster, W. D. *The early history of scientific medicine in Uganda*. Nairobi: East African Literature Bureau, 1970.

Foy, H. & Kendall, A. G. Haemoglobinopathies. In L. C. Vogel *et al.* (Eds.), *Health and disease in Kenya*. Nairobi: East African Literature Bureau, 1974.

Fratkin, E. M. Herbal medicine and concepts of disease in Samburu. University of Nairobi Institute of African Studies Seminar Paper No. 65, October, 1975. (Mimeographed)

Gatere, S. G. Traditional healing methods in psychiatry, Kenya. In A. Kiev, W. Muya, & N. Sartorius (Eds.), *The future of mental health services* (pp. 93–97). Amsterdam: Excerpta Medica, 1980.

Gebbie, D. A. M. Obstetrics and gynaecology. In L. C. Vogel, A. Muller, R. Odingo, Z. Onyango, & A. De Geus (Eds.), *Health and disease in Kenya* (pp. 483–498). Nairobi: East African Literature Bureau, 1974.

Gelfand, M. *Medicine and magic of the Mashona.* Cape Town, 1956.

Gelfand, M. *Witchdoctor: traditional medicine man of Rhodesia.* London: Harvill Press, 1964.

Gerlach, L. Some basic Digo conceptions of health and disease. In *One-day symposium on attitudes to health and disease among some East African tribes.* Kampala, Uganda: East African Institute of Social Research, 1959.

Gershenberg, I. The distribution of medical services in Uganda. *Social Science and Medicine,* 1972, 6, 353–372.

Gesler, W. M. Barriers between people and health care practitioners in Calabar, Nigeria. *Southeastern Geographer,* 1979a, 19, 27–41.

Gesler, W. M. Illness and health practitioner use in Calabar, Nigeria. *Social Science and Medicine,* 1979b, 13D, 223–226.

Giel, R. & van Luijk, J. N. Psychiatric morbidity in a small Ethiopian town. *British Journal of Psychiatry,* 1969, 115, 149–162.

Gilks, J. L. The Medical Department and the health organization in Kenya, 1909–1933. *East African Medical Journal,* 1933, 9, 340–354.

Girt, J. L. Distance to general medical practice and its effect on revealed ill-health in a rural environment. *Canadian Geographer,* 1973, 17, 154–166.

Gish, O. The political economy of primary care and 'health for the people': An historical explanation. *Issues* (African Studies Association), 1979, 9, No. 3, 6–13.

Githagui, N. *Characteristics and practices of traditional birth attendants (TBAs)—a study of three rural Kenyan Communities.* Nairobi: African Medical and Research Foundation, June, 1985. Mimeo.

Githagui, N. *Traditional healing practices in Kenya—a study of four Kenyan Communities.* Nairobi: African Medical and Research Foundation, September, 1985. Mimeo.

Gluckman, M. *Closed systems and open minds.* Edinburgh: Oliver and Boyd, 1944.

Gonzalez, N. Health behavior in cross-cultural perspective. *Human Organization,* 1966, 25, 122–125.

Good, B. The heart of what's the matter: The semantics of illness in Iran. *Culture, Medicine and Psychiatry,* 1977, 1, 25–58.

Good, C. M. Salt, trade, and disease: Aspects of development in Africa's northern Great Lakes Region. *International Journal of African Historical Studies,* 1972, 5, 543–586.

Good, C. M. Traditional medicine: An agenda for medical geography. *Social Science and Medicine,* 1977, 11, 705–713.

Good, C. M. Man, milieu, and the disease factor: Tick-borne relapsing fever in East Africa. In G. W. Hartwig & K. D. Patterson (Eds.), *Disease in African history.* Durham, NC: Duke University Press, 1978.

Good, C. M. A comparison of rural and urban ethnomedicine among the Kamba of Kenya. In P. R. Ulin & Segal, M. (Eds.), *Traditional Health Care Delivery in Contemporary Africa.* Africa Series 35, Maxwell School of Citizenship and Public Affairs, Syracuse University, Syracuse, NY, 1980a.

Good, C. M. Ethnomedical systems in Africa and the LDC's: Key issues for the geographer. In M. Meade (Ed.), *Conceptual and methodological issues in medical geography.* Studies in Geography No. 6, Department of Geography, University of North Carolina at Chapel Hill, 1980b.

Good, C. M. Community health in tropical Africa: Is medical pluralism a hindrance or a resource? In R. Akhtar (Ed.), *Health and disease in tropical Africa: Geographical and medical viewpoints.* London: Gordon and Breach Science Publishers, Ltd., 1987.

Good, C. M., Hunter, J. M., Katz, S. H., & Katz, S. S. The interface of dual systems of health care in the developing world: Toward health policy initiatives in Africa. *Social Science and Medicine,* 1979, 13D, 141–154.

Good, C. M. & Kimani, V. N. Urban traditional medicine: A Nairobi case-study. *East African Medical Journal*, 1980, 57, 301–316.

Good, C. M., Kimani, V. N. and Lawry, J. M. Gukunura mundu mugo: The initiation of a Kikuyu medicine man. *Anthropos*, 1980, 75, 87–116.

Government of Kenya. The Witchcraft Act, Cap. 67, 1925; Rev. 1962.

Green, E. C. Roles for African traditional healers in mental health care. *Medical Anthropology*, 1980, 4, 489–522.

Green, E. C. & Makhubu, L. Traditional healers in Swaziland: Toward improved cooperation between the traditional and modern health sectors. *Social Science and Medicine*, 1984, 18, 1071–1079.

Grossman, L. S. *Peasants, subsistence ecology, and development in the highlands of Papua New Guinea*. Princeton, NJ: Princeton University Press, 1984.

Haaga, J. *et al.* Child malnutrition in rural Kenya: A geographical and agricultural classification. *Ecology of Food and Nutrition*, 1986, 18, 297–308.

Haddock, D. R. W. & Chiduo, A. D. Uvulectomy in coastal Tanzania. *Central African Medical Journal*, 1965, 11, 331–334.

Hake, A. *African metropolis: Nairobi's self-help city*. London: Chatto and Windus, Ltd., 1977.

Halliman, D. M. & Morgan, W. T. W. The city of Nairobi. In W. T. W. Morgan (Ed.), *Nairobi: City and region*. Oxford: Oxford University Press, 1967.

Hand, W. D. (Ed.). *American folk medicine—a symposium*. Berkeley: University of California Press, 1976.

Harley, G. W. *Native African medicine with special reference to its practices in the Mano tribe of Liberia*. Cambridge, MA: Harvard University Press, 1941.

Harlow, V., Chilver, E. M., & Smith, A. (Eds.). *History of East Africa*. Vol. 2. Oxford: Clarendon Press, 1965.

Hartwig, C. W. Church-state relations in Kenya: Health issues. *Social Science and Medicine*, 1979, 13C, 121–127.

Hartwig, G. W. & Patterson, K. D. (Eds.). *Disease in African history*. Durham, NC: Duke University Press, 1978.

Hautvast, J. G. A. & Hautvast-Mertens. Analysis of a Bantu medical system: A Nyakyusa case-study (Tanzania). *Tropical and Geographical Medicine*, 1972, 24, 406–414.

Highton, R. B. Schistosomiasis. In L. C. Vogel *et al.* (Eds.), *Health and disease in Kenya*. Nairobi: East African Literature Bureau, 1974.

Hobley, C. W. *Bantu beliefs and magic*. London: Frank Cass, 1967. (Original work published 1910)

Hopkins, J. M. Theological students and witchcraft beliefs. *Journal of Religion in Africa*, 1980, 11, 56–66.

Hughes, C. C. & Hunter, J. M. Disease and "development" in Africa. *Social Science and Medicine*, 1970, 3, 443–493.

Hulke, M. *The encyclopedia of alternative medicine and self-help*. New York: Schocken Books, 1979.

Hunter, J. M. River blindness in Nangodi, northern Ghana: A hypothesis of cyclical advance and retreat. *Geographical Review*, 1966, 56, 398–416.

Hunter, J. M. On the merits of holism in understanding societal health needs. *Centennial Review*, 1973, 17, 1–19.

Hunter, J. M. The challenge of medical geography. In J. M. Hunter (Ed.), *The geography of health and disease*. Studies in Geography No. 6, Department of Geography, University of North Carolina at Chapel Hill, NC, 1974.

Hunter, J. M. & Arbona, S. Field testing along a disease gradient: Some geographical dimensions of tuberculosis in Puerto Rico. *Social Science and Medicine*, 1985, 21, 1023–1042.

Ijaduola, G. T. A. Hazards of uvulectomy in Nigeria. *East African Medical Journal*, 1982, 59, 771–774.

Imperato, P. J. *African folk medicine. Practices and beliefs of the Bambara and other peoples.* Baltimore, MD: York Press, 1977.

International Development Research Center. *Traditional medicine in Zaire. Present and potential contribution to the health services.* IDRC-137e. Ottawa: International Development Research Center, 1980.

Jacobs, D. R. *The culture themes and puberty rites of the Akamba.* Unpublished Ph.D. dissertation, New York University, 1961.

Janzen, J. M. *The quest for therapy in lower Zaire.* Berkeley: University of California Press, 1978.

Janzen, J. M. Changing concepts of African therapeutics. In Brian M. du Toit & I. H. Abdalla (Eds.), *African healing strategies.* Buffalo, NY: Trado-Medic Books, 1984.

Jarvis, J. F. & Mwathi, S. N. Uvulotomy among East African tribes. *Journal of Larynsolosy and Otolosy,* 1959, 73, 436–438.

Joseph, A. E. & Phillips, D. R. *Accessibility and utilization: Geographical perspectives on health care delivery.* New York: Harper & Row, 1984.

Kato, L. L. Functional psychosis and witchcraft fears. *Law and Society Review,* 1970, 4, 385–406.

Katz, S. S. & Katz, S. H. The evolving role of traditional medicine in Kenya. *African Urban Studies,* 1981, 9, 1–12.

Katz, S. S., Katz, S. G., & Kimani, V. N. The making of an urban mganga: New trends in traditional medicine in urban Kenya. *Medical Anthropology,* 1982 (Spring), 91–112.

Kiev, A. (Ed.). *Magic, faith and healing: Studies in primitive psychiatry today.* London: Collier-Macmillan Ltd., 1964.

Kimani, V. N. Attempts to coordinate the work of traditional and modern doctors in Nairobi in 1980. *Social Science and Medicine,* 1981, 15B, 421–422.

Kikhela, N., Bibeau, G. & Corin, E. Africa's two medical systems: Options for planners. *World Health Forum,* 1981, 2, 96–99.

Kjekshus, H. *Ecology control and economic development in East African history.* London: Heinemann, 1977.

Kleinman, A. Concepts and a model for comparison of medical systems as cultural systems. *Social Science and Medicine,* 1978a, 12, 85–93.

Kleinman, A. International health care from an ethnomedical perspective: Critique and recommendations for change. *Medical Anthropology,* 1978b (Spring), 71–94.

Kleinman, A. *Patients and healers in the context of culture.* Berkeley: University of California Press, 1980.

Kloos, H. Disease concepts and medical practices in relation to malaria among fever cases in Addis Ababa. *Ethnomedizin,* 1973, 2, 229–253.

Kloos, H. Preliminary studies of medicinal plants and plant products in markets of central Ethiopia. *Ethnomedizin,* 1976/1977, 4, 63–103.

Kloos, H. Medicine vendors and their products in the Ethiopian Highlands and Rift Valley. *Ethiopianist Notes* (African Studies Center, Michigan State University), 1978, 2, 47–69.

Kloos, H. *et al.* Utilization of pharmacies and pharmaceutical drugs in Addis Ababa, Ethiopia. *Social Science and Medicine,* 1986, 22, 653–672.

Knight, C. G. *Ecology and change: Rural modernization in an African community.* New York: Academic Press, 1974.

Kokwaro, J. O. *Medicinal plants of East Africa.* Nairobi: East African Literature Bureau, 1976.

Koumare, M. Traditional medicine and psychiatry in Africa. In R. H. Bannerman, J. Burton, & C. Wen-Chieh (Eds.), *Traditional medicine and health care coverage* (pp. 25–36). Geneva: World Health Organization, 1983.

Krapf, J. L. *Travels, researches and missionary labours during an eighteen years' residence in eastern Africa.* London: Trubner, 1860.

Kronholz, J. African healing arts treat saddest of ills, an ache in the heart. *The Wall Street Journal*, July 29, 1982, pp. 1 and 10.

Kunitz, S. J. Underdevelopment, demographic change, and health care on the Navajo Indian Reservation. *Social Science and Medicine*, 1981, 15A, 175–192.

Laman, K. E. *Dictionnaire Kikongo-Français*. Brussels: Institute Royale Coloniale Belge, 1936.

Lambo, T. A. Psychotherapy in Africa. *Human Nature*, 1978, 1, 32–39.

Lasker, J. Choosing among therapies: Illness behavior in the Ivory Coast. *Social Science and Medicine*, 1981, 15A, 157–168.

Last, M. & Chavunduka, G. L. (Eds.). *The professionalisation of African medicine*. Manchester: Manchester University Press, 1986.

Leeson, J. Paths to medical care in Lusaka, Zambia. Paper for International Conference of Africanists, Dakar, Senegal, 1968. (Mimeographed) Cited in Chavunduka (1978).

Leith, J. H. *An introduction to the reformed tradition* (rev. ed.). Atlanta, GA: John Knox Press, 1981.

Leslie, C. Medical pluralism in world perspective. *Social Science and Medicine*, 1980, 14B, 191–195.

Lindblom, G. *The Akamba of British East Africa*. 2nd ed. New York: Negro Universities Press, 1969.

Linsell, C. A. Cancer. In L. C. Vogel *et al.* (Eds.), *Health and disease in Kenya*. Nairobi: East African Literature, 1974.

Lisk, D. R. Some aspects of drug compliance in African epileptics: Experience from Kenyatta National Hospital. *East African Medical Journal*, 1984, 61, 177–183.

Luborsky, L., Singer, B. & Luborsky, L. Comparative studies in psychotherapies. *Archives of General Psychiatry*, 1975, 32, 995–1003.

McDonald, C. A. Political-economic structures—approaches to traditional and modern medical systems. *Social Science and Medicine*, 1981, 15A, 101–108.

McElroy, A. & Townsend, P. K. *Medical anthropology in ecological perspective*. North Scituate, MA: Duxbury Press, 1979.

McEvoy, J. The bus-stop dispenser. *East African Medical Journal*, 1976, 53, 193–194.

McEvoy, P. J. & McEvoy, H. F. Management of psychiatric problems in a Kenyan mission station. *British Medical Journal*, 1976, 1, 1454–1456.

Maclean, U. *Magical medicine. A Nigerian case-study*. Baltimore: Penguin Books, 1971.

Mati, J. K., Anderson, G. E., Carty, M. J., & McGlashan, H. E. A second look at the problems of primary infertility in Kenya. *East African Medical Journal*, 1973, 50, 94–97.

May, J. *Studies in disease ecology*. New York: Hafner Publishing Co., 1961.

Mayer, J. D. Relations between the two traditions of medical geography. *Progress in Human Geography*, 1982, 6, 216–230.

Mbiti, J. S. *African religions and philosophy*. Garden City, NY: Anchor Books, 1970.

Mbiti, J. S. *New Testament eschatology in an African background*. London: Oxford University Press, 1971.

Mbizvo, M. T., Chimbira, T. H. K., & Mkwanzi, J. B. Aetiological factors of male sterility in Zimbabwe. *Central African Journal of Medicine*, 1984, 30, 233–238.

Mburu, F. M. *A socioeconomic epidemiological study: Traditional and modern medicine among the Akamba ethnic group of upland Machakos—Kenya*. Unpublished M.A. thesis, Department of Sociology, Makerere University, Kampala, Uganda, 1973.

Mburu, F. M. The duality of traditional and western medicine in Africa: Mystics, myths and reality. In P. Singer (Ed.), *Traditional healing: new science or new colonialism?* Buffalo, NY: Conch Magazine, Ltd., 1977.

Mburu, F. M. Rhetoric-implementation gap in health policy and health services delivery for a rural population in a developing country. *Social Science and Medicine*, 1979, 13A, 577–583.

Mburu, F. M. Socio-political imperatives in the history of health development in Kenya. *Social Science and Medicine*, 1981, 15A, 521–527.

Mburu, F. M., Smith, M. C., & Sharpe, T. R. The determinants of health services utilization in a rural community in Kenya. *Social Science and Medicine*, 1978, 12, 211–217.

Mead, S. The WHO Reason-for-Encounter classification. *WHO Chronicle*, 1983, 37, 159–162.

Meade, M. Medical geography as human ecology. *Geographical Review*, 1977, 67, 379–393.

Meade, M. Some characteristics and services of Chinese medical practitioners in Malaysia. *Medical Journal of Malaysia*, 1976, 32, 14–16.

Meade, M. Geographical analysis of disease and care. *Ann. Rev. Public Health*, 1986, 7, 313–335.

Miller, N. *Traditional medicine in East Africa*. American Universities Field Staff Reports, No. 22, Africa. Hanover, NH: American Universities Field Staff, 1980.

Molnos, A. *Cultural source materials for population planning in East Africa*, 4 vols. Nairobi: East African Literature Bureau, 1973.

Moore, G. D. Essential drugs for Kenya's rural population. *World Health Forum*, 1982, 3, 196–199.

Moore, O. K. Divination—a new perspective. In A. P. Vayda (Ed.), *Environment and cultural behavior*. Garden City, NY: Natural History Press, 1969, p. 123.

Moore, T. R. Land use and erosion in the Machakos Hills. *Annals, Association of American Geographers*, 1979, 69, 419–431.

Morgan, W. T. W. & Shaffer, N. M. *Population of Kenya. Density and distribution*. Nairobi: Oxford University Press, 1966.

Muchunu, J. A Kenyan remembers. *World Vision*, December 1984–January 1985, 28, 10–11.

Muhangi, J. The nature of psychiatric disorders among a rural population. In T. Asuni (Ed.), *Proceedings of the Third Pan-African Psychiatric Workshop*. Nigeria, 1970.

Mullings, L. *Therapy, ideology, and social change*. Mental healing in urban Ghana. Berkeley: University of California Press, 1984.

Munro, J. F. *Colonial rule and the Kamba. Social change in the Kenya highlands, 1889–1939*. Oxford: Clarendon Press, 1975.

Murphy, J. H. B. The "Kithito" at Mivukoni, Mumoni District, Kenya Colony. *Man*, 1926, No. 135–136, 207.

Mutungi, O. K. *The legal aspects of witchcraft in East Africa*. Nairobi: East African Literature Bureau, 1977.

Mwalali, P. N. The effectiveness of the training of traditional birth attendants in a rural area, Machakos, Kenya, East Africa. Unpublished abstract submitted to National Council for International Health, Washington, DC, 1983.

Nair, K. R., Manji, F. & Gitonga, J. N. The occurrence and distribution of fluoride in groundwaters of Kenya. *East African Medical Journal*, 1984, 61, 503–512.

Nairobi Times. Dar officials blame cholera on traditional healers, January, 15, 1978.

Nchinda, T. C. Traditional and western medicine in Africa: Collaboration or confrontation? *Tropical Doctor*, July 1976, 133–135.

Ndetei, D. M. & Muhangi, J. The prevalence and clinical presentation of psychiatric illness in a rural setting in Kenya. In A. Kiev, W. J. Muya, & N. Sartorious (Eds.), *The future of mental health services*. Amsterdam: Excerpta Medica, 1980.

Ndeti, K. *Elements of Akamba life*. Nairobi: East African Publishing House, 1972.

Ndeti, K. The Kamba of central Kenya. In A. Molnos (Ed.), *Cultural source materials for population planning in East Africa*. Vol. 3. Beliefs and practices. Nairobi: East African Publishing House, 1973.

Ndeti, K. The relevance of African traditional medicine in modern medical training and practice. In F. Grollig & H. B. Haley (Eds.), *Medical anthropology*. The Hague: Mouton, 1976.

Neitschmann, B. *Between land and water: The subsistence ecology of the Miskito Indians, eastern Nicaragua.* New York: Seminar Press, 1973.

Nelson, N. Some aspects of informal social organization of female migrants in a Nairobi squatter neighborhood: Mathare Valley. Paper 1: Strategies of female migration to Mathare Valley. Unpublished paper, February 1975. (Mimeographed)

Nelson, N. Female-centered families: Changing patterns of marriage and family among *buzaa* brewers of Mathare Valley. *African Urban Studies,* 1978-1979 (Winter), 85-104.

Neumann, A. K. & Lauro, P. Ethnomedicine and biomedicine linking. *Social Science and Medicine,* 1982, 16, 1817-1824.

Nordberg, E. M. The true disease pattern in East Africa, Part I. *East African Medical Journal,* 1983, 60, 446-452.

Nottingham, J. C. Sorcery among the Akamba of Kenya. *Journal of African Administration,* 1959, 11, 2-14.

Nsanze, H. Problems and approaches in the surveillance and control of sexually transmitted agents associated with pelvic inflammatory disease in Africa. *American Journal of Obstetrics and Gynecology,* 1980, 138, No. 7, Pt. 2, 1088-1090.

Obudho, R. A. *Urbanization in Kenya.* Lanham, MD: University Press of America, 1983.

Oendo, A. *A brief Kikamba lexicon of health terms.* Nairobi: African Medical and Research Foundation, April 1982. (Mimeographed)

Ogot, B. A. (Ed.) *Zamani: A survey of east African history.* Nairobi: Longman Kenya/East African Publishing House, 1974.

Okwu, A. S. O. Life, death, reincarnation, and traditional healing in Africa. *Issues* (African Studies Association), 1979, 9, 19-24.

O'Leary, M. Responses to drought in Kitui District, Kenya. *Disasters,* 1980, 4, 315-327.

Oliver, R. & Mathew, G. *History of East Africa,* Vol. 1. Oxford: Clarendon Press, 1963.

Oliver, S. C. Individuality, freedom of choice, and cultural flexibility of the Kamba. *American Anthropologist,* 1965, 67, 421-428.

Omondi, H. Back-street doctors causing a sick situation. *Sunday Nation* (Nairobi), November 6, 1977.

Omran, A. R. Epidemiologic transition in the United States. *Population Bulletin,* 1977, 32, 1-42.

Onoge, O. F. Capitalism and public health: A neglected theme in the medical anthropology of Africa. In S. R. Ingman & A. E. Thomas (Eds.), *Topias and utopias in health.* The Hague: Mouton, 1975.

Onyango, R. J. African human trypanosomiasis (sleeping sickness). In L. C. Vogel, A. Muller, R. Odingo, Z. Onyango, & A. DeGeus (Eds.), *Health and disease in Kenya* (pp. 319-330). Nairobi: East African Literature Bureau, 1974.

Oomen, L. Disease pattern in Ukambani, Kenya. *East African Medical Journal,* 1976, 53, 341-349.

Orley, J. *Culture and mental illness.* Nairobi: East African Publishing House, 1970.

Owako, F. N. Machakos land and population problems. In S. H. Ominde (Ed.), *Studies in east African geography and development.* London: Heinemann, 1971.

Owano, N. The Hosken Report: Unflinching look at female circumcision. *Sunday Nation* (Nairobi), May 28, 1978, p. 15.

Owili, P. M. Editorial—sexually transmitted diseases. *East African Medical Journal,* 1983, 60, 281.

Parkin, D. (Ed.) *Town and country in central and eastern Africa.* London: International African Institute, 1975.

Parrinder, G. *Witchcraft: European and African.* London: Faber and Faber, 1963.

Pearce, T. O. Integrating Western orthodox and indigenous medicine. *Social Science and Medicine,* 1982, 16, 1611-1617.

Pfifferling, J.-H. Some issues in the consideration of non-Western and Western folk practices as epidemiologic data. *Social Science and Medicine*, 1975, 9, 655–658.

Pillsbury, B. L. K. *Traditional health care in the Near East*. Washington, DC: United States Agency for International Development, 1978.

Pillsbury, B. L. K. Policy and evaluation perspectives on traditional health care practitioners in national health care systems. *Social Science and Medicine*, 1982, 16, 1825–1834.

Polgar, S. Health and human behavior: areas of interest common in the social and medical sciences. *Current Anthropology*, 1962, 3, 159–205.

Popline (World Population News Service), Vol. 7, Nos. 9 and 10, 1985.

Population Reference Bureau. World Population Data Sheet, 1984. Washington, DC: Population Reference Bureau, Inc., 1984.

Porter, P. W. *Food and development in the semi-arid zone of East Africa*. Foreign and Comparative Studies/African Series No. 32. Maxwell School of Citizenship and Public Affairs, Syracuse University. Syracuse, NY, 1979.

Press, I. Urban folk medicine: A functional overview. *American Anthropologist*, 1978, 80, 71–84.

Press, I. Problems in the definition and classification of medical systems. *Social Science and Medicine*, 1980, 14B, 45–57.

Rappoport, H. The integration of scientific and traditional healing: The problem of demystification. In P. R. Ulin & M. H. Segal (Eds.). *Traditional health care delivery in contemporary Africa*. Africa Series No. 35, Maxwell School of Citizenship and Public Affairs, Syracuse University. Syracuse, NY, 1980.

Rappoport, H. & Rappoport, M. The integration of scientific and traditional healing—a proposed model. *American Psychologist*, 1981, 36, 774–781.

Republic of Kenya. *Kenya Population Census, 1969*, Vol. 1. Statistics Division, Ministry of Finance and Economic Planning, November, 1970. (Published in Nairobi by the Government Printer.)

Republic of Kenya. *Development plan for the period 1979 to 1983*. Nairobi: Government Printer, 1978.

Republic of Kenya. *Kenya Population Census, 1979*, Vol. 1. Central Bureau of Statistics, Ministry of Economic Planning and Development, June, 1981. (Published in Nairobi by the Government Printer.)

Republic of Kenya. *Development plan for the period 1984 to 1988*. Nairobi: Government Printer, 1983.

Republic of Kenya. *Economic management for economic growth*. Sessional Paper No. 1 of 1986. Nairobi: Government Printer, 1986.

Richardson, B. C. *Caribbean migrants*. Knoxville, TN: University of Tennessee Press, 1983.

Roberts, J. M. D. Malaria. In L. C. Vogel, A. Muller, R. Odingo, Z. Onyango, A. DeGeus, (Eds.), *Health and disease in Kenya*. Nairobi: East African Literature Bureau, 1974, pp. 305–317.

Romanucci-Ross, L. Melanesian medicine: Beyond culture to method. In P. Morley and R. Wallis (Eds.), *Culture and curing*. Pittsburgh: University of Pittsburgh Press, 1978.

Ross, M. *The political integration of squatters*. Evanston, IL: Northwestern University Press, 1973.

Ross, M. *Grass roots in an African city: Political behavior in Nairobi*. Cambridge, Mass.: Massachusetts Institute of Technology Press, 1975.

Rotberg, R. I. *Christian missionaries and the creation of Northern Rhodesia, 1880–1924*. Princeton, NJ: Princeton University Press, 1965.

Roundy, R. W. Altitudinal mobility and disease hazards for Ethiopian populations. *Economic Geography*, 1976, 52, 103–115.

Schulpen, T. W. J. & Swinkels, W. J. Machakos Project studies. XIX. The utilization of health services in a rural area of Kenya. *Tropical and Geographical Medicine*, 1980, 32, 340–349.

Schwartz, L. The hierarchy of resort in curative practices: The Admiralty Islands, Melanesia. *Journal of Health and Social Behavior*, 1969, 10, 201–209.

Scrimshaw, S. C. M. A technical manual for private and indigenous health care assessment: Guidelines for health sector analysis. University of California School of Public Health, 1979. (Mimeographed)

Shaffer, R. *Beyond the dispensary (on giving balance to primary health care).* Nairobi: African Medical and Research Foundation, January, 1984. Reprinted 1986.

Sherris, J. D. & Fox, G. Infertility and sexually transmitted disease: A public health challenge. *Population Reports*, 1983, 11, Series L (No. 4).

Sindiga, I. The persistence of high fertility in Kenya. *Social Science and Medicine*, 1985, 20, 71–84.

Singer, P. *Traditional healing: New science or new colonialism?* Buffalo, NY: Conch Magazine, Ltd., 1977.

Slooff, R. & Schulpen, T. W. J. Machakos Project studies. Agents affecting health of mother and child in a rural area of Kenya. VI. The social and hygienic environment. *Tropical and Geographical Medicine*, 1978, 30, 257–270.

Spear, T. *Kenya's past.* London: Longman Group, Ltd., 1981.

Spencer, Paul. *Nomads in alliance.* London: Oxford University Press, 1973.

Spring, A. Traditional and biomedical health care systems in northwest Zambia: A case study of the Luvale. In P. Ulin & M. Segall (Eds.), *Traditional health care delivery in contemporary Africa.* Foreign and Comparative Studies/African Series 35. Syracuse, NY: Maxwell School of Citizenship and Public Affair's, Syracuse University, 1980.

Stanbridge, R. When traditional and Western medicine meet. *New African*, July, 1985, p. 6.

St. John, R. K. & Brown, S. T. (Eds.). International Symposium on Pelvic Inflammatory Disease. *American Journal of Obstetrics and Gynecology*, 1980, 138 (Suppl.).

Stock, R. F. *Health care behaviour in a rural Nigerian setting with particular reference to the utilization of western type health care facilities.* Unpublished Ph.D. dissertation, Department of Geography, University of Liverpool, 1980.

Stock, R. Traditional healers in rural Hausaland. *Geojournal*, 1981, 5, 363–368.

Stock, R. Distance and the utilization of health facilities in rural Nigeria. *Social Science and Medicine*, 1982a, 17, 563–570.

Stock, R. Spirits as disease agents: A northern Nigeria example. (Proceedings). Attidel Primo Seminario Internazionale di Geografia Medica (Rome, November 4–7, 1982). Penigia, Italy: Editrice RUX, 1982b, pp. 403–413.

Stock, R. On the diversity of healers: The Hausa example. Unpublished paper presented at the Annual International Health Conference of the National Council for International Health, Washington, DC, June 13–15, 1983.

Sunday Nation (Nairobi). Backstreet doctors causing a sick situation. November 6, 1977.

Survey of Kenya. *National atlas of Kenya.* 3rd ed. Nairobi: Survey of Kenya, 1970.

Swantz, L. W. The role of the medicine man among the Zaramo of Dar-es-Salaam. Unpublished Ph.D. thesis, University of Dar-es-Salaam, Tanzania, 1974.

Swantz, M. L. *Ritual and symbol in transitional Zaramo society.* Studia Missionalia Upsaliensia XVI, Gleerup, 1970.

Swantz, M. L. Community and healing among the Zaramo in Tanzania. *Social Science and Medicine*, 1979, 13B, 169–173.

Swift, C. R. & Asuni, T. *Mental health and disease in Africa.* Edinburgh: Churchill Livingstone, 1975.

Tauxier, L. *La religion Bambara.* Paris: Librairie Orientaliste, Paul Geuthner, 1927.

Temu, A. J. *British Protestant missions.* London: Longman, 1972.

Thomas, A. E. *Adaptation to modern medicine in lowland Machakos, Kenya: a controlled comparison of two Kamba communities.* Unpublished Ph.D. dissertation, Department of Anthropology, Stanford University, 1971.

Thomas, A. E. Oaths, ordeals, and the Kenyan courts: A policy analysis. *Human Organization,* 1974, 33, 59–70.

Thomas, A. E. Health care in Ukambani Kenya: A socialist critique. In S. R. Ingman & A. E. Thomas (Eds.), *Topias and utopias in health: Policy studies.* The Hague: Mouton, 1975.

Tignor, R. L. *The colonial transformation of Kenya.* Princeton, NJ: Princeton University Press, 1976.

Traditional Medicine Research Unit. Summary report. Muhimbili Government Medical Center. Dar-es-Salaam, Tanzania, 1978. (Mimeographed)

Turshen, M. *The political ecology of disease in Tanzania.* New Brunswick, NJ: Rutgers University Press, 1984.

Twumasi, P. A. Colonialism and international health; a study in social change in Ghana. *Social Science and Medicine,* 1981, 15B, 147–151.

United States Agency for International Development. *Kenya: country development strategy statement.* Washington, DC: Author, 1980.

Unschuld, P. U. The issue of structured coexistence of scientific and alternative medical systems: A comparison of East and West German legislation. *Social Science and Medicine,* 1980, 14B, 15–24.

Van Etten, G. M. *Rural health development in Tanzania.* Assen/Amsterdam: Van Gorcum, 1976.

Van der Geest, S. The illegal distribution of Western medicine in developing countries: Pharmacists, drug pedlars, injection doctors and others: A bibliographic exploration. *Medical Anthropology,* 1982 (Fall), 197–219.

Van Luijk, J. N. Traditional medicine among the Kamba of Machakos District, Kenya. Part II. Draft report, Department of Tropical Hygiene, Royal Tropical Institute, Amsterdam, March 1982.

Van Luijk, J. N. Utilization of modern and traditional medical care by the Kamba of Machakos, Kenya. Part II. Survey on perceived morbidity and medical care: Static and dynamic analysis. Draft report, Department of Tropical Hygiene, Royal Tropical Institute, Amsterdam, January 1983.

Van Luijk, J. N. The utilization of modern and traditional medical care. In J. K. van Ginneken & A. S. Muller (Eds.), *Maternal and child health in rural Kenya* (pp. 281–308). London: Croom Helm, 1984.

Van Steenbergen, W. M., Kusin, J. A., & Rens, M. M. van. Lactation performance of Akamba mothers Kenya. Breast feeding behavior, breast milk yield, and composition. *Journal of Tropical Pediatrics and Environmental Child Health,* 1980; cited in S. Onchere & R. Slooff, Nutrition and disease in Machakos District, Kenya, pp. 41–45, in R. Chambers, R. Longhurst, & A. Pacey (Eds.), *Seasonal Dimensions to Rural Poverty.* London: Frances Pinter, Ltd., 1981.

Van Zwanenberg, R. History and theory of urban poverty in Nairobi: The problem of slum development. *Journal of East African Research and Development,* 1972, 2.

Varner, M. S. & McCandless, A. M. (Eds.). *Proceedings of a symposium on culture and health: Implications for health policy in rural South Carolina.* Center for Metropolitan Affairs and Public Policy, College of Charleston, Charleston, SC, October 1979.

Vecsey, G. Spiritual healing gaining ground with Catholics and Episcopalians. *New York Times,* June 18, 1978, 20.

Verhagen, A. R. H. B. Gonorrhea. In L. Vogel, A. Muller, R. Odingo, Z. Onyango, & A. De-Geus (Eds.), *Health and disease in Kenya.* Nairobi: East African Literature Bureau, 1974. pp. 375–380.

Voorhoeve, A. M., Kars, C., & Van Ginneken, J. K. Machakos Project studies. Agents affecting health of mother and child in a rural area of Kenya. XXI. Antenatal and delivery care. *Tropical and Geographical Medicine*, 1982, 34, 91-101.

Vuori, H. The World Health Organization and traditional medicine. *Community Medicine*, 1982, 4, 129-137.

Warren, D. M. *Disease, medicine and religion among the Techiman-Bono of Ghana: A study of culture change.* Unpublished Ph.D. dissertation, Indiana University, 1974.

Warren, D. M. The role of emic analyses in medical anthropology: The case of the Bono of Ghana. In Z. Ademuwagun *et al.* (Eds.), *African therapeutic systems*. Waltham, MA: African Studies Association, 1979.

Warren, D. M., Bova, G. S., Tregoning, M. A., & Kliewer, M. Ghanaian national policy toward indigenous healers: The case of the Primary Health Training for Indigenous Healers (PRHETIH) Program. *Social Science and Medicine*, 1982, 16, 1873-1881.

Wasunna, A. E. O. & Wasunna, M. Drugs in wrong hands: The trafficking of non-addictive drugs through unqualified and unauthorized persons in Kenya. In A. F. Bagshawe *et al.* (Eds.), *The use and abuse of drugs and chemicals in tropical Africa*. Nairobi: East African Literature Bureau, 1974.

Watts, M. *Silent violence: Food, famine and peasantry in nothern Nigeria*. Berkeley: University of California Press, 1983.

Waweru and associates. Planning and engineering analyses. Nairobi sites. Low cost housing and squatter upgrading study. Progress Report No. 7. Prepared for Ministry of Housing and Social Services, Government of Kenya, and the World Bank. Nairobi, November 1976. (Mimeographed)

Weekly Review (Nairobi). Need for medical manpower acute. February 6, 1981, 8-9.

Weisz, J. R. East African medical attitudes. *Social Science and Medicine*, 1972, 6, 323.

Whiteley, W. H. (Ed.). *Language in Kenya*. Nairobi: Oxford University Press, 1974.

Whyte, S. R. Penicillin, battery acid and sacrifice. *Social Science and Medicine*, 1982, 16, 2055-2064.

Willis, R. G. Magic and "medicine" in Ufipa. In P. Morley & R. Wallis (Eds.), *Culture and curing*. Pittsburgh: University of Pittsburgh Press, 1978.

World Health Organization, Regional Committee for Africa. "Traditional medicine and its role in the development of health services in Africa." AFR/RC26/TO/1, June 23, 1976. Brazzaville: Author.

World Health Organization. *Alma-Ata 1978. Primary health care: Report of the International Conference on Primary Health Care, Alma-Ata, USSR, 6-12 September, 1978*. Geneva: World Health Organization, 1978.

World Health Organization. The extension of health service coverage with traditional birth attendants: A decade of progress. *WHO Chronicle*, 1982, 36, 92-96.

Worsley, P. Non-Western medical systems. *Annual Review of Anthropology*, 1982, 11, 315-348.

Yoder, P. S. (Ed.). *African health and healing systems: Proceedings of a symposium*. University of California at Los Angeles, African Studies Association, Office of International Health and Economic Development, Charles R. Drew Postgraduate Medical School, 1982a.

Yoder, P. S. Biomedical and ethnomedical practice in rural Zaire. *Social Science and Medicine*, 1982b, 16, 1851-1857.

Index